A RESEARCH JOURNAL
DEVOTED TO THE
HISTORY OF SCIENCE
AND ITS
CULTURAL INFLUENCES

EDITOR
ARNOLD THACKRAY

MANAGING EDITOR
FRANCES COULBORN KOHLER

EDITORIAL COORDINATOR
BRUCE LEWENSTEIN

PUBLICATION AND EDITORIAL OFFICE
DEPARTMENT OF HISTORY AND SOCIOLOGY OF SCIENCE
UNIVERSITY OF PENNSYLVANIA
215 SOUTH 34TH STREET/D6
PHILADELPHIA, PENNSYLVANIA 19104, USA

SUGGESTIONS FOR CONTRIBUTORS TO OSIRIS

1. **Manuscripts** (original plus two copies) may be submitted to any member of the Editorial Board, or to the Editorial Office, 215 South 34th Street, Philadelphia, PA 19104, U.S.A. Please include an **abstract** of approximately 150 words. Contributors are advised to retain a copy for reference. If return of submitted material is desired, include return postage or international reply coupons.

2. Manuscripts should be **typewritten** or processed on a **letter-quality** printer and **double-spaced** throughout, including quotations and notes, on paper of standard size or weight. Margins should be wider than usual to allow space for instructions to the typesetter. The right-hand margin should be left ragged (not justified) to maintain even spacing and readability.

3. OSIRIS normally uses double-blind refereeing; authors should therefore identify themselves only on a detachable cover sheet.

4. Bibliographic information should be given in **footnotes** (not parenthetically in the text), typed separately from the main body of the manuscript, **double-** or even **triple-spaced,** numbered consecutively throughout the article, and keyed to reference numbers typed above the line in the text.

   a. References to **books** should include author's full name; complete title of the book, underlined (italics); place of publication and publisher's name for books published after 1900; date of publication, including the original date when a reprint is being cited; page numbers cited. *Example:*

   Joseph Needham, *Science and Civilisation in China,* 5 vols., Vol. I: *Introductory Orientations* (Cambridge: Cambridge Univ. Press, 1954), p. 7.

   b. References to articles in **periodicals** should include author's name; title of article, in quotes; title of periodical, underlined; year; volume number, Arabic and underlined; number of issue if pagination requires it; page numbers of article; number of particular page cited. Journal titles are spelled out in full on first citation and abbreviated subsequently. *Example:*

   John C. Greene, "Reflections on the Progress of Darwin Studies," *Journal of the History of Biology,* 1975, 8:243–273, on p. 270; Dov Ospovat, "God and Natural Selection: The Darwinian Idea of Design," *J. Hist. Biol.,* 1980, 13:169–174, on p. 171.

   c. When first citing a reference, please give the title in full. For succeeding citations, please use an abbreviated version of the title with the author's last name. *Example:*

   Greene, "Reflections," p. 250.

5. Please mark clearly for the typesetter all unusual alphabets, special characters, mathematics, and chemical formulae, and include all diacritical marks.

6. A small number of **figures** may be used to illustrate an article. Line drawings should be directly reproducible; glossy prints should be furnished for all halftone illustrations.

7. Manuscripts should be submitted to OSIRIS with the understanding that upon publication **copyright** will be transferred to the History of Science Society. That understanding precludes OSIRIS from considering material that has been submitted or accepted for publication elsewhere.

OSIRIS (ISSN 0369-7827) is published once a year.

A subscription to OSIRIS is $29 for institutions. Individual subscriptions are $24 (hardcover) and $15 (paperback).

Address editorial correspondence, advertising inquiries, single-issue orders and new subscriptions to HSS Publications Office, 215 South 34th Street D6, Philadelphia, PA 19104, U.S.A. Address renewal orders, claims for missing issues and changes of address to HSS Business Office, P.O. Box 529, Canton, MA 02021, U.S.A. Claims for issues not received should be sent within four months of publication of the issue in question.

OSIRIS will be indexed in major scientific and historical indexing services.

Second-class postage paid at Canton, MA, and additional mailing offices.

Typeset and printed at the Sheridan Press, Inc.

Hardcover edition, ISBN 0-934235-02-3
Paperback edition, ISBN 0-934235-03-1

A RESEARCH JOURNAL
DEVOTED TO THE HISTORY OF SCIENCE
AND ITS CULTURAL INFLUENCES
SECOND SERIES   VOLUME 1   1985

HISTORICAL WRITING ON AMERICAN SCIENCE

EDITED BY

SALLY GREGORY KOHLSTEDT

AND

MARGARET W. ROSSITER

To the scholars and teachers who have supported the study of science in America, especially A. Hunter Dupree, John C. Greene, Brooke Hindle, Nathan Reingold, and Charles Rosenberg

HISTORICAL WRITING ON AMERICAN SCIENCE

Foreword

THE HISTORY OF SCIENCE SOCIETY is proud to present Volume 1 of
the new series of *Osiris*.

The revival of *Osiris* has been made possible by the Fund Drive so ably led
by our thirty-third President, Gerald Holton. That Fund Drive in its turn has
built on and turned to advantage the deepening awareness in American society
that we live in a time of extraordinarily rapid change, change fueled by advances
in the sciences and in the science-based technologies. The sciences in their
modern professional form are now entering their second century in America. It
was in the 1880s that those sciences—echoing earlier developments in Europe—
were reorganized on the basis of specialized learned societies, university departments, graduate study and academic research, and linkages to industrial and
technological concerns. One hundred years ago, the United States was junior
partner to Europe in both the intellectual power and the social organization of
science. However, the twentieth century has seen the rise to dominance of
American science, and an associated development of the history of science itself
as a professional and scholarly pursuit. It is therefore especially fitting that
Volume 1 of the new series of *Osiris* should be devoted to historical writing on
American science.

Working with skill and sagacity, Sally Gregory Kohlstedt and Margaret W.
Rossiter have assembled a set of essays offering exciting "perspectives and prospects" for a field of scholarship that has triumphantly come of age in the past
decade. The essays in this first thematic volume of *Osiris* do not simply capture
the vitality that characterizes the study of their national science by historians in
the United States. These essays also stand as testimony to the level of analysis
that American historians of science bring to the wider world of scholarship, in
studies of Aristotle, Galileo, Newton, Darwin, and the "great tradition." Indeed the community of historians of science in the United States and worldwide
is now so large and securely established that plans are already well articulated
for further guest-edited volumes of *Osiris* on themes that will display, while advancing, scholarship across a wide intellectual spectrum.

Thematic volumes of *Osiris* will interdigitate with volumes (like Volume 2,
now in press) that provide an eclectic mix of articles reporting research at
greater length, but on the same level of scholarly excellence, as in *Isis* articles.
Volumes in this second genre will continue George Sarton's original design for
Osiris. The contributing editors of *Osiris* thus look forward to corresponding
with and receiving submissions from authors from around the globe. When
George Sarton launched *Osiris* in 1936, it was with the hope and conviction that
the history of science community needed to expand the range of its discourse,
while remaining true to the standards represented in *Isis*. Now, almost half a
century later, his hope and conviction may again serve us as we begin on the
second series of *Osiris*.

This first volume of the new *Osiris* will be published on the occasion of the

XVIIth International Congress of the History of Science, in Berkeley, California, August 1985. To our colleagues gathered on that occasion we offer greetings and an invitation to cooperate in securing the continuing vitality of *Osiris*. We also offer this first volume as a token of the strength of the history of science in our day and of the rich themes for research that may be found in *Historical Writing on American Science*.

ARNOLD THACKRAY
MEMORIAL DAY, 1985

Preface

THE APPEARANCE OF THIS VOLUME less than two years after its conception testifies to the cooperation of a wide-ranging group of historians who offered advice and counsel, referees who made invaluable suggestions, contributors who cooperated with revisions, and editorial staff who pushed the work to completion. Thanks to all of them, this volume is far broader and richer than we envisioned as we created preliminary lists of potential authors and sketched out content.

The volume is collaborative in many ways. It grew out of the activities of a group of junior historians of American science in the late 1970s. When plans for a special conference faltered, the group's then six members settled on a newsletter to keep up with growing activities in the field. Since the fall of 1980, *History of Science in America: News and Views,* edited by Clark Elliott, has appeared at least twice a year. When in 1983 the History of Science Society determined to revive *Osiris* and to consider publishing issues on specific topics, a historiographical volume on American science seemed a natural possibility. Several Americanists were ready to assess the current state of various specialties and to indicate what "needs and opportunities" remained after more than a decade of significant activity. The results might also become, it was felt, a tribute to the pioneering work and active advocacy of our predecessors, who, whether retired or not, are still active in the field.

Luckily there were enough historians of American science to make it possible to select suitable topics and authors quickly. Contributors come from a diverse set of backgrounds, including science, American history, history of science, and library and archival work. Most are currently in some branch of academia, but others are at government agencies, archives, learned societies and industry, reflecting the varied roles of science in American culture.

Although some of those invited could not contribute because of prior obligations, we were fortunate in having fifteen essays arrive in time for systematic review. The subject matter partly reflects the editors' beliefs as to which are the central aspects of the history of science in America. The final essays, however, also reflect in important and often creative ways the particular interests and expertise of their authors. Despite our efforts, gaps remain. Anticipated essays on women in science and on scientific biography were not submitted, with the result that neither topic was sufficiently covered by the other authors, who had been told that the topics were being included in separate essays. In addition, because one volume cannot cover all areas, a number of scientific specialties and themes received tangential rather than exclusive attention.

Our contributors operated under a tight schedule that allowed only six months to write their original essays and three months for revisions. Nonetheless they met deadlines, responded to stylistic suggestions, and produced new sections to fill lacunae and achieve overall coherence. Some also served as reviewers for other essays. Through it all, they remained helpful and supportive of the enterprise, for which they have our sincere gratitude.

We also called upon many other scholars to suggest possible authors of essays and to assess early drafts. In fact, a complete list of those who assisted in this editorial work would include a significant proportion of the scholars in the field. Here we can only offer our special thanks to those who with the editors evaluated essays, particularly Maynard Britchford, Hamilton Cravens, Frederick Churchill, Rayna Green, John C. Greene, Carol Gruber, David Hollinger, Daniel Jones, Daniel Kevles, Kenneth Ludmerer, Pamela Mack, Stephen McCluskey, Ronald Rainger, Leonard Reich, Nathan Reingold, Alex Roland, Martin Rudwick, Bruce Sinclair, Michael Sokal, Jeffrey Sturchio, Deborah Warner, Spencer Weart, and Ellis Yochelson.

The Committee on Publications of the History of Science Society, then under the leadership of Barbara Rosenkrantz, understood and supported the early aspirations for this volume. Syracuse University's Dean of the Maxwell School, Guthrie Birkhead, provided institutional resources during the solicitation, circulation and editing of manuscripts. Arnold Thackray, Frances Kohler, and their staff in the History of Science Society's Publications Office shaped the manuscript into its final form, making the challenges of editing less difficult and more satisfying. Understanding family members sustained us with quiet confidence, for which Sally, especially, offers affectionate thanks to David, Kris, and Kurt. Our mutual obligations as coeditors are too intricate to be easily elaborated, but the shared responsibility provided each of us the rewards that come from debate and from cooperation.

S.G.K.
M.W.R.

Introduction

By Margaret W. Rossiter and Sally Gregory Kohlstedt

T HERE IS MUCH EVIDENCE that scholarly work on the history of American science is increasing at a rapid and perhaps even exponential rate: books, articles, and dissertations abound and the number of reference aids is steadily growing. To date there has been no attempt to analyze the entire field. This volume of essays developed from the convictions that such an analysis was in order and that both those new to the study of American science and those already involved in it will benefit from an introduction to and overview of what exists and what remains to be done.

The studies that make up the volume are grouped in four sections. The first revisits "classic themes," examining the latest work and interpretations of topics and themes treated in historical writing in the 1960s and earlier. This section reveals that, as one might have expected, some subjects have grown and certain interpretations or emphases have shifted. The second set of essays discusses historical writing on six scientific specialties or disciplines. This approach is new because much work on national science has been chronological or thematic rather than disciplinary in orientation. Taken collectively, these essays reveal the premises held by those in disciplinary fields and the many gaps in current research. Indeed, the authors in this section call for more work, especially for innovative interpretations. The third section deals with approaches and areas of study that are, in the mid 1980s, gaining or requiring significant attention. Essays here both examine the connections of the history of science with other disciplines and investigate its relationships with industry, public policy makers, and the military. One final, comprehensive essay analyzes the sources and resources available for further historical work. Altogether, these essays indicate that the history of American science is rapidly outgrowing its former preoccupation with the nineteenth century and is both broadening and deepening its domain.

SECTION I: CLASSIC THEMES

This first set of essays reveals that the history of American science is now approaching a certain maturity as older topics are reappraised and some that loomed large fifteen years ago, like American indifference to basic research and the Lazzaroni, are now barely mentioned. Currently being reassessed is the relationship of science to other aspects of American culture and work, particularly religion and medicine. In addition, so much energy has gone into monographs on institutions, controversies, episodes, and even particular aspects of such subjects that it is time to pull all this together, compare scholars' conclusions, and see what remains to be done. "A great deal," is the answer, to judge by all the essays.

In the first essay Sally Gregory Kohlstedt discusses recent work on scientific institutions in the United States. She feels that though historians have done a respectable amount on the origins of institutions, far fewer authors have pursued these organizations into their later and perhaps equally complex years. She suggests, moveover, that much of the work fails to consider the essential role of such institutions for science. A number of historians have analyzed the organizations' internal tensions, as between popularization (or financial survival) and research (or contribution to ideas), changing membership, hidden agendas, and such. Still, the story of one organization can reveal only so much, and even that may be idiosyncratic. Needed are comparative studies that will crack the old chestnut of what is "uniquely American" and what is not. The increasing availability of institutional records, as organizations have installed archivists and curators, makes such work both feasible and attractive in the years ahead.

John Harley Warner, in an essay on the place of science in American medicine, faces the interesting paradox that, precisely because it is no longer assumed, as it was assumed twenty years ago, that improvements in science have underlain the rise of the modern American medical profession, understanding the power of the image and mystique of science is even more essential. Historians have begun to reevaluate the claims made for the new "scientific medicine" toward the turn of the century. In the last decade they have analyzed such rhetorical claims and found evidence of social control, but in the process they have left, Warner feels, the content of the sciences behind. Moreover, they have interpreted "medical science" too narrowly as merely laboratory science and continued to ignore both the clinical sciences and the ecologically oriented sciences, such as medical meteorology, that earlier historians of medicine also tended to pass over. Warner thus urges future historians to consider that "science" has many meanings and should not be limited to a few stereotypically familiar examples.

For Ronald Numbers the history of science and religion in the United States has been marked and marred by military metaphors ever since the subject began about 1870 with Andrew D. White's well-known works. In fact he finds that rather than being a frequent "battlefield," relations between science and religion in America were notable for their harmony. Puritans in America encountered Copernicanism in the seventeenth century with relative equanimity, and the antebellum clergy accommodated new scientific findings to their religious beliefs rapidly and with minimal discomfort. Many devout Americans, despite a few well-known skirmishes, even reconciled themselves to theories of biological evolution relatively quickly. It is the nature of such harmony rather than any warfare that needs exploring. In general Numbers calls for, as do several other authors, more clarity of thinking and modesty of claims as to who was converted to what, where and when.

Sharon Gibbs Thibodeau revisits the whole realm of science and the federal government made evident in the 1950s by A. Hunter Dupree's classic work on that subject. She finds that a certain amount of work has appeared on these themes since 1957 but that, now as then, the whole period since 1940 awaits its historian. Some topics have been explored by academic historians, others by former participants in scientific events or institutions, and still others by critics and journalists, who generally do not use the archives that might be available to them. She reminds us that the "official record" is not as dry and dull as reputed and that the whole topic of the impact of federal funding on science requires assessment.

SECTION II: SCIENCE IN SPECIALTIES

The second set of essays, on scientific specialties, presents for the first time in a collection on American science a detailed analysis of particular disciplines. The contributors here argue that they are at a mapping stage where there are more topics to pursue than persons to do them. Even as they bemoan their thin ranks, the authors openly recruit future historians into their specialties. They use work to date to demonstrate that research is sufficiently advanced for trends to be discerned. Older generalizations about American science are being re-examined using techniques borrowed from other sciences and from the social sciences.

In a direct assault on what has passed for scholarship in the history of American geology to date (compelling because echoed in subsequent essays), Mott Greene outlines three forms of what he and others call practitioner history of science (here, history of geology by geologists). This genre is composed of celebratory articles written for obituaries, centennials, and the like; review articles that usually take a contrived whiggish path to the present; and polemic accounts, often disguised as efforts to "put the record straight." When practitioners' histories are available, they thus offer advantages to but also create problems for disciplinary historians. Whether for modern periods or earlier ones, trained scientists often have more expertise on technical issues than do historians or historians of science. But they generally (though exceptions abound) lack a sense of the larger picture of what is important and significant. Thus they can present research on a range of topics from what Greene considers to be of "numbing triviality" in the history of geology to what Hamilton Cravens more generously estimates to be of value in the recent history of the social sciences.

Marc Rothenberg similarly finds that the history of astronomy in America is underdeveloped, with too few scholars in it and too few topics being worked on. He hopes for an eventual synthesis but admits it seems remote at present. Much of what is written is intended to highlight "major achievements," while whole subject areas, such as the role of instrumentation, so central to astronomers, lack systematic critical analysis. He finds too that most work on the discipline focuses on the period between 1870 and 1920. Earlier topics are rarely examined by historians of astronomy, because, as John Warner found with his physicians, historians have generally prejudged antebellum science as insignificant. Rothenberg also touches on the theme (as does Albert Moyer in his essay on history of physics) of the role of elites and democratic ideals in the history of American science. Strains between amateur and professional astronomers existed, but so did a certain complementarity of effort. In general Rothenberg indicates that the facts about the history of astronomy in America, especially in certain periods, are increasingly clear but interpretations are still too few for much sustained debate.

In many ways, however, the history of astronomy, one of the smallest of the sciences, is better developed than is that of chemistry, among the largest in the

United States. John Servos finds this field neglected in recent decades, for reasons that are not entirely clear. Like Mott Greene on geology, he writes that one could dig in almost anywhere and make a contribution. In particular, disputes and debates are often fruitful topics to pursue, for as Greene puts it in his essay, anything scientists find worth fighting over, historians should find worth writing about. Nevertheless, Servos finds comparative questions particularly interesting: for example, why did organic chemistry and physical chemistry develop at different rates? Like several other authors, he urges future contributors to go beyond highly visible topics, such as Du Pont, which is currently receiving much attention, and get on to subjects that might better typify the institutions and research concerns of American chemists. Perhaps because this field is so underworked, Servos has only praise for the few historians active in the history of chemistry. Getting the facts out provides plenty of work for everyone.

Historians of biology have, according to Jane Maienschein, only recently explored in any detail the external history of their science. Although various individuals were publishing in the 1950s on the history of American biology, the field has been dominated to date by internalists who focus on particular persons and topics. Only now are younger historians of biology discovering the classic studies on the social history of American natural history and biology and claiming them as worthy. They are beginning to open more diverse areas for research and concerning themselves with methodology. Maienschein documents the considerable emphasis given to T. H. Morgan and describes current historical research on Woods Hole laboratories at the turn of the century. She also finds increasing interest in the emergence of subspecialties, such as genetics from embryology around 1900.

Quite promising, according to Albert Moyer, is current work on the history of American physics. There are more good people working on the subject than ever before, although, he explains, much of their effort has concentrated on the atomic bomb and high-energy physics to the detriment of other topics and areas. To offset this tendency, he sees significant research going on now in the history of solid-state physics, with glimmerings in the history of physics education. Although Daniel Kevles's lengthy survey provides one of the few overviews of a scientific field, Moyer thinks that still more synthetic work should be done. He also calls for more diversity in defining physics and its boundaries, for, as Warner writes about medical sciences, the naive assumption that one area represents the whole field can lead to simplistic misperceptions that are no longer tolerable.

Hamilton Cravens observes that work on the history of the social sciences has found recent recognition and legitimacy, particularly as practitioners appropriate the standards of research and interpretation of the historical profession. Nevertheless, he admits, the field is still chiefly a collection of topics, albeit important ones such as mental hospitals, child development, and eugenics. Synthetic overviews are needed in the traditional areas of anthropology, psychology, and sociology as well as in social sciences that have been virtually ignored. As more people study the social sciences, especially those equipped to do archival research, a host of new topics will open for investigation. This is important because, Cravens feels, the social sciences lie close to the center of modern life.

SECTION III: NEW AREAS

The third set of essays surveys new areas or approaches to the history of science in America. Prominent among these and relatively unexplored (by historians, if not by anthropologists) is the area of ethnoscience and archaeoscience that developed rapidly in the 1970s. Clara Sue Kidwell describes recent efforts in this area. Because written records, on which history typically depends, are generally lacking for most native peoples of North and Central America, some scientists and scholars have used other tools to investigate past Indian practices and beliefs. These have included both archaeoastronomy (since its application to Stonehenge in the 1960s), which is the extrapolation from past building sites and locations to the probable accuracy of astronomical observations, and ethnobotany, which uses, for example, modern linguistics and pharmacology to identify the names, composition, and efficacy of various drugs used by native peoples. Yet despite such technical advances, Kidwell indicates that these are relatively naive methodologies that merely judge the accuracy of pretelescopic Indian astronomy and determine the identity of the drugs but do not penetrate or explore thought patterns of early native Americans, who stressed unity with nature, approached through rituals and ceremonies. History of science will be enhanced by considering these subjects and their often technical methodologies and by assessing anew native American conceptual structures and practices.

George Wise sees technology used to quite different ends in modern science policy. He discusses the differing views of the science and technology relationship held by policymakers (with budgets to defend) and historians of science and technology (largely without them). Wise focuses on the period after 1945, when leaders of science agencies in the United States, particularly Vannevar Bush in his report *The Endless Frontier*, propagandized an older idea that only "basic" science led to applied science, future technology, improved living standards, and military defense. If this seemed a truism in 1945 after the wonders of penicillin were highly publicized, and later that year when the atomic bomb was dropped, by the 1950s and 1960s this "assembly line" model began to seem atypical and untenable. Historians who had studied more and other examples thoroughly began to devise other theories and models with which to explain the relationship. By the 1970s historians of technology and science were using such analogies as "marriage" and "mirror-image twins" to explain the interactions of science and technology. These views have not yet, Wise indicates, coalesced into a post-assembly-line viewpoint, and even if they are more correct historically, it is far from clear that current policymakers have accepted them. The old ideas, despite their questionable accuracy, had provided effective justification for science appropriations for at least two decades and are therefore not about to be abandoned without being replaced by an equally persuasive argument. Historians, meanwhile, are turning their attention to other questions, adding to the growing number of impressive books on particular episodes and even, as in the case of Thomas Hughes, on entire industries.

Alex Roland provides a wide-ranging survey of the relations between war, science, and technology from the days of the American colonies to the Vietnam War. He suggests, for example, that overall the impact of science and technology on war has been overstated by historians in many fields. A close look at specific cases, for example, quickly reveals that the *Monitor* and the *Merrimac* did not decide the Civil War and that the atomic bomb was not essential to the outcome of World War II. The relations between science and technology were more long lasting and complex and went in both directions. The connection between science and technology did affect the practice of war, certainly, but when all adversaries had access to specialized knowledge, the impact was neutralized. More significantly, war shaped future science and technology generally via the institutions created during or immediately after war years, such as the land-grant colleges, the National Academy of Sciences (and later the National Research Council), and the National Science Foundation. Even the Army Corps of Engineers, which sounds like a military organization, has had more impact on internal domestic improvements than on the battlefield. Roland also presents an abundance of military histories, often descriptive and of limited scope but helpful at times, that historians of science or technology have not used. In short, there is a whole field awaiting good critical analysis. The important issues need to be defined and more works written that relate to central themes.

Margaret Rossiter provides a sweeping overview of important topics for the post–World War II era, which she claims is the most important period in the history of American science but one so unexamined that only a mapping is possible. She does not go into science itself, which other authors have treated in the disciplinary chapters, but concentrates on public policy, with topics that range from education to economics and from health to space. Like other authors she finds that certain events and episodes have been studied but that overviews and syntheses do not yet exist. Historians of science could have a special role to play in working on this period, she feels, since those government officials who have kept "big science" funded and functioning have left many records that purport to explain and justify to Congress and the public just what their agency is doing. To get at these often voluminous materials, however, historians will need good archives and should expect to be persistent and imaginative in their search. Rossiter is less optimistic about oral history, seeing both difficulties and advantages in that approach. One particular problem in using nearly contemporary studies may be getting permission to cite unflattering sources, for, she asks, if political scientists, journalists, statisticians, and sociologists obtain information by guaranteeing their informants confidentiality and anonymity, should historians necessarily follow? Would it still be history if they did?

SECTION IV: ACCESS TO THE SOURCES

Clark Elliott's essay offers a final summary of significant reference materials and resource centers, in which he describes the bibliographies, reference books, and manuscript finding aids that scholars now have and soon will need. In recent years considerable energy has gone into preparing reference works, both those which provide broad coverage, such as his own directory of nineteenth-century scientists, and others that touch only in part on the history of American science. Many of these volumes will be unfamiliar to practicing historians of science. Elliott points out that though most reference works codify and make accessible biographical, institutional, and other data, their implications are much broader, and he urges attention to methodological issues. He is convinced that the examination and analysis of reference works and aids to documentation are so important that such evaluation should become a speciality in its own right; that is, historical works and historians themselves will become the subject of documentary analysis.

CONCLUSION

When all is read and analyzed, it is evident that there has been a great deal of work in the field of the history of American science since George Daniels hosted a meeting on the subject at Northwestern University in 1970 (and where the two editors of this volume met, in fact, as graduate students). At that time the word "archaeoastronomy" was little known, many archives were not yet opened, and few scholarly inquiries went beyond the 1870s. Fifteen years later so many themes have been explored and monographs written that even larger projects now seem very possible. Because ambitions have broadened and consciousness has been raised, new deficencies are being exposed: in most areas, the coverage has been uneven, gaps remain, and differences in interpretation are too few to merit much discussion. Several essayists lament, as well, the lack of synthetic works.

Nonetheless, there seems to be room for optimism, perhaps for the first time. When Hunter Dupree published his classic essay "American Science: A Field Finds Itself" in 1966, the most he could anticipate was a "corporal's guard" of historians and archivists to tend the records and write occasional reports on the history of American science. Now twenty years later the numbers have risen to the point where, to extend the military metaphor, squadrons or battalions of specialists are emerging. Even if their representatives in this volume seem to be complaining in varying voices about particular weaknesses, such commentary is simply the sound of specialists stretching against their current confinements. Now the mood is no longer one of caution and gradualism but of optimism, as key topics have been identified and the community of interested scholars is expanding, drawing in fresh doctoral candidates as well as enthusiastic internalists who also now trek to the Rockefeller Archive Center and similar sources. There is less hesitation to tackle difficult and complex modern issues. Researchers have more options and more materials with which to work. It is to be hoped that this volume will speed historians on to accomplish aspirations expressed here, and more, in the decade ahead.

APPENDIX: SELECTED READINGS

The following readings represent significant thinking about the state of the history of American science. The individual works, written by observers and advocates of the field over the past half century, are arranged here in chronological order to suggest the changing orientation of those contributing to the subject.

Shryock, Richard. "The Need for Studies in the History of American Science," *Isis,* 1944, *35*:10–13.

Schlesinger, Arthur M. "An American Historian Looks at Science and Technology," *Isis,* 1946, *36*:162–166.

Fulton, John. "The Impact of Science on American History," *Isis,* 1951, *42*:176–191.

Bell, Whitfield. *Early American Science: Needs and Opportunities for Study* (Williamsburg: Institute of Early American History and Culture, 1955).

Cohen, I. Bernard. "Some Reflections on the State of Science in America during the

Nineteenth Century," *Proceedings of the National Academy of Sciences,* 1959, *45*:666–677.

Dupree, A. Hunter. "History of American Science—A Field Finds Itself," *American Historical Review,* 1966, *71*:863–874.

Hindle, Brooke. *Technology in Early America: Needs and Opportunities for Study* (Chapel Hill: Univ. North Carolina Press, 1966).

Van Tassel, David, and Michael G. Hall, eds. *Science and Society in the United States* (Homewood, Ill.: Dorsey, 1966).

Greene, John C. "American Science Comes of Age, 1780–1820," *Journal of American History,* 1968, *55*:22–41.

Daniels, George, ed. *Nineteenth-Century American Science: A Reappraisal* (Evanston, Ill.: Northwestern Univ. Press, 1972).

Bender, Thomas. "Science and the Culture of American Communities: The Nineteenth Century," *History of Education Quarterly,* 1976, *16*:63–77.

Reingold, Nathan. "Reflections on 200 Years of Science in the United States," *Nature,* 1976, *262*:9–13.

Kohlstedt, Sally Gregory. "Reassessing Science in Antebellum America," *American Quarterly,* 1977, *29*:444–453.

Rossiter, Margaret W. "American Science in the 1970s," *Reviews in American History,* 1980, *8*:547–552.

Reingold, Nathan. "Clio as Physicist and Machinist," *Reviews in American History,* 1982, *8*:264–280.

Thackray, Arnold. "On American Science," *Isis,* 1982, *73*:7–10.

Beardsley, Edward H. "The History of American Science and Medicine," in *Information Sources in the History of Science and Medicine* (Boston: Butterworth Scientific, 1983).

Rosenberg, Charles. "Science in American Society: A Generation of Historical Debate," *Isis,* 1983, *74*:356–367.

Institutional History

By Sally Gregory Kohlstedt*

THE STUDY OF INSTITUTIONS is a venerable tradition in the history of
science in the United States. The explanation lies in part in the cultural
patterning of Americans and in part in the assumptions of modern scientists.
From the founding of the new nation, Americans have been preoccupied with
institution building, for reasons that are at once personal, practical, patriotic,
and at the outset intellectually presumptive. Ideas about the value of collective
efforts permeated the republic, which sought to cultivate cultural as well as civic
values in its citizens. An inevitable theme in literature, the arts, and sciences
was that mutual efforts would advance knowledge and general social purposes
more rapidly and more comprehensively than the enterprise of individuals
working alone. Voluntarism was the technique, with governmental support initially sporadic and specific rather than sustaining with regard to activities intended to advance knowledge and expertise. Science in America came of age in
this atmosphere, and practitioners here took their cues from local contemporaries even as they adopted immigrant ideas and experiences, initially from
Britain and France, although very shortly from Germany as well. The tradition
of assuming that institutions shaped the nature of scientific inquiry became pervasive in the lore and history of science in the United States.[1] For this reason,
historians, whether enthusiastic about or skeptical of the purposes and power
of institutions, can seldom avoid institutional developments in studying the history of modern science.

Because many of the early leaders established and then sustained organizations from which their lives and work seemed almost indistinguishable, the line

* History Department, Syracuse University, Syracuse, New York 13210.

The research of this essay was sponsored in part by the Syracuse University Fund for Research
and the National Science Foundation, Grant No. 3535502. An early version was presented in 1984
at the University of West Virginia, which helped sponsor the research.

[1] This theme is developed in Sally Gregory Kohlstedt, "George Brown Goode: The Naturalists'
Tradition in the History of American Science" (Paper presented at the annual meeting of the American Historical Association, 28 Dec. 1984). The term institution, which had earlier religious roots,
became more widely used in secular settings after the establishment of the popular Royal Institution
in Great Britain in 1800. See *Oxford English Dictionary*, (1933), Vol. V, p. 354, s.v. "institution."
Current synonyms include institute, society, academy, association, and organization as well as such
functional terms as library, museum, laboratory, observatory. Surprisingly little has been done to
investigate the ways in which institutional norms might inhibit rather than enhance a researcher's
creative efforts. Such conservative tendencies within the Geological Society of London, however,
are discussed in Rachel Laudan, "Ideas and Organizations in British Geology: A Case Study in
Institutional History," *Isis*, 1977, *68*:527–538.

between biography and institutional history has not been precisely drawn in the history of science. The nineteenth-century naturalist and museum builder Louis Agassiz, according to his biographer, prefigured "the career of types of personality whose original intellectual eminence gave them structure and power to shape institutional change." Other examples abound, such as Joseph Henry, who embodied and articulated the philosophy that guided the Smithsonian Institution while he was its first secretary, a philosophy that combined a commitment to increase abstract knowledge with a sense of responsibility to disseminate that knowledge.[2] Similar observations might be made on Alexander Dallas Bache of the United States Coast Survey, Simon Newcomb at the Naval Observatory, John Wesley Powell with the United States Geological Survey, and a score of others.[3] Whether closely linked or not, both biography and institutional studies offer convenient foci for historical work, as evidenced by the use of those two categories for the *Isis Cumulative Index*. The study of individuals and individual institutions provide, moreover, an intellectual coherence: they have a specific origin and identity, and, often equally important, their activities are likely to be systematically recorded and preserved in a major repository.[4]

Although some institutions have been designed to advance what we identify today as the biological, physical, and social sciences, many have expressed broader aims. This essay, surrounded by others on specific disciplines and topics, will suggest the themes and questions provided by historians who have closely investigated how science was conducted in learned societies and museums, colleges and universities, professional and popular periodicals, and foundations and research facilities. Historians of American science have long identified such institutions as central to their inquiry into the nature of science and its relationship to broader social, political, and economic phenomena, particularly in documenting the powerful influence exerted by urbanization and the communications revolution, the growth of federal government, and the resources generated for research by agricultural and industrial concerns.[5]

---

[2] Edward Lurie, *Louis Agassiz: A Life in Science* (Chicago: Univ. Chicago Press, 1960), p. xi; and Arthur P. Molella and Nathan Reingold, "Theorists and Ingenious Mechanics: Joseph Henry Defines Science," *Science Studies*, 1973, *3*:323–351. See also Nathan Reingold et al., eds., *The Papers of Joseph Henry*, 4 vols. to date (Washington, D.C.: Smithsonian Institution Press, 1972–present).

[3] See Nathan Reingold, "Alexander Dallas Bache: Science and Technology in the American Idiom," *Technology and Culture*, 1970, *11*:163–177; Arthur Norberg, "Simon Newcomb's Early Astronomical Career," *Isis*, 1978, *69*:209–225; and William Culp Darrah, *Powell of the Colorado* (Princeton: Princeton Univ. Press, 1951). An excellent resource on biography, with appendixes indicating career and institutional affiliations, is Clark A. Elliott, *Biographical Dictionary of American Science: The Seventeenth Through the Nineteenth Centuries* (Westport, Conn.: Greenwood, 1979).

[4] A particularly thoughtful piece on biography is Thomas L. Hankins, "In Defence of Biography: The Use of Biography in the History of Science," *British Journal for the History of Science*, 1979, *17*:1–16. The standard directory of institutions is Ralph S. Bates, *Scientific Societies in the United States*, 3rd ed. (Cambridge: MIT Press, 1965). More information on larger institutions currently in operation may be found in Joseph C. Kiger, *Research Institutions and Research Societies* (Westport, Conn.: Greenwood, 1982).

[5] Historians of the early modern period are once again emphasizing the close relationship between the revolution in scientific thought and organized academies or societies; see, e.g., Bruce T. Moran, "German Prince-Practitioners: Aspects in the Development of Courtly Science, Technology, and Procedures in The Renaissance," *Technol. Cult.*, 1981, *22*:253–274. See also, for the early twentieth century an interpretive work on capitalism and engineers in David F. Noble, *America By Design: Science, Technology, and the Rise of Corporate Capitalism* (New York: Knopf, 1977).

OVERVIEW

Institutions have grown at an accelerating rate, becoming both more specialized and more complex in their purposes and structure. Perhaps because early institutions have been less difficult to describe, leaving from the difficulty of limited sources for some, colonial and antebellum organizations have received a large share of attention from historians seeking the foundations of scientific development. As the numbers and diversity of institutions increase in the late nineteenth and twentieth centuries, historians seem less able to encompass larger patterns and rely even more on single case studies as a complement to biographical and thematic research.

During the colonial period, the organizing efforts of a Cotton Mather, a Benjamin Franklin, and an Alexander Garden were typically short-lived and of limited influence when compared to their intellectual productivity. For this reason historians necessarily concentrate on the individuals and their relatively independent results.[6] However, by the nineteenth century the "nation of joiners," as Alexis de Tocqueville called the young country in his widely cited commentary *Democracy in America,* believed that collective associations could help them realize social, political and intellectual goals. High ideals about American opportunities were widespread, and institutions were conceived of as vehicles to achieve such goals as a moral (slave-free, religious, peace-loving) society. For those seeking an intellectually advanced culture, technical and scientific knowledge was an accepted counterpart to literature and the arts. The multiple interests were evident in a proliferation of societies and publications, a phenomenon that has drawn quite varied interpretation.[7] It was also during the nineteenth century that American scientists coordinated and publicly justified their work as part of an effort to establish places in which to pursue it. Contemporary accounts describe institutions in detail because reporters themselves were self-consciously building new organizations and vehicles for publication as they sought to earn livings in government, academic, or learned settings. Not surprisingly, their reflections on the past, typically eulogies written for colleagues and anniversary commemorations of societies, stressed the kind of activities which they themselves valued. Concentrating on achievements rather than process and problems, these laudatory accounts provide helpful materials but lack critical explanations for historians today.

[6] The most prominent American scientists are included in the *Dictionary of Scientific Biography,* ed. Charles Gillispie, 16 vols. (New York: Scribners, 1970–1980). On Franklin's scientific work see I. Bernard Cohen, *Franklin and Newton: An Inquiry into Speculative Newtonian Experimental Science and Franklin's Work in Electricity as an Example Thereof* (Cambridge, Mass.: Harvard Univ. Press, 1966).

[7] Irving Bartlett, *The American Mind in the Mid-Nineteenth Century* (Northbrook, Ill.: AHM Press, 1967), places these activities, including science, in the context of a democratic spirit without losing sight of the strains and ambiguities of the period. Peter Dobkin Hall, *The Organization of American Culture, 1700–1900: Private Institutions, Elites, and the Origins of American Nationality* (New York: New York Univ. Press, 1982), instead stresses a mid-century crisis of authority and subsequent efforts to reestablish a coherent and cohesive link between knowledge and elites. George Frederickson, *The Inner Civil War: Northern Intellectuals and the Crisis of the Union* (New York: Harper & Row, 1965), infers from the rapid demise of many institutions that the antebellum mood was anti-institutional; an equally plausible explanation for the rapid turnover but steady growth of institutions, however, is that organizers were experimental, even pragmatic, in their efforts to find techniques to deal with their increasingly complex urban and industrial society.

Detailed descriptions of various institutions by participants, interwoven with accounts of their lives, reached a new level in the historical surveys of George Brown Goode in the 1880s. Initially his accounts detailed the lives of major figures in American natural history as well as in physics, astronomy, and other sciences. Goode's career at the Smithsonian Institution, which included supervising its display at the Philadelphia Centennial Exposition in 1876, made him particularly sensitive to the role of institutions in sponsoring as well as in diffusing scientific knowledge. The Smithsonian, after all, assumed the connotations of permanence and authority implied in the title of "institution." An advocate of institutional expansion, Goode had broad ideas about the functions of the American Historical Association as well. As a member of its Council, he helped arrange for the AHA offices to be located at the Smithsonian "castle," and he used that organization and others as a forum for his ideas about American science. Although a naturalist, in his historical writings he emphasized the evolution of scientific and educational institutions much as the professional historians at Johns Hopkins University concentrated on similar themes in their studies of political and social institutions. His wide-ranging overview also emphasized the relationship between government and science and characterized the emerging federalism as important in the process of modern scientific investigation. His analysis of nineteenth-century science stressed the importance but also the limitations of individual and private efforts, and his historical work reflected the naturalists' traditional enthusiasm for biography.[8] After Goode there was a relative hiatus of work on history of American science, although certain basic reference works such as Max Meisel's bibliography on natural history demonstrated the possibilities which existed for such study.

In the past three decades there has been rapidly growing enthusiasm for the study of science in America, and results suggest that while the traditional methods have not been much discussed, their influence persists. Only recently, moreover, have bibliographies, narratives, biographies, institutional studies, or thematic inquiries on the history of science in the United States reflected trends in social and cultural history. A few historians have begun to look at the amateur tradition of American science as well as at professionals; at the dynamics within the scientific community; and at the rhetoric and ideology of scientists, their promoters, and their detractors. Most of the current work, however, merely sketches the terrain, using institutions as markers and identifying significant social forms upon which more interpretative studies may be based. What follows is a topographical reconnaissance that highlights the institutions that have received attention from scholars. Possibilities for research are suggested throughout, while the conclusion focuses on methods and approaches.

[8] Goode's major essays are reprinted in Smithsonian Institution, *Memorial of George Brown Goode, Together with a Selection of his Papers on Museums and on the History of Science in America,* (Annual Report for 1897, Part 2) (Washington, D.C., 1901). Other significant reference works included George P. Merrill, *Contributions to a History of American State Geological and Natural History Surveys* (U.S. National Museum Bulletin 109) (Washington, D.C.: National Museum, 1920), and Max Meisel, *A Bibliography of American Natural History; The Pioneer Century, 1769–1865,* 3 vols. (Brooklyn: The Premier Pub. Co., 1924–1929). Aside from sketches by contemporaries, there is relatively little work on Goode; see Carroll Lindsay, "George Brown Goode," in *Keepers of the Past,* ed. Clifford L. Lord (Chapel Hill: Univ. of North Carolina Press, 1965), pp. 127–140, and discussion in Edward P. Alexander, *Museum Masters: Their Museums and Their Influence* (Nashville: American Association for State and Local History, 1983), pp. 277–310.

LEARNED AND PROFESSIONAL ORGANIZATIONS

The insistence that institutions are the essential foundation for scientific work begins, with negative evidence, in the histories of the colonial period. Raymond Stearns's *Science in the British Colonies of America,* winning recognition for the field by achieving a National Book Award in 1971, stresses the limits of local communication and the importance of transatlantic correspondence networks that sustained those who studied the natural sciences in the distant and intellectually challenging New World. Because there were so few interested naturalists in the colonies, results of investigation here were presented through the British Royal Society and might eventually find their way into the society's *Transactions.* But the very efforts to organize alluded to above—of Cotton Mather in Boston, Benjamin Franklin in Philadelphia, and Alexander Garden in Charleston—despite their limited success demonstrated provincial aspirations. The arrival of visitors became an excuse to meet and to create, even temporarily, a community among like-minded naturalists and philosophers. This phenomenon would be repeated on subsequent frontiers across the country, although studies of regional science have been relatively rare and the subject invites investigation.[9]

The apparently pervasive incentive to share information, books, philosophical apparatus, and duplicate specimens led to more formal learned societies, most of which were sufficiently ecumenical in their purposes to embrace a wide spectrum of intellectual interests. By the time of the Revolution, Brooke Hindle argues, the library companies and philosophical societies appearing in Boston, New York, Philadelphia, and Charleston were as dedicated to science as to the arts.[10] Patriotic zeal and very real economic competition also provided incentives to build libraries, natural history collections, and physical science apparatus which would establish literary and scientific studies on the eastern shore of the Atlantic. Individual efforts figure prominently in Stearns's and Hindle's accounts, and both historians justify institution building in the personal and patriotic terms put forward by contemporaries.

John C. Greene's richly detailed *American Science in the Age of Jefferson* sustains the approach developed by Stearns and Hindle. His first five chapters develop what he terms the "traditional view" that science requires an institutional base in order for individuals to achieve sustained research results. Responding to scholars who have distinguished among class interests, political affiliation, and even the religious orientation of scientists, Greene concludes that there was a rather "astonishing unity of attitude and purpose" in spite of such differences. The institutions built in the early nineteenth century, he contends, were remarkably similar and rivaled each other on regional rather than on

---

[9] Raymond Stearns, *Science in the British Colonies of America* (Urbana: Univ. Illinois Press, 1970), p. xiv. Stearns uses the American fellows of the Royal Society as the basis for his collective analysis and argues that most of the promoters of investigation were savants in London, with a few on the Continent. His emphasis on the Anglo-American connection is appropriate but deserves reinvestigation with the use of newer approaches. A recent promising beginning on regional science is represented in the First Bernard-Millington Symposium on Southern Science and Medicine, "Science in the Old South," held in March of 1982 and whose proceedings are available in typescript from the Center for the Study of the South, University of Mississippi, Oxford, Miss.

[10] Brooke Hindle, *The Pursuit of Science in Revolutionary America, 1735–1789* (Chapel Hill: Univ. North Carolina Press, 1956).

philosophical grounds. Explaining the subsequent troubles of many short-lived groups, however, Greene approaches the older theme that the general public was not interested in scientific work and that such indifference hindered those building new institutions during the first half of the nineteenth century.[11]

In a now classic (and perhaps for that reason contested) essay, Richard Shryock maintained that this lack of sufficient public support undermined early scientific efforts in America, forcing the establishment of multiple private institutions as well as the concentration on applied science that seemed to characterize antebellum science. Recent respondents have explored how science and applied sciences were linked in the minds and efforts of early practitioners and document how much, in fact, was accomplished. The question of whether multiple and competing organizations reflected weakness or strength, however, remains unsettled.[12]

The questions concerning the strength and significance of scientific study as well as the search for explanations of the visibility and authority of science have, however, revived and revised older studies of institutions. Earlier scholars, using Max Meisel's classic bibliography from the 1920s, provided substantial evidence of efforts to establish scientific societies and journals—while demonstrating the short-lived nature and tentative successes of even the best-founded. Recent accounts, however, no longer assume that the transitional nature of such groups demonstrates a larger anti-institutional bias or indifference among contemporaries. Nor do they follow the themes of revisionists writing on reform movements, mental asylums, and public schools, and argue that the learned societies struggled because they countered democratic ideals by establishing bastions of privilege and agencies of social control.[13] Most frequently these new studies assert that the ambitions of more ephemeral institutions simply exceeded their means of support. In this context the rapid growth, intermittent activities and publications, and networks of communication are important for the process of establishing science, whatever a particular institution's problems might be. Certainly data in the essays in a volume sponsored by the American Academy of Arts and Sciences, *The Pursuit of Knowledge in Early America,* indicate that individuals frequently moved from institution to institution, taking with them into regional, national, specialized, or career organizations many practices and incentives derived from apparently defunct institutions.[14]

[11] John C. Greene, *American Science in the Age of Jefferson* (Ames: Iowa State Univ. Press, 1984). An earlier treatment is Greene, "Science, Learning, and Utility: Patterns of Organization in the Early American Republic," in *The Pursuit of Knowlege in the Early American Republic: American Learned and Scientific Societies from Colonial Times to the Civil War,* ed. Alexandra Oleson and Sanborn C. Brown (Baltimore: Johns Hopkins Univ. Press, 1976).

[12] See Richard H. Shryock, "American Indifference to Basic Science During the Nineteenth Century," *Archives internationales d'histoire de science,* 1948, *28*:50–65; and I. B. Cohen, *Science and American Society in the First Century of the Republic* (Columbus: Ohio State Univ. Press, 1961); see also Nathan Reingold, "American Indifference to Basic Research: A Reappraisal," *Nineteenth-century American Science: A Reappraisal* (Evanston: Northwestern Univ. Press, 1972); Reingold ed., *Science in Nineteenth-century America: A Documentary History* (New York: Hill & Wang, 1964); and Reingold and Arthur Molella, "Theorists and Ingenious Mechanics: Joseph Henry Defines Science," *Science Studies,* 1973, *3*:323–351.

[13] For revisionist thinking see the examples given in Joseph Kett, "On Revisionism," *The History of Education Quarterly,* 1979, *19*:229–235; for a hint of this attitude see the discussion of agricultural experimentalists in Charles Rosenberg, *No Other Gods: On Science and American Social Thought* (Baltimore: Johns Hopkins Univ. Press, 1976), pp. 135–152.

[14] See *Pursuit of Knowledge,* ed. Oleson and Brown (cit. n. 11); and Walter B. Hendrickson,

As historians study individuals and institutions in detail, they also discover that contemporaries had a very complex view of theoretical and utilitarian science.[15] Particularly revealing is Bruce Sinclair's suggestively entitled *Philadelphia's Philosopher Mechanics,* which demonstrates the orientation of newer institutional studies. Sinclair explains his concentration on the Franklin Institute in institutional terms: "Institutions are particularly valuable subjects for analysis. They reflect the rhetoric which surrounds self-conscious social inventions; and their successes and failures reveal what lies behind pronouncements."[16] In other words, institutions expose much about a culture, but the assessments by participants must be approached cautiously. Rhetoric may suggest—but may also mask—the intentions of institution builders. Sinclair's emphasis is on the *process* of institutional development. The Franklin Institute was directed by influential manufacturers and political leaders, but that fact did not make it a static organization compelled by its founders' vision. Rather, it existed in a climate of competing aspirations and expectations. Philadelphia's working classes had independent ideas about the aims and practical programs of the educators, and when not consulted directly, they expressed their opinions by their attendance. Moreover, the research activities sponsored at the Franklin Institute provide evidence about the relationship between practical and abstract knowledge (and the aspiration reflected in the term philosopher mechanics) during precisely those years when the term technology was coined. Sinclair therefore uses the microscopic approach, focusing closely on one carefully selected organization during a relatively short period of time, in order to explore a number of issues in the history of science and technology.

Nineteenth-century scientists watched with interest the professionalizing activities in such neighboring fields as law, medicine, and education. Critical of what they termed amateur science or even pseudoscience and simultaneously eager to establish a public forum and a coordinated voice for lobbying, scientists organized the American Association for the Advancement of Science, which they anticipated would contain and constrain a "national scientific community." The results did not meet the private hopes of the best placed and most productive scientists, who organized the National Academy of Sciences during the course of the Civil War and, in the following decades, other more specialized societies. A history of the National Academy has recently appeared but its significance for its members, who were publicly and privately powerful in Washington intellectual circles, has yet to be explored in detail.[17]

"Science and Culture in the American Middle West," *Isis,* 1973, *64:*326–340, rpt. in *Science in America Since 1820,* ed. Nathan Reingold (New York: Science History, 1976). A number of recent analyses of British organizations offer insights useful to students of American scientific institutions; see esp. Jack Morrell and Arnold Thackray, *Gentlemen of Science: Early Years of the British Association for the Advancement of Science* (Oxford: Clarendon, 1981); and S. F. Cannon, *Science in Culture: The Early Victorian Period* (New York: Science History, 1978).

[15] On empiricist, "Baconian" science see George Daniels, *American Science in the Age of Jackson* (New York: Columbia Univ. Press, 1968).

[16] Bruce Sinclair, *Philadelphia's Philosopher Mechanics: A History of the Franklin Institute, 1824–1865* (Baltimore: Johns Hopkins Univ. Press, 1974), p. ix.

[17] Sally Gregory Kohlstedt, *The Formation of the American Scientific Community: The American Association for the Advancement of Science, 1848–1860* (Urbana: Univ. Illinois Press, 1976). See George H. Daniels, "The Process of Professionalization in American Science: The Emergent Period, 1820–1860," *Isis,* 1967, *58:*151–166; Nathan Reingold, "Definitions and Speculations: The Professionalization of Science in America in the Nineteenth Century," in *The Pursuit of Knowledge,* ed.

The claims that scientists made for their methodology and objectivity increasingly began to interest reformers and intellectuals seeking to study human society more systematically. The American Social Science Association exemplifies the institutionalization of this effort. Thomas L. Haskell traces the dynamic process of institutional development, concentrating on how changing norms and powerful individuals account for the direction taken by the ASSA. Historians of science have not yet analyzed in any comprehensive way the rise of specialized societies. The AAAS provided a convenient mechanism for some specialties to formulate new societies through its sectional meetings, but it may have also retarded that movement toward autonomy. Ralph Bates assumes a positive influence, but the later years of the AAAS have not yet been analyzed in any detail, nor have there been adequate historical analyses of other umbrella organizations—such as the American Council of Learned Societies, the Association of the Academies of Sciences, and the American Association of Museums—to establish their effect. Residual efforts to maintain coordination across disciplinary lines and among those inside and outside research universities in such groups as the American Society of Naturalists remind historians of the cross-currents affecting the course of events—too often described as a straight and narrow channel.[18]

Although the distinction between professional and amateur became more precise, in some disciplines there was also an accommodation between vocational and avocational practitioners, particularly when data could be gathered on a massive scale by observant amateurs, as in ornithology and, for a time, in astronomy.[19] The complex means by which institutions resolved the "threats" of amateurism have received little attention in comparison with the often narrowly defined studies of the progress of specialties, which depend heavily on the records of a few prominent officers. Nonetheless, general discussions tend to agree that the specialization (or fragmentation) of knowledge within science "required a new set of institutional arrangements in post-Appomattox America."[20]

Institutional strength may, or may not, be an indicator of the strength of studies in a discipline or locality. Some scholars have suggested that professional groups are even an explanation of strength. Daniel Kevles points out in *The Physicists*, for example, that the relative differences in the size and strength of the fields of geology and physics at the turn of the century may be attributed

Oleson and Brown (cit. n. 11), pp. 33–69; Rexford Canning Cochrane, *The National Academy of Sciences: The First Hundred Years, 1863–1963* (Washington: National Academy of Sciences, 1978); and A. Hunter Dupree, "The National Academy of Sciences and American Definitions of Science," in *The Pursuit of Knowledge*, ed. Oleson and Brown, pp. 33–69.

[18] Thomas Haskell, *The Emergence of Professional Social Science: The American Social Science Association and the Nineteenth-Century Crisis of Authority* (Urbana: Univ. Illinois Press, 1977). Also see Mary O. Furner, *Advocacy and Objectivity: A Crisis in the Professionalization of American Social Science, 1865–1905* (Lexington: Univ. Kentucky Press, 1975); and for some discussion of the ASN, Hamilton Cravens, *The Triumph of Evolution: American Scientists and the Heredity-Environment Controversy, 1900–1941* (Philadelphia: Univ. Pennsylvania Press, 1978), pp. 28–29.

[19] See, e.g., Marianne Gosztonyi Ainley, "The Contribution of the Amateur to North American Ornithology: A Historical Perspective," *The Living Bird*, 1979–1980, *18*:161–177; John Lankford, "Amateurs Versus Professionals: The Controversy over Telescope Size in Late Victorian Science," *Isis*, 1981, *72*:11–28; and Marc Rothenberg, "Organization and Control: Professionals and Amateurs in American Astronomy, 1899–1918," *Social Studies of Science*, 1981, *11*:305–325.

[20] Hamilton Cravens, "American Science Comes of Age: An Institutional Perspective, 1850–1930," *American Studies*, 1976, *17*:49–70; and Daniel J. Kevles, Jeffrey L. Sturchio, and P. Thomas Carroll, "The Sciences in America, circa 1880," *Science*, 1980, *209*:27–32.

to their institutional bases, which he takes to include not simply the voluntary and professional associations but also government sponsorship and facilities in universities and other research settings.[21] Whether historians assess growth in fields over time or look across several disciplines at any given point in time, membership and leadership lists, budgets, and recorded activities provide useful and reasonably comparable data.

Admitting that institutions are vehicles for change permits closer investigation of *how* scientists and the public have used them. Certainly organizations provide a forum for "public scientists" to argue the cause of a particular scientific project before a general audience, and these same organizations sometimes acknowledge that the public has the right to respond as well. The Seismological Society of America, according to Arnold Meltsner, disseminated public information about predictive data both because the members believed the public was not sufficiently informed of local risks and because they hoped to gain more public financial support for research. Studies of specialized groups are also likely to reveal substantial discontinuities between local affiliates and national organizations.[22]

Similarly the issues of roles and ranking have often depended on analysis of institutions. Initially discussed in connection with the catch-all terms of *amateur* and *professional,* roles and ranking are now recognized as having more subtle refinements.[23] The tug-of-war for power (or autonomy) was an indication of the ideology and the strength of groups that competed within the same organization or sought to establish the claims of their institution over that of another. Amateurism persisted and established its place in America by the end of the nineteenth century in connection with educational and leisured activities. Despite disapproval of the experts, many practitioners maintained their interests in such "pseudosciences" as phrenology, spiritualism, and animal magnetism.[24] Moreover, studies of those who worked hard to establish credentials and other attributes of professionalism did not always meet their own high ideals of disinterested research and open access, particularly when dealing with women and minorities. Margaret Rossiter's broad-based analysis of women in all fields of science is already a classic, and Kenneth Manning's excellent study of a single

[21] Daniel Kevles, *The Physicists: The History of a Scientific Community in Modern America* (New York: Knopf, 1978), p. 37. On professionalization more generally, see Burton Bledstein, *The Culture of Professionalization: The Middle Class and the Development of Higher Education in America* (New York: Norton, 1976); and a rather different set of perspectives presented by Thomas Bender, Thomas Haskell, David Hollinger, and Dorothy Ross in *The Authority of Experts: Studies in History and Theory,* ed. Thomas Haskell (Bloomington: Indiana Univ. Press, 1984).

[22] For the term *public scientist,* see Frank M. Turner, "Public Science in Britain, 1880–1919," *Isis,* 1980, 71:589–608. For a political scientist with both intellectual and organizational skills, see Barry D. Karl, *Charles E. Merriam and the Study of Politics* (Chicago: Univ. Chicago, 1974); on the SSA, see Arnold Meltsner, "The Communication of Scientific Information to the Wider Public: The Case of Seismology in California," *Minerva,* 1979, 17:331–354. British commonwealth historians of science are particularly interested in describing the distinction between "provincial" (or colonial) and "cosmopolitan" science; see, e.g., Stephen Hill, "Questioning the Influence of a 'Social System of Science': A Study of Australian Scientists," *Sci. Stud.,* 1974, 4:135–163.

[23] Nathan Reingold, "Definitions and Speculations: The Professionalization of Science in America in the Nineteenth Century," in *Pursuit of Knowledge,* ed. Oleson and Brown (cit. n. 11), pp. 33–69.

[24] See, e.g., R. Laurence Moore, *In Search of White Crows: Spiritualism, Parapsychology, and American Culture* (Oxford Univ. Press, 1977); and Herbert Leventhal, *In the Shadow of the Enlightenment: Occultism and Renaissance Science in Eighteenth-Century America* (New York: New York Univ. Press, 1976).

black man demonstrates very effectively the reality of discrimination and employment segregation.[25]

Once established in the late nineteenth century, professional organizations became a forum for the "public scientists" who sought to influence public opinion.[26] Recognition gave them a role in the building of urban culture as well as a voice in regional and national politics. By the end of the century, museums for natural history and technology were only the most publicly visible component of reconstituted intellectual communities that exercised considerable local influence. This was particularly evident in Washington, D.C., where the number of government employees working on projects related to science had increased substantially, although they have not been counted on this basis.[27] Other cities like New York and Chicago had similar core networks through which recognized scientists coordinated with philanthropists, academics, entrepreneurs, and institutional builders.[28] Learned societies have remained significant, holding annual meetings, publishing premier journals, and providing a forum for certain debates. In the twentieth century, however, their functions have been circumscribed by other publicly visible and financially powerful institutions, both private and governmental.

MUSEUMS

Natural history collections were based in substance and in theory on a long-standing tradition. Eighteenth-century learned societies collected books and objects indiscriminately, unwilling to rely simply on "authority" for explanations of the natural world. By the middle of the following century, the museum components of both general and specialized societies were either dropped entirely as logistically and financially problematical, as at the New York Academy of Sciences, or made the focus of attention, as at the Boston Society of Natural History. Because the correspondence and records indicating the source and geographical origin of specimens are crucial to taxonomists, museum archives have typically been maintained in conjunction with specimens themselves and thus provide a rich and still surprisingly underutilized resource for historians of science.[29] Early museums, particularly that of Charles Willson Peale, have recently

[25] Margaret Rossiter, *Women Scientists in America: Struggles and Strategies to 1940* (Baltimore: Johns Hopkins Univ. Press, 1982); and Kenneth Manning, *Black Apollo of Science: The Life of Ernest Everett Just* (New York: Oxford Univ. Press, 1983); on women see also essays by Sally Gregory Kohlstedt, Donna Harraway, Michele Aldrich, and others in a special issue "Women and Science" in *Signs,* 1978, *4.*

[26] See the public lectures reprinted in Robert H. Kargon, ed., *The Maturing of American Science: A Portrait of Science in Public Life Drawn from the Presidential Addresses of the American Association for the Advancement of Science 1920–1970* (Washington, D.C.: AAAS, 1974).

[27] J. Kirkpatrick Flack, *Desideratum in Washington: The Intellectual Community in the Capital City, 1870–1900* (Cambridge: Schenkman, 1975); and Michael James Lacey, "The Mysteries of Earth-Making Dissolve: A Study of Washington's Intellectual Community and the Origins of Environmentalism in the Late Nineteenth Century" (Ph.D. diss., George Washington Univ., 1979).

[28] Thomas Bender, "Science and the Culture of American Communities: The Nineteenth Century," *Hist. Educ. Quart.,* 1976, *16*:63–77, and Douglas Sloan, "Science in New York City, 1867–1907," *Isis,* 1980, *71*:35–76.

[29] See Sally Gregory Kohlstedt, "From Learned Society to Public Museum: The Boston Society of Natural History," in *The Organization of Knowledge in Modern America, 1860–1920,* ed. Alexandra Oleson and John Voss (Baltimore: Johns Hopkins Univ. Press, 1979), pp. 386–406; Smithsonian Institution, *Guide to the Smithsonian Archives* (1983); Venia Phillips, *Guide to the Microfilm*

An idealized sketch of the Peabody Academy of Science, under Frederic Ward Putnam. From Frank Leslie's Illustrated Newspaper (*4 September 1869*), *courtesy of the Peabody Museum, Salem, Massachusetts.*

received considerable attention, from historians of both science and art. That situation is likely to continue, given the recent availability of Peale manuscripts on microfiche and in print. Peale's popularity (there were over 10,000 visitors to the Philadelphia museum in 1816) and the significance of his collections for early writers, such as the ornithologist Alexander Wilson, are increasingly evident. A number of efforts to establish similar museums, however, including several attempts by Peale's own sons, were more ephemeral. Historians have yet to explain the apparent slide into "humbug" of the various Peale museums, not to mention enterprises from Charleston to St. Louis in the 1840s.[30] Here, too, the distance in mid century between self-defined experts and those involved in "popular science" has yet to be explored.

Despite the richness of their archives, surprisingly few serious historical

Publication of the Minutes and Correspondence of the Academy of Natural Sciences of Philadelphia, 1812–1942 (Philadelphia: ANS, 1967); and for the historical uses of museum records (based on Australian materials but applicable to the American experience as well), see Sally Gregory Kohlstedt, "Historical Records in Australian Museums of Natural History," in *Australian Historical Bibliography* (Bulletin 10 of the Reference Section of Australia 1788–1988: A Bicentennial History) (Sydney: Univ. New South Wales, 1984).

[30] See Charles Coleman Sellers, *Mr. Peale's Museum: Charles Willson Peale and the First Popular Museum of Natural History and Art* (New York: Norton, 1980); for a detailed finding aid see Lillian B. Miller, ed., *The Collected Papers of Charles Willson Peale and His Family: A Guide and Index to the Microfiche Edition* (Millwood, N. Y.: Kraus, 1980); see also Miller, ed. *The Selected Papers of Charles Willson Peale and His Family,* Vol. I (New Haven: Yale Univ. Press, 1983); on other museums of the early period see Walter Muir Whitehead, ed., *A Cabinet of Curiosities: Five Episodes in the Evolution of American Museums* (Charlottesville: Univ. Virginia, 1967), esp. the essay by Louis Leonard Tucker.

studies of natural history museums have been written. The "nation's attic" in Washington has received the most attention, in part because early administrators like Joseph Henry and Spencer F. Baird were so influential in other institutional and scientific developments and in part because the Smithsonian Institution is publicly and politically visible in the capital. For similar reasons of personality, geography, and financial support, the American Museum of Natural History in New York is the subject of several accounts, with particular attention being paid to its political connections and philanthropic supporters.[31] Recent work on museums promises to expand in two distinct directions. The first is leading historians of various disciplines and specialties to work on departments in museums, particularly those that have their roots in these institutions. The second trend is toward examination of the relationship between science and popular culture, and, at various levels, the role of education in this relationship. Despite their apparent divergence, both trends reflect the tendencies that persist in modern museums and provide a continuing source of tension.[32]

Similar issues exist in varying degrees within such publicly sponsored or endowed institutions as botanical gardens, arboretums, zoos, laboratories, and astronomical observatories. Their definitions and directions, too, have changed in the past two hundred years. The manuscripts of such natural science organizations as the New York Botanical Garden in Brooklyn and the Hunt Botanical Institute in Pittsburgh include significant, organized, and still underutilized records, as do those of a number of physics laboratories discussed elsewhere in this volume.[33]

PUBLICATIONS

Like museums, scientific journals struggled to balance the goal of publishing for a learned community with the need to obtain broad financial support. Scientific publication has not as yet received much scholarly attention, despite its impor-

---

[31] Most accounts of the Smithsonian are intended for a popular audience, although some are based on substantial research as well; see, e.g., Paul H. Oehser, *Sons of Science: The Story of the Smithsonian Institution and Its Leaders* (New York: Schuman, 1949); Philip Kopper, *The National Museum of Natural History* (New York: Abrams, 1982); and on architectural themes Kenneth Hafertepe, *America's Castle: The Evolution of the Smithsonian Building and Its Institution, 1840–1878* (Washington, D.C.: Smithsonian Institution Press, 1984). The political and philanthropic dimensions of the Museum of Natural History are the focus of John M. Kennedy, "Philanthropy and Science in New York City: The American Museum of Natural History, 1868–1968" (Ph. D. diss., Yale Univ., 1968); and of the popularly written book by Geoffrey Hellman, *Bankers, Bones and Beetles: The First Century of the American Museum of Natural History* (Garden City, N.Y.: Natural History Press, 1969).

[32] Curtis J. Hinsley, Jr., *Savages and Scientists: The Smithsonian Institution and the Development of American Anthropology, 1846–1910* (Washington, D.C.: Smithsonian Institution Press, 1980); and Joseph T. Gregory, "North American Vertebrate Paleontology, 1776–1976," in *Two Hundred Years of Geology in America: Proceedings of the New Hampshire Bicentennial Conference on the History of Geology,* Cecil Schneer, ed. (Hanover: Univ. New England Press, 1979). On popular culture see Nancy Oestereich Lurie, *A Special Style: The Milwaukee Public Museum, 1882–1982* (Milwaukee: Milwaukee Public Museum, 1984), which offers a thoughtful account of that institution. On some parallel interests of college and public museums see Sally Gregory Kohlstedt, "Henry A. Ward: The Merchant Naturalist and American Museum Development," *Journal of the Society for the Bibliography of Natural History,* 1980, 9:647–661. The persistent tension is reflected in the new journal *Museums Studies,* which contains historical essays in every issue and also documents continuing efforts to apportion staff and financial resources between research and education.

[33] On zoos see Helen L. Horowitz, "Animal and Man in the New York Zoological Park," *New York History,* 1975, 55:426–455.

tance for scientific communication. Compiled before much work had been done on the history of science in America, the important preliminary survey of bibliographer Frank Mott is no longer current and is misleading in some details.[34]

Most work on the subject has tended to look at scientific journals primarily as a function of their editors or publishers rather than to explore how contributors and readers may have influenced the journals' content—admittedly a more elusive problem. Accounts of several long-lived and intellectually comprehensive journals, such as *The American Journal of Science, Scientific American, Popular Scientific Monthly,* and *Science,* suggest that successful publications had editors who were well-connected within the scientific community, which valued the enterprise and in return provided information and even contributed articles.[35] Journals with more specialized readers are being studied as a way to describe and explain the emergence of recent scientific specialties, particularly for the modern period, where citation and other indexes are available. Too often, however, the results have been used as explanations of rather than simply as evidence for the development of a discipline.[36]

Many journals were published by learned societies and by independent publishers who also produced monograph series, textbooks, and more ephemeral pamphlet literature. Moreover, many publications required collaborative effort involving naturalists, illustrators, and patrons.[37] Particularly revealing about the dissemination of scientific knowledge are the textbooks and manuals that appeared throughout the nineteenth century. Most major publishers produced scientific series, with volumes written by educators, including many women from the increasingly feminized profession, as well as by freelance writers and naturalists. The changing demands made on these writers, their point of view on scientific methodology, and the attention they paid (or did not pay) to the issue of religion would certainly reveal much about what was being taught and discussed in schools and local nature clubs.

[34] Frank Luther Mott, *A History of American Magazines,* 5 vols. (Cambridge: Belknap Press of Harvard Univ. Press, 1930–1968).

[35] Those accounts that do exist are often written or sponsored by the publications and seem determined to document the case that publications were central both to the development of particular disciplines and to national scientific efforts. See, e.g., Edward S. Dana et al., *A Century of Science in America, with Special Reference to the American Journal of Science, 1818–1918* (New Haven: Yale Univ. Press, 1918); for the more recent period, see John Hammond Moore, *Wiley: One Hundred and Seventy Five Years of Publishing* (New York: Wiley, 1982), an "official" history written for a general audience. See also Robert Post, "Science, Public Policy, and Popular Precepts: Alexander Dallas Bache and Alfred Beach as Symbolic Adversaries," in *The Sciences in the American Context: New Perspectives,* ed. Nathan Reingold (Washington, D.C.: Smithsonian Institution Press, 1979), pp. 77–98; William E. Leverette, Jr. "E. L. Youmans' Crusade for Scientific Autonomy and Respectability," *American Quarterly,* 1965, *17*:12–32; and Sally Gregory Kohlstedt, "*Science:* The Struggle for Survival, 1880–1894," Michael Sokal, "*Science* and James McKeen Cattell, 1894–1945," and John Walsh, *Science* in Transition, 1948–1962," all in *Science,* 1980, *209*:33–57. A particularly useful effort to identify the structure and appeal of popular science is Donald Zochert, "Science and the Common Man in Ante-Bellum America," *Isis,* 1974, *65*:448–473.

[36] One research technique is demonstrated in Ronald Tobey, "Methodological Appendix," in *Saving the Prairies: The Life Cycle of the Founding School of American Plant Ecology* (Berkeley: Univ. California Press, 1982), pp. 223–247. One of the early important recommendations for using citation indexes came from Derek J. de Solla Price, "Networks of Scientific Papers," *Science,* 1965, *149*:510–515. An example of this technique is Martha Chappell Dean, "The Evolution of Experimental Operant Psychology: Quantitative Analysis of 'Progress' in Behavioral Science" (Ph.D. diss., Syracuse Univ., 1980).

[37] See Charlotte Porter, "The Eagle's Nest: Natural History and American Ideas, 1812–1848," (Univ. Alabama Press, forthcoming; based on Ph.D. diss., Harvard Univ., 1978).

EDUCATION

Developments in academic institutions and changing expectations regarding advanced education in the United States both contributed to the rationale that established university claims for hegemony regarding the discovery and diffusion of knowledge between the Civil War and World War I. Stanley Guralnick has demonstrated how early and systematically science found a place in leading men's colleges, successfully establishing a claim for their aspirations in science.[38] But the task of building up the place of science in the college curriculum, an effort that led to the founding of the Sheffield and Lawrence scientific schools in the middle of the nineteenth century, though significant, merely formed the backdrop for the more dramatic place assumed by science in the increasingly research-oriented universities. For the most part, however, historians focusing on disciplines have concentrated on administrative advocacy and on the initial transfer of the laboratory ideal rather than on the practical implementation of these ideals. Research on the twentieth century concentrates typically on the origins of particular universities or even of individual departments, but the patterns among the universities have yet to be discussed comprehensively. The somewhat different pattern of teaching the applied sciences that occurred in the land-grant colleges and universities has also to date been treated only in histories of individual institutions.[39]

Other evidence suggests that the concern for science education reached beyond higher education, into the academies and teachers' colleges and even into the secondary and primary schools. Deborah Warner has argued that leading female academies, unconstrained by the need to provide a classical precollegiate curriculum, were at least as well equipped with scientific apparatus and instructors as their fraternal counterparts in the middle of the century. Margaret Rossiter has documented the productivity of the (usually female) faculty at the women's colleges, where, despite heavy teaching loads and limited resources, they established a legacy of women pursuing scientific careers. In fact, in the late nineteenth century there were many informal ways to acquire scientific training, including short-term institutes, evening and weekend classes, summer institutions, and various local and national societies. Others have found in the public institutions, particularly in the institutions devoted to teacher training known as normal schools, a curriculum that contained chemistry, botany, and other scientific subjects.[40]

---

[38] Edward Shils, "The Order of Learning in the United States from 1865 to 1920: The Ascendency of the Universities," *Minerva*, 1978, *16*:159–195; Stanley Guralnick, "Sources of Misconception on the Role of Science in the Nineteenth-Century American College," *Isis*, 1974, *65*:352–366; and Guralnick, *Science and the Ante-Bellum American College* (Memoirs of the American Philosophical Society, 109) (Philadelphia: APS, 1975).

[39] On the research ideal, see esp. Lawrence Vesey, *The Emergence of the American University* (Chicago: Univ. Chicago Press, 1965); Hugh Hawkins, *Pioneer: A History of the Johns Hopkins University, 1874–1899* (Ithaca: Cornell Univ. Press, 1960); and Hawkins, *Between Harvard and America: The Educational Leadership of Charles W. Eliot* (New York: Oxford Univ. Press, 1972); on science-based institutions see Robert H. Kargon, "Temple to Science: Cooperative Research and the Birth of the California Institute of Technology," *Historical Studies in the Physical Sciences,* 1977, *8*:3–31; and John W. Servos, "The Industrial Relations of Science: Chemical Engineering at MIT, 1900–1939," *Isis*, 1980, *71*:531–549; on land grant institutions see, e.g., Winton U. Solberg, *The University of Illinois, 1867–1894; an Intellectual and Cultural History* (Urbana: Univ. Illinois Press, 1968).

[40] Deborah Warner, "Science Education for Women in Antebellum America," *Isis*, 1978, *69*:58–

There may well have been political and professional penalties to pay for such a nonstandard education. What happened to this preparation in school settings or in the world of work has remained largely unexplored, with the exception of a thoughtful account of elementary arithmetic.[41] One place to begin research would be with the annual reports of the Commissioner of Education, a position held in the late nineteenth century by philosopher William T. Harris, which were surprisingly attentive to if uneven in their coverage of science education. Participants in the annual meetings of the National Education Association also frequently discussed science education the elementary and secondary levels, often with an eye to its implications for future study in college. In the twentieth century various commissions and educational associations reported on science education. These reports, too, like those just mentioned, outlined primarily critics' and reformers' views, giving some attention to contemporary practices but concentrating particularly on the subjects and methods they believed should comprise scientific education.[42] Such accounts were often sponsored by foundations interested in how their support for education and for science would be channeled, providing a resource for historians on both the practice and changing ideals of scientific education.

PHILANTHROPY

Laments about the lack of support for science can be traced back to the founding years of the republic when intellectuals sought to match what they believed was more generous patronage in Europe. Most advocates of cultural activities also perceived as an advantage of their situation the fact that in the United States no government or church controlled intellectual activity. Pluralism in religious and social matters led them to the voluntary cooperation that seems in retrospect so characteristic of nineteenth-century America. Learned societies brought wealthy sponsors together with talented investigators and writers, but the collaboration proved more effective for specific, short-term projects than long-term activities.

Howard Miller's *Dollars for Research* demonstrates the ingenuity and pervasiveness of fund-raisers from established centers like Boston to the frontiers of settlement. Miller's pioneering book reminded historians that Louis Agassiz arranged to have his Museum of Comparative Zoology housed at Harvard University, persuaded the Massachusetts legislature to provide $100,000, and raised

67; Rossiter, *Women in Science in America*, Ch. 1. On informal ways of acquiring expertise in science see Lois Arnold, *Four Lives in Science: Women's Education in the Nineteenth Century* (New York: Schocken, 1983); Sally Gregory Kohlstedt, "In from the Periphery: American Women in Science, 1830–1880," *Signs*, 1978, *4*:81–96, and Joan N. Burstyn, "Early Women in Education: The Role of the Anderson School of Natural History," *Journal of Education*, 1977, *159*:50–64; and on normal schools, e.g., Louis I. Kuslan, "Elementary Science in Connecticut, 1850–1900," *Science Education*, 1959, *43*:286–289.

[41] Patricia Cline Cohen, *A Calculating People: The Spread of Numeracy in Early America* (Chicago: Univ. Chicago Press, 1983).

[42] The first volume of the *Annual Reports of the Commissioner of Education* is dated 1867/1868; the National Education Association published its proceedings and addresses under various titles from 1858 on. In the 1960s some quasi-historical discussions designed to support efforts for contemporary educational reform made use of these sources. See Paul DeHart Hurd, *Biological Education in American Secondary Schools, 1890–1960* (Washington, D.C.: American Institute of Biological Sciences, 1961); and Theodore R. Sizer, *Secondary Schools at the Turn of the Century* (New Haven: Yale Univ. Press, 1964).

nearly $75,000 from private sponsors. Using rather different methods of appeal
in Cincinnati, Ormsby McKnight Mitchel raised the money for his observatory
entirely from private donors, some of whom pledged as little as a dollar. A
number of historians of science have demonstrated that the federal government
also provided significant but selective support for exploring expeditions and for
projects related to mercantile ventures. Such money, however, like funds provided locally, was subject to the whims of legislators and was most often allocated for precisely focused work.[43]

Although Howard Miller has demonstrated that nineteenth-century capitalists
and others could be persuaded to support science and A. Hunter Dupree has
shown how much science was pursued within government agencies, the scientists themselves advocated more regular and independent sources. As their costs
increased with more expensive equipment and more collaborative research, scientists experimented in the late nineteenth century with a variety of approaches,
alternatively constrained and encouraged by the responses they had from pro-
spective donors. Sponsorship appeared easier to acquire for those at well-
established universities who had in mind, for example, capital equipment which
might be named for a donor. New observatories and science buildings were an
essential and dramatic supplement to student tuition and state government sponsorship, but scientists sought sustained support for their actual work. Another
attractive option, industrial support, also proved to have limitations from the
scientists' point of view, in part because it focused research so closely.[44]

In the late nineteenth century, yet another potential resource, developed by
wealthy entrepreneurs and encouraged by state and national legislation, became
available: the philanthropic foundation. Such foundations emerged as a promising means to provide support while balancing control between donor and re-
cipient. Andrew Carnegie unquestionably established an influential precedent
with his continued financial support of the Carnegie Institution, initially providing small grants to individuals but eventually concentrating on the institution's own departments, laboratories, and observatories. Large research foundations came to wield an influence beyond their monetary contribution because
they trained administrators who became effective managers of science. Thus,
according to Robert Kohler, Warren Weaver made the Rockefeller Foundation
not "simply a passive patron but a promoter of science along particular lines."[45]

[43] For a general overview see the still useful historical study by Robert Bremner, *American Philanthropy* (Chicago: Univ. Chicago Press, 1960); see also Howard Miller, *Dollars for Research; Science and Its Patrons in Nineteenth-Century America* (Seattle: Univ. Washington Press, 1970); on
exploration see William Stanton, *The Great United States Exploring Expedition of 1838–1842*
(Berkeley/Los Angeles: Univ. California Press, 1975), and David B. Tyler, *The Wilkes Expedition:
The First United States Exploring Expedition, 1838–1842* (Philadelphia: American Philosophical Society, 1968); also on early government research, see Bruce Sinclair, *Early Research at the Franklin
Institute: An Investigation into the Causes of Steam Boiler Explosions, 1830–1837* (Philadelphia:
Franklin Institute, 1966); and on mercantile incentives, see Harold L. Burstyn, "Seafaring the the
Emergence of American Science," *The Atlantic World of Robert G. Albion,* ed. Benjamin W. Lar-
rabee (Middleton, Conn.: Wesleyan Univ. Press, 1975), pp. 76–109.

[44] For one unsuccessful effort to broaden industrial support, see Lance E. Davis and Daniel J.
Kevles, "The National Research Fund: A Case Study in the Industrial Support of Academic Science," *Minerva,* 1974, *12:*207–220.

[45] For a lawyer's perspective on the incorporation of foundations see James Willard Hurst's
chapter "Science, Technology and Public Policy," in *Law and Social Order in the United States*
(Ithaca, N.Y.: Cornell Univ. Press, 1977), pp. 157–213. See also Nathan Reingold, "National Science
Policy in a Private Foundation: The Carnegie Institution of Washington, 1902–1920," in *The Orga-*

Independent research laboratories, often connected in some way with foundations and universities, became significant centers for research, suggesting the potential connections among private corporations, academic institutions, and governmental departments. The oceanographic and biological laboratories in Woods Hole, Massachusetts, with significant connections to the Boston Society of Natural History and other interested Bostonians, to Harvard University, and to the United States Fish Commission, provided a significant setting in which scientific specialties developed at the turn of the century. Such research centers as Scripps Institution of Oceanography have recently hired full-time archivists and made their records available. Industries, which tended to maintain independent laboratories of their own and brought in researchers from outside on a consultative basis, have never established substantial and sustained support for scientific work outside their purview, although recently closer connections have been forged with leading universities of technology.[46] It seems important to understand the extent to which such collaboration existed in the past and to understand the inhibitions that now appear to have been released.

According to Stanley Coben, a significant new initiative was demonstrated in the 1920s, when several foundations committed themselves to supporting the research of individual scholars on a large scale. Although initially directed toward younger scholars, these programs soon provided release time for the scientists whose academic teaching and administrative responsibilities distracted them from research. The assumptions undergirding this policy continue to guide policy in both private foundations and in the federal science and humanities foundations established since World War II. Some recent scholarship has evaluated foundations critically, focusing on their efforts to control research, their connections to each other, and their links to higher education. Historians of science might well examine questions like those raised in essays in Robert Arnove's *Philanthropy and Cultural Imperialism: The Foundations at Home and Abroad*.[47]

NEEDS AND OPPORTUNITIES

As historians seek the connections among intellectual, social, and cultural history, institutional studies become one point of convergence. Intellectual historians concur that "intellectual life requires the stimulation and discipline of dense intercommunication or community." The changing location of such "communities of discourse," however, is based on factors often external to the discussions of the intellectuals. To put it another way, in different settings similar goals may require quite different institutions. Thus the explanation of why

---

nization of Knowledge in Modern America, ed. Oleson and Voss (cit. n. 29); and Reingold and Ida H. Reingold, eds., *Science in America: A Documentary History, 1900–1939* (Chicago: Univ. Chicago Press, 1981), esp. pp. 7–55; and Robert Kohler, "The Management of Science: The Experience of Warren Weaver and the Rockefeller Foundation Programme in Molecular Biology," *Minerva*, 1976, 14:303.

[46] For recent sources on research centers, see Kiger, *Research Institutions* (cit n. 4).

[47] Stanley Coben, "American Foundations as Patrons of Science: The Commitment to Individual Research," *The Sciences in the American Context*, ed. Reingold, pp. 229–247; and Robert Arnove, *Philanthropy and Cultural Imperialism: The Foundations at Home and Abroad* (Boston: G. K. Hall, 1980), esp. Barbara Howe, "The Emergence of Scientific Philanthropy, 1900–1920," and Sheila Slaughter and Edward Silva, "Looking Backwards: How Foundations Formulated Ideology in the Progressive Period."

nearly identical, sometimes competing organizations proliferated in the United States is related in part to James Bryce's astute observation a century ago that it is "the only great country in the world which has no capital."[48] Equally important, activities take place in a larger matrix of institutions which are set in a culture that simultaneously values voluntarism and seeks to provide order through organizational mechanisms. There are several promising approaches.

One approach would be to analyze what Diana Crane has called "invisible colleges," that is, the informal but powerful cliques that exert particular influence. Studies of the self-styled Lazzaroni and references to the Cosmos Club in Washington have sometimes implied narrow self-interest.[49] But in fact the connections both among scientists in general and between scientists and others in particular regional settings have been critical in determining the viability of such learned societies as the Academy of Natural Sciences of Philadelphia and such publications as *Science* magazine. What has made some communities more effective than others? How were decisions made when there were a number of options—to concentrate resources on publications or collections or personnel, to organize by self-selection or invitation, to sponsor private laboratories, a university department, or a learned organization, or to turn to public or private sponsors?

While the study of individual institutions yields much information about scientific practitioners, working on a cluster of them yields even more, just as social historians found when using collective techniques to assess general characteristics and differences among workers, ethnic groups, and social classes. The practical problems of data gathering are significant, but the patterns may reveal the common assumptions held among those who are, for example, establishing in parallel fashion oceanographic summer schools, university departments of anthropology, or state agricultural experiment stations. The collective study of institutions can reveal the extent to which a regional or disciplinary "ideology" has been at work and perhaps uncover networks by which common goals have been fostered.

International or intranational contexts are particularly significant. Perhaps historians of American science, who so often encounter what appear to be derivative organizations modeled on apparent counterparts from Britain, France, or Germany are particularly aware of how much is to be learned from comparative studies. Continuities reveal the tenacity of values and organizational patterns while discontinuities suggest ways in which social, economic, geographic and political forces modify initially similar objectives. Such work should be extended into the twentieth century, where science policy, the role and status of academic scientists, the nature of graduate education, and the role of experts in popularization all invite scrutiny. Current institutional research on the context of science in Britain and her former colonies, Europe, and even Asia make comparisons possible.[50] Some of the same sources would yield material on the regional

[48] Haskell, *The Authority of Experts,* pp. 81, 100; for more discussion on the proliferation of learned societies see "Learned and Professional Organizations" above.

[49] Diana Crane, *Invisible Colleges: Diffusion of Knowledge in Scientific Communities* (Chicago: Univ. Chicago Press, 1972); and Mark Beach, "Was There a Scientific Lazzaroni?" in *Nineteenth-Century American Science: A Reappraisal,* ed. George Daniels (Evanston: Northwestern Univ. Press, 1972).

[50] For discussion of the colonial connections, see the papers prepared for a conference "Scientific

connections and exchanges among nations as well. One study that demonstrates the power of international investigation is Thomas Hughes's *Networks of Power*. True to the double meaning of its title, the work traces structural aspects of electrical power systems in four countries and describes the power brokerage of entrepreneurs and engineers who made them possible. Hughes transcends the boundaries of private and public sponsorship to analyze individual and organizational initiatives in the industry, providing a model that might well be used by historians of science dealing with interdependent institutions and interconnected research inquiries.[51]

Connections among different types of institutions have sometimes meant combined resources and have sometimes meant competition. At different periods and in distinct fields the relative importance of meetings, membership, the publication of journals, and the establishment of research facilities may vary substantially. How were choices made by foundering organizations that had attempted too much too soon and consequently found themselves unable to sustain a journal, a collection of specimens, and field researchers? When resources were available, which activities took priority? In the twentieth century regular staff members have come to play an increasingly powerful role in such decision making, by virtue of their familiarity with the institution and their connections with others holding positions of similar responsibility in other institutions. Their loyalties and points of view, however, are shaped by such forces as their training and professional affiliations in other than strictly scientific societies. The process of selecting work and financial priorities within such institutions as museums, for example, has inevitably been complicated as both staff and officers respond to what museum visitors anticipate on display, what philanthropists will fund, and what advertisers will pay for. Institutional archives can reveal how influence is exerted by external sources.

Like biography, institutional study can slip into debunking or heroic modes, minimizing the role of either aspirations or circumstances. David Noble's *America By Design* is a forceful reminder that a matrix of powerful organizations can establish modes of operation and behavior. Arguing from a neo-Marxist perspective, he offers persuasive evidence of collaboration among industry, scientific and engineering organizations, and higher education in the twentieth century. The success of the engineers, however, was apparently possible precisely because their operating ideals fit so well with contemporary ideologies among most powerful industrial and political groups. The limiting factors to hegemony include the layers of multiple expectations and scrutiny articulated by either

Colonialism, 1800–1930: A Cross-Cultural Comparison," held in May 1981, at the University of Melbourne, Australia, now being edited by Nathan Reingold and Marc Rothenberg for the Smithsonian Institution Press. Work on Britain is extensive, and reviews of current work may be found in both the *British Journal for the History of Science* and *Annals of Science*. A new journal deals with the colonial period and after in the South Pacific—*Historical Records of Australian Science*. On France see Mary Jo Nye, "Recent Sources and Problems in the History of French Science, *Historical Studies in the Physical Sciences*, 1983, *13*:401–415; and Robert Fox and George Weisz, ed., *The Origins of Science and Technology in France, 1808–1914* (Cambridge: Univ. Cambridge Press, 1980).

[51] Thomas P. Hughes, *Networks of Power: Electrification in Western Society, 1880–1930* (Baltimore: Johns Hopkins Univ. Press, 1983). For a case study of the parallel dimensions of institution building, see Sally Gregory Kohlstedt, "Australian Museums of Natural History: Public Priorities and Scientific Initiatives in the Nineteenth Century," *Historical Records of Australian Science*, 1983, 5:1–29.

law or tradition; by layers of bureaucracy, including members, sponsors, ac-
crediting agencies, and a government alert to tax-exempt status (among other
matters); and by the publicly informed thoughts of an often skeptical press.

To many scholars, institutions still seem to be bastions of old ideas. Critics
have charged, often fairly, that institutional studies have been too narrowly
constructed.[52] Some of us have argued, however, that they are also likely to
represent ideals in operation. Those historians who can go beyond preoccupation with origins, such as the revolution in higher education or the origins of
specialization, and analyze the ongoing process of institutional redefinition are
the ones who can bridge the distance between intellectual and social history.

Historians who understand that institutions provide vantage points from which
to view the connections between ideology and operation, moreover, realize with
Joseph Henry that the incentives for scientific societies lie in part in human
personality. In an address to a new Washington organization, the Smithsonian's
director pointed out that as "sympathetic" beings, scientists created communities through which they intended to have science "advanced and its results given
to the world."[53] What he knew, but rarely admitted in public, was that such
collaborations were also necessarily the product of political and cultural incentives that pulled on the collective enterprises and reshaped even their stated
purposes. It is the historian who must use the remarkable tendency for institutions to remember and retain a record of their past while recognizing as
well their capacity to forget, when convenient. Institutional history at its best
will reveal the ways in which learned societies, museums, universities, laboratories, and research centers provide an essential though sometimes tempestuous
channel for the passage of personal and intellectual aspirations in the world of
modern science.

[52] Reflecting this uneasiness with institutionally oriented work is Charles Rosenberg, "American
Science: A Generation of Historical Debate," *Isis*, 1983, 74:356–367.
[53] "Address," *Bulletin of the Philosophical Society of Washington*, 1871, p. 11.

Science in Medicine

By John Harley Warner*

"THE HISTORIAN OF MEDICINE who imagines that he is *ipso facto* a historian of science, is laboring under a gross delusion," George Sarton charged half a century ago. Sarton was miffed because while he had no institute for the history of science to direct, the Institute of the History of Medicine, with Henry Sigerist at its head, had just been established at Johns Hopkins. Considered as history of science, Sarton continued in his open letter to Sigerist, the history of medicine was "like the play of Hamlet with Hamlet left out." Sigerist retorted that "a cat is misleading considered as a dog" and that Sarton "could just as well have said that Othello is incomplete because there is no Hamlet in it."[1]

Sigerist, who saw medical history as a medical discipline, not a branch of the history of science, did not doubt that science, like economics, religion, social structure, and politics, was one important aspect of medical history. Historians of medicine in subsequent years shared Sigerist's view, though many of them placed more emphasis on science and less on society than he did. But the historiographic trends of the past two decades have made it decreasingly self-evident that the historian who writes about medicine particularly cares about the history of science. While this troubles some historians of medicine, it is nonetheless clear that a growing proportion of those who study the history of health care have more to talk about with historians of the city, women, social welfare, political culture, demography, or labor than they do with historians of the sciences.

Since the 1960s, recruitment of scholars to what is conventionally termed the new social history has increasingly diverted attention from analysis of the content and internal logic of medicine. In proselytizing the social history mission, the programmatic rhetoric of this movement often stigmatized close study of the development of medical science as antiquarian and more than likely positivistic. Especially among younger historians, the subject of science as a system of ideas and practices in medicine became an unpopular subject for historical inquiry.

Yet the same historiographic impulse greatly broadened the range of questions that historians asked about science in medicine. As historical study of the cognitive development of medical science diminished, scrutiny of the cultural role

* Wellcome Institute for the History of Medicine, London, NW1 2BP, England.

This essay was supported in part by NIH Grant LM 03910-02 from the National Library of Medicine, a research award from an Arthur Vining Davis Foundation grant to the Department of Social Medicine and Health Policy, Harvard Medical School, and a NATO Postdoctoral Fellowship from the National Science Foundation.

[1] George Sarton, "The History of Science versus the History of Medicine," *Isis*, 1935, *23*:319–320; Henry E. Sigerist, "The History of Medicine *and* the History of Science," *Bulletin of the Institute of the History of Medicine*, 1936, 4:6. A recent word of reconciliation between the fields is Gert H. Brieger, "The History of Medicine and the History of Science," *Isis*, 1981, 72:537–540.

and imagery of science in medicine was activated. While examining only shallowly the scientific concepts and methods of medicine, practitioners of the newer historiography have asked questions about other aspects of science's place in medicine, such as its function as an ideology, a source of cultural authority, an agent of professional legitimation, and a tool for attaining social, economic, and political objectives.

I wish to suggest that during the past decade the most vibrant theme dealing with science in the emerging historiography of American medicine has been the assessment of the *meaning* of science in medicine. The social history agenda that has nudged science as a body of knowledge to the periphery has at the same time drawn the meaning of science in medicine to the center of historical attention. Few studies developing this theme have taken as self-consciously expansive a view of cultural context as might inform an anthropologist's analysis of meaning. But they have revealed specific aspects of the meaning of science in medicine, exploring for example its function as a source of cultural and clinical power, its role in Americans' imposition of order on morbid reality, and its exhibition of societal values and needs. Recent essays surveying the current state of the historiography of medicine have extensively reviewed the varying approaches favored by historians with different training, the demographic changes in the field, and the ways that writing on the history of medicine has been shaped by scholarship in, for example, history, anthropology, and sociology.[2] My narrower focus here is on work of the past decade or so that has elucidated the place, function, and nature of science in American medicine and on the need and means to develop a more ample and balanced history of the meanings of that science.

CHANGING PERSPECTIVES ON SCIENCE IN MEDICINE

Underlying much of the writing in the past decade has been a new attitude toward the role and worth of science in modern medicine. From the 1960s the notion that the progressive infusion of scientific knowledge and methods into medicine inevitably improved patient care has steadily lost ground, with challenges coming from several fronts. Antipsychiatry was the most vivid emblem

[2] Gert H. Brieger, "History of Medicine," in *A Guide to the Culture of Science, Technology, and Medicine*, ed. Paul T. Durbin (New York: Free Press, 1980), pp. 121–194; John Burnham, "Will Medical History Join the American Mainstream?" *Reviews in American History*, 1978, 6:43–49; Karl Figlio, "The Historiography of Medicine: An Invitation to the Human Sciences," *Comparative Studies in History and Society*, 1977, 19:262–286; Daniel M. Fox, "Recent Marxist Interpretations of the History of Medicine in the United States," *Clio Medica*, 1982, 16:225–231; Gerald Grob, "The Social History of Medicine and Disease in America: Problems and Possibilities," in *The Medicine Show: Patients, Physicians, and the Perplexities of the Health Revolution in Modern Society*, ed. Patricia Branca (New York: Science History, 1977), pp. 1–19; Ronald L. Numbers, "The History of American Medicine: A Field in Ferment," *Rev. Amer. Hist.*, 1982, 10 (4):245–263; Susan Reverby and David Rosner, "Beyond 'the Great Doctors,'" in *Health Care in America: Essays in Social History*, ed. Reverby and Rosner (Philadelphia: Temple Univ. Press, 1979), pp. 3–18; Charles Rosenberg, "Science in American Society: A Generation of Historical Debate," *Isis*, 1983, 74:356–367; Charles Webster, "The Historiography of Medicine," L. J. Jordanova, "The Social Sciences and History of Science and Medicine," Margaret Pelling, "Medicine since 1500," and Edward H. Beardsley, "The History of American Science and Medicine," all in *Information Sources in the History of Science and Medicine*, ed. Pietro Corsi and Paul Weindling (London: Butterworth Scientific, 1983), pp. 29–43, 81–96, 379–407, 411–435; Arnold Thackray, "History of Science in the 1980s: Science, Technology, and Medicine," *Journal of Interdisciplinary History*, 1981, 12:299–314.

of a reevaluation downward of medical authority, but the same tendency was expressed in a resurgence of self-help medicine and in calls from the women's movement for women to regain control of their own bodies. In public sight, too, was anxiety about the social implications of technological medicine's rising cost. American society could not afford to grant all its members unlimited access to the fruits of biomedical science, and debates about how expenditure for health care should be contained emphasized that medical knowledge and services are political commodities. Biomedical authority became one target of assaults upon the American health care system's social and economic ills, and some critics, notoriously Ivan Illich, asserted that scientific medicine was hazardous to the people's health.[3]

At the same time, a critique of reductionist medicine emerged within the medical community. Clinicians and medical ethicists, disturbed by what they regarded as the dehumanization of modern medicine, suggested that science and the technology it informed stripped clinical medicine of important dimensions of healing. By the 1970s, medical humanities publications, programs for practicing physicians, and courses in medical schools, all of which promoted medical ethics, philosophy, and history, proliferated. Typical of this humanist concern was physician-ethicist Stanley Reiser's argument, embedded in an historical narrative, that while the rise of medical technology has improved care in some respects, it has vitiated the doctor-patient relationship and obscured the patient as person. Concurrently, some physicians expressed their resentment of the material support enjoyed by research laboratories geared more to satisfying intellectual curiosity than to providing clinically usable products and their concern about the subordination of clinical judgment to the decrees of the diagnostic laboratory isolated from sick individuals. Neither manifestation of this clinical skepticism of the laboratory was novel, but both were infused with new energy by clinicians like Alvan Feinstein, who wrote thoughtfully about the limits and legitimacy of clinical inquiry. Feinstein proposed, in *Clinical Judgment,* that overreliance on the scientific medicine of the laboratory could actually distort the clinician's judgment, to the detriment of the patient.[4]

Historical research generated yet another challenge to the putative boon to health bestowed by the biomedical sciences' growth. Thomas McKeown used studies of England and Wales to support his claim that prior to the 1930s and 1940s, advances in medicine did little to increase life expectancy or diminish mortality. The marked growth of population since the eighteenth century had more to do with better diet and hygiene than with curative medicine, he argued, and the role of medical science in improving health was much less significant

[3] Ivan Illich, *Medical Nemesis: The Expropriation of Health* (London: Calder & Boyars, 1975). See also John C. Burnham, "American Medicine's Golden Age: What Happened to It?" *Science,* 1982, *215*:1474–1479; and Linda H. Aikeen and Howard E. Freeman, "Medical Sociology and Science and Technology in Medicine," in *The Culture of Science,* ed. Durbin, pp. 527–580. Daniel M. Fox discusses the historiographic consequences of a move away from the idea of progress among historians of medicine in "The Decline of Historicism: The Case of Compulsory Health Insurance in the United States," *Bulletin of the History of Medicine,* 1983, *57*:596–610.

[4] Stanley Joel Reiser, *Medicine and the Reign of Technology* (Cambridge: Cambridge Univ. Press, 1978); Alvan R. Feinstein, *Clinical Judgment* (Baltimore: Williams & Wilkins, 1967). A recent evaluation of the role of the humanities in tempering the medical student's socialization as a scientist is Jerome J. Bylebyl, ed., *Teaching the History of Medicine at a Medical Center* (Baltimore: Johns Hopkins Univ. Press, 1982).

than hitherto supposed. Social and behavioral factors contributed more to health than did the remarkable increase of scientific knowledge, and McKeown believes that this past pattern holds important implications for future policy. While historians have bickered much about McKeown's assertions, Judith Leavitt gave his thesis one of its first American tests in her study of public health in Milwaukee, and her findings tend to confirm his historical conclusions.[5]

The common message of these diverse critiques was that there is no simple correlation between increased scientific knowledge in medicine and better health care. Recognition of this point, bolstered by increased attention to the political, economic, and social uses and abuses of biomedical authority, promoted dissatisfaction among historians with accounts of the function of science in medicine that considered only its clinical utility and its place in the history of ideas. It suggested, for example, that in order to explain the appeal of science to physicians, something more than its technical value should be sought, especially prior to the middle third of the twentieth century. To be sure, some historians have rightly continued to emphasize and display the clinical applications of the basic sciences. But the intriguing and unmistakable fact remains that the growth of the American medical profession's prestige and power from the late nineteenth century, a phenomenon often attributed to the advances in laboratory science, preceded rather than followed any substantial demonstration of the power of science to make physicians better healers.

One historiographic project informed by this critical skepticism endeavored to show how science in medicine was made to serve unsavory social ends. This aim was nowhere more starkly displayed than in some of the feminist histories published in the early 1970s, which depicted medical science as a sword wielded by male physicians in the social control of women. The authority of biomedical science, the argument went, was used to oppress women by defining the social limits of what was natural and healthy. Physicians' advice, according to one appraisal, "was not science after all, but only the ideology of a masculinist society, dressed up as objective truth."[6] Science in medicine was also depicted in these writings as a convention physicians used to sustain male dominance and to justify cruel treatments that exercised male hostility toward women.[7]

Less ideologically committed scholars recognized that such conspiracy models, which turned upon a psychosexual dynamic and reduced women to vic-

[5] Thomas McKeown, *The Modern Rise of Population* (New York: Academic Press, 1976); McKeown, *The Role of Medicine: Dream, Mirage, or Nemesis?* (Princeton, N.J.: Princeton Univ. Press, 1979); Judith Walzer Leavitt, *The Healthiest City: Milwaukee and the Politics of Health Reform* (Princeton, N.J.: Princeton Univ. Press, 1982). For a sample of the response to McKeown's work see *Health and Society: The Milbank Memorial Fund Quarterly*, 1977, 55:361–428. On McKeown's preoccupation with current health care issues in the selection of topics for historical study, see his "A Sociological Approach to the History of Medicine," *Medical History*, 1970, 14:342–351.

[6] Barbara Ehrenreich and Deidre English, *For Her Own Good: 150 Years of the Experts' Advice to Women* (Garden City, N.Y.: Doubleday, Anchor Press, 1978), p. 4; see also Ehrenreich and English, *Complaints and Disorders: The Sexual Politics of Sickness* (Old Westbury, N.Y.: Feminist Press, 1973); and Linda Gordon, *Woman's Body, Woman's Right: A Social History of Birth Control in America* (New York: Grossman, 1976).

[7] G. J. Barker-Benfield, *The Horrors of the Half-Known Life: Male Attitudes toward Women and Sexuality in Nineteenth-Century America* (New York: Harper & Row, 1976); Jane B. Donegan, *Women and Men Midwives: Medicine, Morality, and Misogyny in Early America* (Westport, Conn.: Greenwood, 1978); and Ann Douglas Wood, " 'The Fashionable Diseases': Women's Complaints and Their Treatment in Nineteenth-Century America," *J. Interdis. Hist.* 1973, 4:25–52.

tims, were simplistic. At the same time, they convincingly displayed in subtler writings how biomedical theories about women were used to legitimate the existing social order. They further showed how some women physicians exploited the same theories to justify their own professional role.[8] The ways that scientific ideas in medicine functioned to explain and maintain the order of society have also been explicated in the case of physicians' views of blacks. Medical rationales for racism in the antebellum South have provided an especially fertile context for exploring the sociopolitical uses of biomedical concepts.[9] The use of psychiatric authority in sanctioning the imposition of a particular value system upon society has also been much analyzed and disputed.[10]

A parallel and in some ways more problematic theme has been the way science legitimated and augmented the medical practitioner's authority. While the physician sometimes used this authority as an agent of his gender, class, and race, he employed it more self-consciously in defending the prerogatives of his vocation. The invocation of science in a variety of healers' claims to legitimacy is readily discerned at any period in American history. Cotton Mather, for example, relied on seventeenth-century atomism in his vindication of the minister's role in curing disease.[11] But most attention to the use of science to elevate the medical practitioner's authority has focused on the decades from the 1870s to the 1910s, when experimental laboratory science made vigorous claims to medical relevance. Historians' inordinate preoccupation with this period in part reflects their perception that it was during these decades that the structure, values,

[8] Virginia G. Drachman, *Hospital with a Heart: Women Doctors and the Paradox of Separatism at the New England Hospital, 1862–1969* (Ithaca, N.Y.: Cornell Univ. Press, 1984); John S. Haller, Jr., and Robin M. Haller, *The Physician and Sexuality in Victorian America* (Urbana: Univ. Illinois Press, 1974); Regina Morantz, "The Lady and Her Physician," in *Clio's Consciousness Raised: New Perspectives on the History of Women*, ed. Mary S. Hartman and Lois Banner (New York: Harper & Row, 1974), pp. 38–53; Carroll Smith-Rosenberg, "Puberty to Menopause: The Cycle of Femininity in Nineteenth-Century America," *Feminist Studies*, 1973, *1*:58–73; Smith-Rosenberg and Charles Rosenberg, "The Female Animal: Medical and Biological Views of Woman and Her Role in Nineteenth-Century America," *Journal of American History*, 1973, *60*:332–356; Mary Roth Walsh, *"Doctors Wanted: No Women Need Apply": Sexual Barriers in the Medical Profession, 1835–1975* (New Haven, Conn.: Yale Univ. Press, 1977); Richard W. Wertz and Dorothy C. Wertz, *Lying-In: A History of Childbirth in America* (New York: Schocken, 1979).

[9] James O. Breeden, "States-Rights Medicine in the Old South," *Bulletin of the New York Academy of Medicine*, 1976, *52*:348–372; John S. Haller, Jr., "The Negro and the Southern Physician: A Study of Medical and Racial Attitudes, 1800–1860," *Med. Hist.*, 1972, *16*:238–253. See also Kenneth F. Kiple and Virginia Himmelsteib King, *Another Dimension to the Black Diaspora: Diet, Disease, and Racism* (Cambridge: Cambridge Univ. Press, 1981); and Todd L. Savitt, *Medicine and Slavery: The Diseases and Health Care of Blacks in Antebellum Virginia* (Urbana: Univ. Illinois Press, 1978).

[10] For a review of this literature see Gerald N. Grob, "Rediscovering Asylums: The Unhistorical History of the Mental Hospital," in *The Therapeutic Revolution: Essays in the Social History of American Medicine*, ed. Morris J. Vogel and Charles E. Rosenberg (Philadelphia: Univ. Pennsylvania Press, 1979), pp. 135–157; Andrew Scull, "Humanitarianism or Control: Observations on the Historiography of Anglo-American Psychiatry," *Rice University Studies*, 1981, *67*:21–41; and Peter L. Tyor and Jamil S. Zainaldin, "Asylum and Society: An Approach to Institutional Change," *Journal of Social History*, 1979, *13*:23–48.

[11] Margaret Humphreys Warner, "Vindicating the Minister's Medical Role: Cotton Mather's Concept of the *Nishmath-Chajim* and the Spiritualization of Medicine," *Journal of the History of Medicine*, 1981, *36*:278–295. On the cases of pediatricians, obstetricians, and patent medicine makers, see Rima D. Apple, " 'To Be Used Only Under the Direction of a Physician': Commercial Infant Feeding and Medical Practice, 1870–1940," *Bull. Hist. Med.*, 1980, *54*:402–417; Judith Walzer Leavitt, " 'Science' Enters the Birthing Room: Obstetrics in America since the Eighteenth Century," *J. Amer. Hist.*, 1983, *70*:281–304; and Sarah Stage, *Female Complaints: Lydia Pinkham and the Business of Women's Medicine* (New York: Norton, 1979).

and shortcomings of the health care system we live with today took rough shape.

The theme of science as an ideological platform for raising the physician's prestige has been developed most fully in accounts of the process ordinarily called professionalization. The marked elevation in the American medical profession's status from the late nineteenth century has been linked from that time until the present to the concomitant growth of laboratory science. Certainly most historians recognize that some substantial measure of the profession's betterment in the public's eyes must be ascribed to its increased ability to explain and predict the course of disease, its appropriation of credit for the very real advances in public health, and its demonstrated power to prolong life through developments in, for example, pediatric surgery. Ronald Numbers has recently cautioned against trying to account for a greater increase in professional power than actually occurred, suggesting that it is wrong to search for the magic lamp that fulfilled physicians' professional agenda in return for only a politically adroit rub and wish. After all, he points out, orthodox physicians were unable to squelch such competing sects as osteopathy, chiropractic, and Christian Science, which persisted and thrived in the twentieth century. Nevertheless, orthodox medical practitioners' immense success at professionalization has properly encouraged historians to look to more than increased efficacy for an explanation.[12]

What distinguishes recent analyses of this process from earlier ones is their endorsement of the proposition that science's catalytic power might have come more from its use as a cultural than as a technical resource. Gerald Geison has bravely suggested that the clinical utility of experimental physiology has been meager. Yet its professional value has been considerable, he proposes, "for the experimental sciences, like Latin in an earlier era, have given medicine a new and now culturally compelling basis for consolidating its status as an autonomous 'learned profession,' with all of the corporate and material advantages that such status implies." S. E. D. Shortt has similarly argued that the key to understanding the relationship between medical science and medical professionalization is recognizing the extent to which physicians used "not the content, but the rhetoric of science." The cultural authority physicians derived from science and used in expanding professional power is also a persistent theme in sociologist Paul Starr's historical analysis of the American medical profession's development, the most synthetic of recent studies.[13]

Barbara Gutmann Rosenkrantz has gone furthest in showing that not simply the incorporation of science into American medicine but the changing meaning of that science transformed physicians' professional identity (as well as professional status and power) in the late nineteenth century. She has analyzed the

---

[12] Ronald L. Numbers, "The Fall and Rise of the American Medical Profession," in *The Professions in America,* ed. Nathan Hatch (South Bend, Ind.: Univ. Notre Dame Press, forthcoming).

[13] Gerald L. Geison, " 'Divided We Stand': Physiologists and Clinicians in the American Context," in *The Therapeutic Revolution,* ed. Vogel and Rosenberg, pp. 67–90, on p. 85; S. E. D. Shortt, "Physicians, Science, and Status: Issues in the Professionalization of Anglo-American Medicine in the Nineteenth Century," *Med. Hist.,* 1983, *27*:51–68, on p. 60; also see Shortt, "The New Social History of Medicine: Some Implications for Research," *Archivaria,* 1980, *10*:5–22; Paul Starr, *The Social Transformation of American Medicine* (New York: Basic, 1982). An earlier, deterministic conception of the relationship between late-nineteenth-century science and professionalization is exemplified by William G. Rothstein, *American Physicians in the Nineteenth Century: From Sects to Science* (Baltimore: Johns Hopkins Univ. Press, 1972).

transition "from the old view of science as explanatory language of natural order" to "a new conception of science as the powerful intervening instrument which conferred expert status on those who mastered the skills of the trade" and observed that in the process a new professional ethos emerged in which "accountability to science replaced relations with patients." Science was far more to American physicians than a convenient tool for the acquisition of economic and social advantage, Rosenkrantz argues; its acceptance as the core of what it meant to be a physician cannot be reduced to such marketplace determinism. "Science," she noted, "endowed the professions not only with a new authority, but more significantly, with a new morality." Rosenkrantz has also examined public health, charting the ways in which developments within bacteriology and immunology reoriented professional identity by narrowing the definition of expertise, laying the basis for the "New Public Health Movement."[14] Her writings emphasize that what was new in late nineteenth-century medicine was not science itself, but science redefined and taken up in new ways.

Medical education was the principal conduit for infusing laboratory science and its ideals into the medical ethos. Accordingly, it is especially in studies of educational reform that revisionists have sought the ideological uses of science. A leading question about the changes that made laboratory science a central part of medical education in the early twentieth century has been: Who profited from reform? Not the public, have answered many who regard these educational reorientations as a means of restricting entry into medicine and establishing medical elitism.[15] In a book that created a greater brouhaha among historians of medicine than any other in the past decade, E. Richard Brown reduced the ascendancy of laboratory science in American medicine to terms amenable to a doctrinaire Marxist analysis and presented "modern scientific medicine" as above all an "ideological weapon." The American medical elite first embraced the ideal of "scientific medicine" as a way to limit the production of doctors and elevate the profession's economic and social standing, Brown argued. Their strategy worked because corporate capitalists recognized that scientific medicine would serve their interests, and, acting especially through the Rockefeller and Carnegie Foundations, provided the money required to enact the expensive reform of medical schools. Although infidelity to his sources ultimately leads Brown's work to collapse, the publicity his thesis received brought the issue of the ideological uses of medical science to the forefront of medical historians' attention.[16]

[14] Barbara G. Rosenkrantz, "The Search for Professional Order in 19th Century American Medicine," *Proceedings of the XIVth International Congress of the History of Science* (Tokyo: Science Council of Japan, 1975), no. 4, pp. 113–124, quoting pp. 118, 122, 121; Barbara Gutmann Rosenkrantz, "Cart before Horse: Theory, Practice, and Professional Image in American Public Health, 1870–1920," *J. Hist. Med.*, 1974, 29:55–73; also Rosenkrantz, *Public Health and the State: Changing Views in Massachusetts, 1842–1936* (Cambridge, Mass.: Harvard Univ. Press, 1972).

[15] Howard Berliner, "A Larger Perspective on the Flexner Report," *International Journal of Health Services,* 1975, 5:573–592; Gerald E. Markowitz and David Karl Rosner, "Doctors in Crisis: A Study of the Use of Medical Education Reform to Establish Modern Professional Elitism in Medicine," *American Quarterly,* 1973, 25:83–107.

[16] E. Richard Brown, *Rockefeller Medicine Men: Medicine and Capitalism in America* (Berkeley/ Los Angeles: Univ. California Press, 1979), on p. 10. An alternative interpretation that identifies evangelical protestantism as a motivation behind one part of the Rockefeller Foundation's support of medical reform is John Ettling, *The Germ of Laziness: Rockefeller Philanthropy and Public Health in the New South* (Cambridge, Mass.: Harvard Univ. Press, 1981). The animated discussion of Brown's study included Howard S. Berliner, review of *Rockefeller Medicine Men* and other books,

Whereas historians have looked hard at what the growth of the laboratory sciences from the late nineteenth century meant for medical education, the inverse has received much more cursory examination. Yet a few studies have begun to reveal how the ideology of scientific medicine fundamentally shaped the way the biological sciences developed in America. Reformed medical schools did more than train "scientific" doctors; they also gave a home to those who had been lured to basic research. The institutional and ideological changes in American medical education presented a ripe context for discipline building. In the richest study to date of the development of any biomedical discipline, Robert Kohler shows in *From Medical Chemistry to Biochemistry* that only by understanding the intimate relation between biochemists and medical school reform can the formers' singular success at discipline building in the United States (compared with Germany and Britain) be understood. Basic scientists offered physicians instruction, legitimating knowledge, and clinically usable techniques and in return received financial and institutional support.[17] Although the spontaneous biological experiments performed by disease have often been crucial to the elucidation of scientific questions, the conceptual profits the sciences have derived from medicine have been even less systematically explored than the institutional benefits.[18]

Basic scientists and medical education reformers sometimes appeared to uphold a shared ideal of medical science, even though it had different personal and professional meanings for them. But the last decade's writings have begun to underscore the ambivalence toward experimental medical science felt by many lay and medical Americans. In condemning the moral and epistemological legitimacy of knowledge gained by animal experimentation, antivivisectionists, some of them physicians, offered perhaps the most blatant expression of dissent from the emerging model of medical science. The antivivisection movement targeted both the methods of science in medicine and the application of its products. Several historians have traced the growth of antivivisectionist sentiment after the 1870s in America, but it still awaits probing analysis comparable to that given the British movement.[19]

Bull. Hist. Med., 1980, *54*:131–134; Lloyd G. Stevenson, "A Second Opinion," *ibid.*, pp. 134–140; E. Richard Brown, "Response to Lloyd Stevenson's 'Second Opinion,' " *ibid.*, pp. 589–591; Daniel M. Fox, "Rockefeller Medicine Men Again: Ideology vs. Methodology," *ibid.*, pp. 591–593; Harold Y. Vanderpool, "Letter to the Editor," *ibid.*, 1981, *55*:434–437; and Saul Benison, "Ideology *über alles*: An Essay Review," *J. Hist. Med.*, 1982, *37*:83–90.

[17] Geison, " 'Divided We Stand,' "; Robert E. Kohler, *From Medical Chemistry to Biochemistry: The Making of a Biomedical Discipline* (Cambridge: Cambridge Univ. Press, 1982); and Philip J. Pauly, "The Appearance of Academic Biology in Late Nineteenth Century America," *Journal of the History of Biology*, 1984, *17*:369–397.

[18] Medical contributions to basic science during recent decades that await historical scrutiny are sketched in Henry K. Beecher, ed., *Disease and the Advancement of Basic Science* (Cambridge, Mass.: Harvard Univ. Press, 1960). I am grateful to an anonymous reviewer for guiding me to this source.

[19] Richard D. French, *Antivivisection and Medical Science in Victorian Society* (Princeton, N.J.: Princeton Univ. Press, 1975); William Gary Roberts, "Man before Beast: The Response of Organized Medicine to the American Antivivisection Movement" (A.B. thesis, Harvard Univ., 1979); and James Turner, *Reckoning with the Beast: Animals, Pain, and Humanity in the Victorian Mind* (Baltimore: Johns Hopkins Univ. Press, 1980). This national imbalance will be largely redressed by Susan Eyrich Lederer, "An Ethical Problem: The Controversy over Human and Animal Experimentation in Late Nineteenth-Century America" (Ph.D. diss., Univ. Wisconsin–Madison, in progress); see also Lederer, "The Inevitable Tendency of Vivisection: Hideyo Noguchi's Luetin Experiments and the Antivivisectionists," *Isis*, 1985, *76*:31–48.

A more subtle vein of resistance to the ascendancy of experimental science in medicine that has recently drawn scrutiny is the enduring ambivalence of many practicing physicians toward the laboratory's workers, methods, and products. One tack has been to analyze closely the ideal of medical science to which intellectually elite physicians in the mid-nineteenth century gave their allegiance and to show how its methodological and epistemological character was in some ways antithetical to that of laboratory science. Thinking physicians who opposed the rise of laboratory science, such a view urges, did not necessarily oppose science in medicine but instead objected to the new definition of what constituted science, rooted more in experimental laboratory physiology than in empirical clinical observation.[20]

The divergent perceptions of medical science held by laboratory scientists and by clinicians after the 1870s, and the tensions that resulted, have been more extensively analyzed. Russell Maulitz's " 'Physician versus Bacteriologist': The Ideology of Science in Clinical Medicine" represents this genre at its best. Bacteriology, he notes, attracted clinicians through its promise to place powerful new diagnostic and therapeutic tools in their hands. But at the same time it threatened them with a vision of the laboratory rather than the bedside as the hub about which scientific medicine would revolve. Other studies have focused on many clinicians' leading question about research in such laboratory sciences as physiology and bacteriology—What difference does it make at the bedside?—and their discomfort with the answers. Such studies of clinicians' ambivalence toward laboratory science, among the most provocative of recent work on the relation between science and clinical medicine, have been fueled in part by some modern clinicians' skepticism of laboratory research and tests and by the common suggestion that the stress on laboratory experience in medical schools' curricula should be reduced.[21]

Defining the relationship between science, profession, and practitioner was especially arduous in nursing, as some of the flourishing work on nursing history has started to disclose. In the field's formalization in the late nineteenth century, the feminine virtues celebrated by Florence Nightingale—compassion, self-sacrifice, orderliness, spirituality—were far more central to the nurse's identity than an allegiance to science. As Susan Reverby has shown, when a small group of American nursing reformers endeavored in the early twentieth century "to move nursing from sentiment to science," this view that the nurse's abilities emanated naturally from her gender increased resistance to their task. Reverby has further observed that different elite reformers saw the potential of science for professionalizing nursing in different ways. For some reformers, science was an activity, a patient-oriented style of clinical research that would advance nursing

---

[20] John Harley Warner, " 'The Nature-Trusting Heresy': American Physicians and the Concept of the Healing Power of Nature in the 1850's and 1860's," *Perspectives in American History,* 1977–1978, *11*:291–324.

[21] Peter C. English, *Shock, Physiological Surgery, and George Washington Crile: Medical Innovation in the Progressive Era* (Westport, Conn.: Greenwood, 1980); Geison, " 'Divided We Stand' "; and Russell C. Maulitz, " 'Physician versus Bacteriologist': The Ideology of Science in Clinical Medicine," in *The Therapeutic Revolution,* ed. Vogel and Rosenberg (cit. n. 10), pp. 91–107. On expressions of this tension in discussions of medical curriculum reform, see Ronald L. Numbers, ed., *The Education of American Physicians: Historical Essays* (Berkeley/Los Angeles: Univ. California Press, 1980).

practice; others "saw science as the objective authority to be called upon to improve nursing's status."[22]

The great imperative of social history to move away from studying only established elites has led a number of scholars to ask how patients and alternative healers viewed orthodox medicine. A few historians have looked at self-help healing and folk practices of Americans who did not necessarily dissent from orthodox medicine's belief system. On the whole, such work has focused on sources of health care rather than attitudes toward organized natural knowledge.[23] But far more attention has been directed toward the beliefs and practices of certain intellectually marginal groups who typically saw themselves arrayed against the medical orthodoxy and its science. William Rothstein's influential *American Physicians in the Nineteenth Century* was a harbinger of such studies. Though decidedly positivistic in regarding the orthodox profession as but one among a variety of medical sects until it was uplifted by science and its presumed power to guide "valid therapy," the book nonetheless emphasized the worth of studying nonorthodox systems of healing. Histories of diverse groups that held alternative conceptions of nature to that of regular medicine—such as Adventists, Grahamites, homeopaths, osteopaths, and sundry species of health reformers—have proliferated.[24] Still, most studies make little distinction between attitudes toward orthodox science and attitudes toward orthodox practice.

The transformation of the general hospital, like professionalization and medical education reform, is a phenomenon once explained by the growth of medical science that has drawn revisionist attention. A charitable refuge for the sick poor through the first three quarters of the nineteenth century, by the early twentieth century the hospital had become a center for medical research and care for people of all classes. Medical science, recent studies have emphasized, did not, as hitherto assumed, compel this change and therefore should not be seen as the sole, or even the chief, factor in accounts of the hospital's ascendancy. In his study of Boston hospitals, for example, Morris Vogel argued that explanations for the hospitalization of the middle classes and the increasing centrality of the hospital to medical care must be sought largely in social, political, and

[22] Susan Reverby, "A Sensibility for Science: Perspectives from 20th Century American Nursing," paper given at the Sixth Berkshire Conference on the History of Women, Smith College, Northampton, Massachusetts, 3 June 1984. Also see, e.g., Janet Wilson James, "Isabel Hampton and the Professionalization of Nursing in the 1890s," in *The Therapeutic Revolution*, ed. Vogel and Rosenberg, pp. 201–244; Ellen Condliffe Lagemann, ed., *Nursing History: New Perspectives, New Possibilities* (New York: Teachers College Press, 1983); Barbara Melosh, *The Physician's Hand: Work Culture and Conflict in American Nursing* (Philadelphia: Temple Univ. Press, 1982); and Charles Rosenberg, "Florence Nightingale on Contagion: The Hospital as a Moral Universe," in *Healing and History: Essays for George Rosen*, ed. C. Rosenberg (New York: Science History Publications, 1979), pp. 116–136.

[23] Guenter B. Risse, Ronald L. Numbers, and Judith Walzer Leavitt, eds., *Medicine without Doctors: Home Health Care in American History* (New York: Science History, 1977).

[24] Rothstein, *American Physicians* (cit. n. 13). See also Harris L. Coulter, *Divided Legacy: A History of the Schism in Medical Thought* (Washington, D.C.: McGrath, 1973), Vol. III; Norman Gevitz, *The D.O.'s: Osteopathic Medicine in America* (Baltimore: Johns Hopkins Univ. Press, 1982); Russell W. Gibbons, "Physician-Chiropractors: Medical Presence in the Evolution of Chiropractic," *Bull. Hist. Med.*, 1981, 55:233–245; Martin Kaufman, *Homeopathy in America: The Rise and Fall of a Medical Heresy* (Baltimore: Johns Hopkins Press, 1971); Stephen Nissenbaum, *Sex, Diet, and Debility in Jacksonian America: Sylvester Graham and Health Reform* (Westport, Conn.: Greenwood, 1980); Ronald L. Numbers, *Prophetess of Health: A Study of Ellen G. White* (New York: Harper & Row, 1976); James C. Whorton, *Crusaders for Fitness: The History of American Health Reformers* (Princeton, N.J.: Princeton Univ. Press, 1982).

economic permutations in the community. He did acknowledge that science and the technical tools it guided altered the purposes of hospitals and catalyzed the shift in control within them from lay trustees to physicians. Nevertheless, the ideological role of science in supplanting Christian stewardship as the hospital's justification is intercalated deeper into Vogel's account than are actual shifts in scientific belief and practice or their consequences. This marginality of the content of science in hospital history took its most extreme form in David Rosner's account of Brooklyn hospitals during the Progressive Era. Rosner's book is state-of-the-art social history and one of the best contributions to American medical history of the past decade, and it is precisely for this reason that his decision virtually to ignore medical science is so historiographically telling.[25]

A few historians more interested in the medical implications of the transfiguration of the hospital have begun to look closely at how some physicians committed to the ideal of scientific medicine saw hospitals as the workshops where an avowedly clinical science would be fashioned. Especially in the early twentieth century, certain clinical investigators sought to define the clinic rather than the laboratory as the primary locus of scientific medicine and to use the clinical science pursued there to distinguish themselves from both general medical practitioners and benchmen. The writings of clinician-historians like Maulitz and Kenneth Ludmerer have brought a new analytical sophistication to the history of clinical science as an ideology and activity and corrected the too-common equation of modern medical science with the laboratory. Ludmerer's essays have traced the incorporation of a Germanic ideal of scientific research into American clinical investigation and analyzed the place clinical science held in the reformed curricula of medical schools.[26] The study of clinical specialization, like that of the creation of basic scientific disciplines, has also received a new impulse from the tendency to see scientific and technical innovations as important but inadequate to account fully for changes in professional structure.[27]

The ongoing reassessment of what physicians gained from science has begun to have a liberating effect on study not only of the profession's structures and institutions, but also of medical practice. Nowhere is this more evident than in

[25] Morris J. Vogel, *The Invention of the Modern Hospital: Boston, 1870–1930* (Chicago: Univ. Chicago Press, 1980); David K. Rosner, *A Once Charitable Enterprise: Hospitals and Health Care in Brooklyn and New York, 1885–1915* (Cambridge: Cambridge Univ. Press, 1982). See also Charles E. Rosenberg, "And Heal the Sick: The Hospital and the Patient in 19th Century America," *Journal of Social History,* 1977, *10*:428–447; and Rosenberg, "Inward Vision and Outward Glance: The Shaping of the American Hospital, 1880–1914," *Bull. Hist. Med.,* 1979, *53*:346–391.

[26] Kenneth M. Ludmerer, "The Plight of Clinical Teaching in America," *Bull. Hist. Med.,* 1983, *57*:218–229; Ludmerer, "Reform at Harvard Medical School, 1869–1909," *ibid.,* 1981, *55*:343–370; Ludmerer, "Reform of Medical Education at Washington University," *J. Hist. Med.,* 1980, *35*:149–173; Ludmerer, "The Rise of the Teaching Hospital in America," *ibid.,* 1983, *38*:389–414. On clinical research, see Saul Benison, "Poliomyelitis and the Rockefeller Institute," *ibid.,* 1974, *29*:74–92; and A. McGehee Harvey, *Science at the Bedside: Clinical Research in American Medicine, 1905–1945* (Baltimore: Johns Hopkins Univ. Press, 1981). A study of the infusion of the experimental laboratory's way of thinking into British clinical thinking that should stand as a model for studies on the American context is Christopher Lawrence, "Moderns and Ancients: The 'New Cardiology' in Britain, 1880–1930," in *The Emergence of Modern Cardiology,* ed. W. F. Bynum and C. Lawrence, Suppl. 5 to *Med. Hist.* (London: Wellcome Institute for the History of Medicine, in press).

[27] Bonnie Ellen Blustein, "A New York Medical Man: William Alexander Hammond, M.D. (1828–1900), Neurologist" (Ph.D. diss., Univ. Pennsylvania, 1979); and Blustein, "New York Neurologists and the Specialization of American Medicine," *Bull. Hist. Med.,* 1979, *53*:170–183. Rosemary Stevens's *American Medicine and the Public Interest* (New Haven, Conn.: Yale Univ. Press, 1971) remains the most comprehensive study of specialization in American medicine.

the history of medical therapeutics. A bicentennial survey of American medicine typified an older historiography in dismissing extensive attention to the practice of the American physician before the 1870s as unrewarding because, as the authors noted, "lacking a scientific base, his therapy for the more serious maladies was founded upon false hypotheses and fanciful systems and was for the most part ineffective."[28] Such a statement is ahistorical in ignoring what nineteenth-century physicians regarded as the solid scientific foundation of their practices.

More than this, to say that therapy was inefficacious because it was not scientific simply makes no sense. Thus Charles Rosenberg has approached nineteenth-century orthodox therapeutics with the initial proposition that therapy worked, though not perhaps as judged by criteria of efficacy satisfying to a twentieth-century pharmacologist; the historian's task is to understand what made it meaningful. Proceeding from this premise, historians can give due weight to the power of early nineteenth-century medical science to manage sickness by explaining it, and they can evaluate science as one component of a belief system that practitioners and their patients largely shared. A similar approach holds much promise for understanding the alternative systems of medical belief and practice to which many Americans turned. While a relativism that fails to see the very real progress that scientific advances have brought to health care is the chief risk in such an approach (and one to which Rosenberg has not fallen prey), taking medical science prior to the era of the laboratory seriously is the only way historians can hope to grasp its meaning.[29]

PROGRESS AND POVERTY

Some excellent historical studies of such topics as biomedical investigation, research tools, and the reception of scientific innovations have appeared, but on the whole these subjects have occupied a narrower band in the spectrum of work in the field than in earlier decades.[30] Perhaps the most glaring deficit in the best of the past decade's work that has touched on science in American medicine is its slight attention to the content of that science. Those who have embraced the newer historiography have tended to look *at* science more than *into* it; they have

---

[28] James Bordley III and A. McGehee Harvey, *Two Centuries of American Medicine, 1776–1976* (Philadelphia: Saunders, 1976), p. vii.

[29] Charles E. Rosenberg, "The Therapeutic Revolution: Medicine, Meaning, and Social Change in Nineteenth-Century America," *Perspectives in Biology and Medicine,* 1977, 20:485–506.

[30] On biomedical investigation see, e.g., Edward C. Atwater, " 'Squeezing Mother Nature': Experimental Physiology in the United States Before 1870," *Bull. Hist. Med.,* 1978, 52:313–335; Harry F. Dowling, *Fighting Infection: Conquests in the Twentieth Century* (Cambridge, Mass.: Harvard Univ. Press, 1977); J. Worth Estes, *Hall Jackson and the Purple Foxglove: Medical Practice and Research in Revolutionary America, 1760–1820* (Hanover, N.H.: Univ. Press of New England, 1979); John Parascandola, "John J. Abel and the Early Development of Pharmacology at the Johns Hopkins University," *Bull. Hist. Med.,* 1982, 56:512–527; Lewis Rubin, "Leo Loeb's Role in the Development of Tissue Culture," *Clio Med.,* 1977, 12:33–56; and Margaret Warner, "Hunting the Yellow Fever Germ: The Principle and Practice of Etiological Proof in Late Nineteenth-Century America," *Bull. Hist. Med.,* in press. On research tools see, e.g., on statistics, James H. Cassedy, *American Medicine and Statistical Thinking, 1800–1860* (Cambridge, Mass.: Harvard Univ. Press, 1984); and Gerald N. Grob, *Edward Jarvis and the Medical World of Nineteenth-Century America* (Knoxville: Univ. Tennessee Press, 1978). On innovations see, e.g., Thomas Gariepy, "The Acceptance of Antiseptic Surgery in the United States" (M.A. thesis, Univ. Notre Dame, 1976); Dale C. Smith, "Gerhard's Distinction between Typhoid and Typhus and Its Reception in America, 1833–1860," *Bull. Hist. Med.* 1980, 54:368–385; and Patricia Spain Ward, "The American Reception of Salvarsan," *J. Hist. Med.,* 1981, 36:44–62.

said much about its imagery, social uses, and ideological character but relatively little about the changing structure of scientific thinking and technique in medicine or the ways that scientific developments reshaped medical theories and were applied at the bedside. The expression of this tendency in an able collection of essays on the history of Wisconsin medicine recently prompted reviewer Erwin Ackerknecht to comment: "As I wrote one of the first monographs in the social history of medicine fifty years ago, may I be allowed to plead for the rediscovery of the medical history of medicine? It exists too."[31] Ackerknecht's dismay cannot be dismissed as a mere reaction against a social history approach to medicine and should be taken seriously as a caution that important currents of medical history are being too thoroughly channeled off from the mainstream.

This circumstance has seriously disturbed some medical historians and has elicited complaints about an emerging history of medicine without medicine, science, or doctors, complaints that echo those over a history of science without science.[32] In fact, while the questions asked by social historians of health care and more traditional medical historians differ, neither approach intrinsically has more or less to do with science than the other. Like earlier works in medical history that ordinarily gave scant attention to the ideological uses and ambiguities of science, newer approaches informed by the protocols of social history that ignore the internal content of medical science are incomplete. But defensiveness at both historiographic poles has led to invidious claims about what constitutes legitimate medical history, with scant recognition of the value of both perspectives. As regrettable as the baiting rhetoric from some social historians who deprecate the shortcomings of traditional, iatrocentric medical history have been the countering barbs that historians without formal medical and scientific training are doubtfully competent to contribute to the history of medicine. The issue of who is not qualified to write medical history has absorbed much energy and has strongly shaped the allocation of professional rewards; but intellectually the issue is a red herring, and those trolling in these waters have bigger fish to fry.

To one interested in understanding the meaning of science in American medicine, historiographic reductionism has been the most deplorable consequence of the field's divisiveness. If asked, most historians would certainly agree that any given scientific concept in medicine might simultaneously be useful both in furthering research and practice and in advancing socioeconomic and political goals, but in their work many tend to act as if they must choose between these functions.[33] Those who have hold of the elephant's trunk and know it to be round are loath to tolerate protests from those grasping the ear that it is flat. It

---

[31] Erwin H. Ackerknecht, review of *Wisconsin Medicine: Historical Perspectives*, ed. Ronald L. Numbers and Judith Walzer Leavitt (1981), *Med. Hist.*, 1982, 26:476.

[32] [Leonard Wilson], "Medical History without Medicine," *J. Hist. Med.*, 1980, 35:5–7; also see Wilson, "The Education of Historians of Medicine," *ibid.*, 1982, 37:265–268; and Wilson, "Schizophrenia in Learned Societies: Professionalism vs. Scholarship," *ibid.*, 1981, 36:5–8. On the analogous concern in the history of science, see Charles C. Gillispie's Sarton lecture as reported in William J. Broad, "History of Science Losing Its Science," *Science*, 1980, 207:389. One recent attempt to lay the ground for a rapprochement between physician- and lay-historians is S. E. D. Shortt, "Clinical Practice and the Social History of Medicine: A Theoretical Accord," *Bull. Hist. Med.*, 1981, 55:533–542.

[33] Steven Shapin noted a similar either/or mentality among historians of science in "History of Science and Its Sociological Reconstructions," *History of Science*, 1982, 20:157–211, on p. 187.

is just as futile to seek the full meaning of a scientific concept in medicine by looking only at its use in social control, advancing professionalization, or furthering physicians' sociopolitical goals as it is by examining only its implications for medical practice or the research front. Historiographic reductionism has been useful in drawing attention to the significant role of social context in giving medical science part of its meaning, but it can take understanding only so far.

Even though the most conspicuous shortcoming of the social history that has illuminated the meaning of science is inattention to the content of that science, it is not the most disabling. Rather, I would suggest that the most serious curb on our bettering our comprehension of the meaning of science in American medicine is one that the new history has inherited from the old, namely, a stilted model of what constitutes medical science. It is ironic that social historians who proclaim that writing history from the bottom up is a hallmark of their endeavor have clung to a conception of science that is often staunchly elitist. In cultivating a history of health care that is broader than the iatrocentric, elitist focus of traditional medical history, they have by and large retained a narrow vision of science defined by those at the top. The same historians have also tended tacitly to identify the term "science" with what is now the emblem of modern medical research, the experimental laboratory. The perspective on science in American medicine that many social history studies have taken has been both elitist and presentistic.

THE MULTIPLE MEANINGS OF SCIENCE IN MEDICINE

"Science" in American medicine was never monolithic. It meant different things at various times and to assorted social groups, including the varieties of medical practitioners.[34] No historian would decline to acknowledge this diversity, yet only a small number have made analyzing it prominent in their work. Nevertheless, a self-conscious awareness of the multiplicity of the meanings of science in medicine is one of the most promising guides to research that can illuminate the place and function of science in medicine and of medical science in American culture. It is easy enough to come up with topics for historical reexamination to which this perspective could be applied: clinical, laboratory, and public health perceptions of early bacteriology; business, public, and medical views of twentieth-century pharmacological research; or the divergent medical epistemologies that coexisted within the American medical profession at any time during the nineteenth century. But more profitable than cataloguing my personal research want-list will be identifying some of the ways in which science's meaning varied that should be taken into account in analyzing the meaning of science in medicine.

What constituted medical science changed fundamentally over time. Constructing rationalistic systems of pathology and therapeutics in the late eighteenth century and elucidating clinical natural history by careful observation in the mid-nineteenth century were medical enterprises every bit as scientific to their practitioners as experimenting in the laboratory became after the 1870s.

---

[34] Charles E. Rosenberg makes this point about science in American society in his writings generally, and especially in "Introduction: Science, Society, and Social Thought," in *No Other Gods: On Science and American Social Thought* (Baltimore: Johns Hopkins Univ. Press, 1976), pp. 1–21.

One form of "scientific medicine." Preserved tumors from the pathological teaching collection of Thomas D. Mütter (1811–1859), professor of surgery at Thomas Jefferson Hospital. Courtesy of the Mütter Museum, College of Physicians, Philadelphia.

Yet too often in historical discourse the terms "scientific medicine" and "medical science" have been narrowly associated with the particular form of laboratory-based science that came to the fore only after the Civil War. More is at stake here than fuzzy vocabulary, for the identification of science with the form it assumed during a particular period of American history has misshaped the questions historians have asked about the role of science in the late nineteenth-century transformation of medicine. It has especially misled historians to ask: Why and how did medicine become scientific in the late nineteenth century? Recognizing that medicine did not simply become more scientific, but that instead what constituted science was redefined, makes it clear that the question glosses over the principal change. More probing formulations might ask how one sort of scientific foundation for medicine began to supplant an established one or what the specific appeals of the laboratory were compared with alternative models of science.

Even at any given time, science in medicine had vastly different meanings for various social aggregates of Americans. The assortment of coexisting medical belief systems that lay Americans subscribed to has drawn considerable historical attention. In contrast, the nondissenting public's perceptions of orthodox medical science have been virtually ignored. Martha Verbrugge's study of middle-class women's attraction to physiology and hygiene in nineteenth-century Boston remains unusual as a close examination of how one segment of the public regarded medical science and the special meaning it held for them. "Understanding of the condition and potential of mankind as revealed by physiological law," Verbrugge observed, was hailed by such women as "the panacea for moral and social decay as well as for personal ill-health."[35] Social historians have recognized the value of studying the medical experiences of women, immigrants, blacks, and other consumers of medical services. But they have tended to analyze the ways these groups were affected by a white male elite interpretation of science rather than to investigate how members of these various groups thought about and judged orthodox medical science.

Indeed, despite much programmatic talk about rewriting the history of health care from the layperson's perspective, there is virtually nothing that explicitly analyzes the meaning of science in medicine from the American patient's point of view.[36] Yet to understand, for example, the success of laboratory science as a professional leaven, it is plainly necessary to consider why the American people bought what many physicians believed laboratory science gave them to sell. Even if the ideal of experimental science won the hearts and minds of physicians, why and to what extent it did the same for the lay man and woman remains unclear, especially given the doubtful ability of scientific medicine in the late nineteenth century to deliver the goods in terms of a demonstrably elevated power to cure.

Further, science held diverse meanings for the variety of physicians who made up the regular medical profession in America at any given time. Orthodox physicians' disagreements about medical science went far deeper than disputes over each other's theories. They often disagreed fundamentally about what kinds of natural knowledge were and were not legitimate and desirable in medicine. Sometimes these differences mirrored shifts in the profession's epistemological orientation from one generation to the next, as in the strident opposition of American clinical empiricists bred in the sensualist tradition of E. B. de Condillac and P. J. G. Cabanis to those older physicians who continued through the first half of the nineteenth century to endorse systems of practice cast in the mold of Enlightenment medical rationalism. Other differences can be attributed to physicians' social and intellectual diversity. It is unreasonable, for instance, to assume that the average rural practitioner and the urban medical professor defined science in the same ways or had the same expectations of it.

Regina Markell Morantz has elegantly illustrated the variations even in apparently similar physicians' postures toward science by contrasting the scientific faiths of two early and prominent women physicians, Elizabeth Blackwell and Mary Putnam Jacobi. Blackwell believed that women's gender made them peculiarly suited to be physicians through their capacities to support, nurture, and minister to their patients' spiritual needs. She grew up in a family committed to abolitionism, women's rights, and Christian perfectionism and saw in medicine a vehicle for the elevation of women and moral reform. Late nineteenth-century medical science appeared to be at odds with these reformist goals, for it threatened to corrupt the female practitioner by diverting her attention from the whole

[35] Martha H. Verbrugge, "The Social Meaning of Personal Health: The Ladies' Physiological Institute of Boston and Vicinity in the 1850s," in *Health Care in America*, ed. Reverby and Rosner (cit. n. 2), pp. 45–66, on p. 50. See also the use of oral history in James H. Jones, *Bad Blood: The Tuskegee Syphilis Experiment* (New York: Free Press, 1981).

[36] By far the most sophisticated of such calls I have read, and one that specifies a concrete research plan to bring this program to fruition, is Roy Porter, "Introduction," in *Health, Healing, and the People*, ed. R. Porter (Cambridge: Cambridge Univ. Press, in press).

patient to germs and diseased organs. Jacobi, in contrast, saw science as an unambiguous good and the main source of medical progress. Science rather than feminine virtues, she believed, should be the primary foundation of the female physician's professional identity. By rejecting the notion that science conveyed the same messages even to two contemporaneous women physicians whose shared but unusual professional status might imply a common attitude toward science, Morantz's study underscores the error of assuming that science held a unified meaning for all the members of a motley profession.[37] By and large, though, historians have failed to explicate this diversity in appraisals of science or to recognize its critical importance in explaining the profession's debates on, for example, medical education, the production of new knowledge, the limitations of medical power, and the proper emphases to be placed on cure and prevention.

The meaning of science in medicine was not monolithic for the individual at any single time either. A common tendency in historical writing has been to focus upon a single function of medical science—be it social control, professionalization, or the improvement of patient care—and to treat that as if it were the only one that mattered. But science conveyed multiple meanings simultaneously. Kohler emphasized this point in his study of the development of biochemistry as a discipline. "One cannot distinguish purely technical aspects of ideas from their role as political strategies in the competition for resources," he argued. "Ideas are judged not only for their truth value but also for their utility in discipline building."[38] That Kohler elected to scrutinize biochemical ideas more as agents in the political economy of the discipline than as components of a particular system of knowledge in no way diminishes the cogency of his assertion that for the biochemist or physician who regarded biochemistry as a medical science, intellectual, clinical, and political significance coexisted. Similarly, recent studies of the theoretical beliefs that antebellum Southern physicians held about their region's medical distinctiveness have emphasized that this concept served their intellectual and political purposes alike. Recognizing the sociopolitical utility of the notion of Southern medical distinctiveness in furthering Southern nationalism does not change the fact that as a scientific concept it was central to the thinking physician's intellectual program for medicine.[39] A historian's decision to analyze only one function of medical science by no means implies that it did not serve the same people in other ways that were equally important.

Calls such as Kohler's to desist from seeing the intellectual and sociopolitical aspects of any endeavor as intrinsically at odds have been more developed and heeded outside the history of medicine than within it. The Renaissance historian

---

[37] Regina Markell Morantz, "Feminism, Professionalism, and Germs: The Thought of Mary Putnam Jacobi and Elizabeth Blackwell," *Amer. Quart.,* 1982, *34*:459–478. An explicit analysis of physicians' differing views of science in the British context is John Harley Warner, "Therapeutic Explanation and the Edinburgh Bloodletting Controversy: Two Perspectives on the Medical Meaning of Science in the Mid-Nineteenth Century," *Med. Hist.,* 1980, *24*:241–258.

[38] Kohler, *From Medical Chemistry to Biochemistry* (cit. n. 17), p. 6.

[39] John Harley Warner, "The Idea of Southern Medical Distinctiveness: Medical Knowledge and Practice in the Old South," in "Science and Medicine in the Old South," ed. Ronald L. Numbers and Todd L. Savitt, forthcoming; and John Harley Warner, "A Southern Medical Reform: The Meaning of the Antebellum Argument for Southern Medical Education," *Bull. Hist. Med.,* 1983, *57*:364–381.

William Bouwsma, for example, argued that tight compartmentalization into historiographic approaches serves historical understanding poorly. Rather than mourning the decline of intellectual history, his field, Bouwsma proclaimed that it has been rightly merged into a new style of historical analysis, which he termed the history of meaning. He would have historians abandon the view of man dealing with his world through the antiquated faculty of "intellect" and instead approach the subjects with which intellectual historians have traditionally worked as a subset of the broader category of "all efforts to discover or to impose meaning on our experience." Bouwsma went on to explain that "these efforts are not the work of the 'intellect' or of any particular area of the personality. They are rather a function of the human organism as a whole; they are carried on both consciously and unconsciously, and they are presupposed by, and merge with, every more specific human activity."[40] It is no accident, Bouwsma maintained, that some of the best recent historical work defies classification as intellectual, social, economic, or political history. The acceptance of a history-of-meanings approach, and with it the freedom to consider broadly the human response to phenomena and ideas, has in his view unfettered the historian from the constraints imposed by a single historiographic stance.

MAKING THE MOST OF MEANINGS

The kind of history of meanings Bouwsma endorsed emphasizes that questions about the meaning of science in medicine will necessarily have multiple answers that little respect neat divisions between intellectual and social explanations or between internal and external approaches. For example, a question that is infrequently asked explicitly but that is central to grasping the meaning of science to American physicians is, What did physicians gain from science? Certainly science yielded a practical reward in helping physicians to understand, explain, and manage disease. Yet science's technical promise hardly explains the full extent to which nineteenth-century physicians pursued not only sciences with recognized clinical relevance, such as anatomy and chemistry, but also branches, such as geology, ethnology, and astronomy, with less obvious pertinence to the patient's bedside. Physicians were among the leaders of scientific cultivation, or the pursuit of science as rational amusement, in many American communities and actively participated in scientific societies that had no medical objectives. To account for this, it is necessary to look beyond utilitarian value to cultural benefits such as those Ian Inkster and Arnold Thackray have so lucidly traced in British physicians' participation in voluntary organizations.[41] Yet in partially answering the question by regarding science as a cultural resource

[40] William J. Bouwsma, "Intellectual History in the 1980s: From History of Ideas to History of Meaning," *J. Interdisc. Hist.*, 1981, *12*:279–291, on p. 283; see also John Higham and Paul K. Conkin, eds., *New Directions in American Intellectual History* (Baltimore: Johns Hopkins Univ. Press, 1979), esp. pp. xi–xix.

[41] Ian Inkster, "Marginal Men: Aspects of the Social Role of the Medical Community in Sheffield, 1790–1850," in *Health Care and Popular Medicine in Nineteenth Century England: Essays in the Social History of Medicine,* ed. John Woodward and David Richards (London: Croom Helm, 1977), pp. 128–163; Arnold Thackray, "Natural Knowledge in Cultural Context: The Manchester Model," *American Historical Review*, 1974, *79*:672–709. On the need for analyses of the cultural and other appeals of science to the practitioner, see John Harley Warner, essay review of Ian Inkster and Jack Morrell, eds., *Metropolis and Province: Science in British Culture, 1780–1850* (1983), *Transactions and Studies of the College of Physicians of Philadelphia*, 5th Ser., 1983, *5*:377–384.

(a task that remains to be taken on in earnest for American physicians), the historian does not have to dismiss or discount the practical, clinical profits. It is legitimate and oftentimes necessary for the historian to tease out for close scrutiny one aspect of the meaning of science for the physician, but he or she should not lose sight of the fact that physicians themselves did not conventionally experience their indebtedness to science in a one-dimensional way.

At the same time, the medical meaning of science must also be judged by what difference scientific beliefs made in actual medical behavior. Ackerknecht urged almost two decades ago that behavior—what medical practitioners did and not just what they said they did—should become a central focus in medical historiography. Much touted in programmatic statements, this advice has been but little put into practice. Within the past decade, however, several historians have analyzed medical case records from private and hospital practice, as well as pharmacy prescription books, to construct operational accounts of what physicians *did* therapeutically.[42] Especially intriguing have been the works of historians motivated to study therapeutic behavior less by an interest in empirical reconstruction than by the proposition that ideas can be fully understood only in their relation to action.

Regina Morantz, working with Sue Zschoche, for example, analyzed the medical prescribing practices of two late nineteenth-century Boston hospitals, one staffed by women physicians and the other by men, to show that male and female physicians practiced obstetrics in much the same way. Rhetoric about a distinctively feminine, mild therapeutic management was an inaccurate description of reality. One of the best studies relating medical ideology to bedside activity has been written by Martin Pernick. Analyzing the types of patients who did and did not receive anesthesia for major limb surgery during the decades after anesthesia was introduced, he demonstrated the practical meaning of scientific categorizations of patients according to such factors as ethnicity, age, and gender. Certainly statistical analyses of patient records must be related to knowledge gained from narrative sources, or they will fall prey to the tendency of cliometrics to become the new antiquarianism. Nevertheless, the further application of this operational approach is among the most promising ways of understanding what scientific beliefs meant to the clinical experiences of practitioners and patients.[43]

Beyond fostering new tacks in evaluating the medical implications of science, a self-conscious emphasis on the multiplicity of science's meanings can bring

[42] Erwin H. Ackerknecht, "A Plea for a 'Behaviorist' Approach in Writing the History of Medicine," *J. Hist. Med.*, 1967, 22:211–214; David L. Cowen, Louis D. King, and Nicholas G. Lordi, "Nineteenth Century Drug Therapy: Computer Analysis of the 1854 Prescription File of a Burlington Pharmacy," *Journal of the Medical Society of New Jersey*, 1981, 78:758–761; and J. Worth Estes, "Therapeutic Practice in Colonial New England," in *Medicine in Colonial Massachusetts, 1620–1820,* ed. Philip Cash, Eric H. Christianson, and J. Worth Estes (Boston: Colonial Society of Massachusetts, 1980), pp. 289–383.

[43] Regina Markell Morantz and Sue Zschoche, "Professionalism, Feminism, and Gender Roles: A Comparative Study of Nineteenth-Century Medical Therapeutics," *J. Amer. Hist.*, 1980, 67:568–588; Martin S. Pernick, "A Calculus of Suffering: Pain, Professionalism, and Anesthesia in Nineteenth Century American Medicine" (Ph.D. diss., Columbia Univ., 1979). Similarly, therapeutic ideas and professional values are interpreted against the backdrop of bedside behavior (informed in part by a computer-aided analysis of hospital and private-practice medical case records) in John Harley Warner, "The Therapeutic Perspective: Medical Knowledge, Practice, and Professional Identity in America, 1820–1885" (Ph.D. diss., Harvard Univ., 1984).

new vigor to some of the most hackneyed themes in the historiography of science in American medicine. For example, it suggests that the long-standing view that American physicians during the first three quarters of the nineteenth century were apathetic about science is much overstated. An elitist perspective of what constituted science has led historians to ignore the scientific interests and activities of the larger proportion of the profession and thereby to underestimate greatly the extent of American medical interest in science. Looking beyond the scattering of American contributions at the cutting edge of medical research to what the bulk of practitioners themselves would have regarded as science reveals vigorous and widespread scientific activity among regular physicians.

Few Americans at mid century had access to the large numbers of hospital patients required to objectify the reigning ideal of scientific medicine, which turned upon the correlation of systematic clinical observations with pathoanatomical findings made at autopsy. Judged by activity of this type, the barrenness of American medical science seems evident. But at the same time a strikingly large proportion of American practitioners regularly observed and kept records of local meteorology, fluctuating disease conditions, and other changes in the natural environment. These physicians, who often did not practice in large cities, unmistakably regarded their study of the natural history of diseases in relation to their physical and social environments as science, yet this activity has attracted scant attention from historians. Similarly, while the prevailing caricature of medical societies during this period as sociopolitical institutions largely refraining from scientific activities of universal interest is accurate, historians have inadequately recognized the extent to which such societies were centers for cultivating local knowledge about disease and therapy and thereby served important scientific functions for their members.[44]

The same departure from an elitist view of science also urges a thoroughgoing reconsideration of the notion of American indifference to basic science in the context of medicine. The simple fact is that a substantial number of American physicians during the nineteenth century pursued natural historical and pathological research on a local level with the leading objective of understanding the patterns of health and disease that immediately confronted them. And while the ability to explain the nature and occurrence of a particular locale's diseases had a practical dimension, physicians' environmental and epidemiological investigations must be regarded as in substantial measure basic research by any categorization applicable to a field like medicine. Most of this research was of little interest outside the region where it was conducted, but to regard only universal knowledge as basic is to ignore the localistic nature of much of nineteenth-century medical science.

Thus the historian of medicine Richard H. Shryock, who wrote the canonical statement of the American indifference argument in 1948, overlooked most of the extensive basic science inquiry among American physicians by not casting his sights low enough. And while Nathan Reingold's proposal that the question of American indifference simply cannot be resolved may be right for the other sciences, this is not true of American physicians' pursuits. To understand the

---

[44] See Ronald L. Numbers and John Harley Warner, "The Maturation of American Medical Science," in *Scientific Colonialism, 1800–1930: A Cross-Cultural Comparison*, ed. Nathan Reingold and Marc Rothenberg (Washington, D.C.: Smithsonian Institution Press, forthcoming).

place of this basic science research in American medicine, it will be necessary to unrivet our gaze from those few outstanding American contributions to medical science that attracted international notice and to examine closely the nature, motivation, and extent of the scientific enterprise that actually occupied the energies of so many ordinary American physicians.[45] Of course, even if American medicine was not distinguished by indifference to basic science, a recognition of medical science's multivocality certainly suggests that science occasionally did speak in a distinctively American dialect worth listening for.

Emphasizing that thinking physicians sometimes held divergent views of what was and was not science also presents a topic like the medical reception of bacteriology under a new aspect. There is no recent systematic study of the reception of bacteriology into American medicine, though many writings have touched upon aspects of this theme. One crucial ingredient in such a study will be an appreciation of the diversity of epistemological commitments among American physicians. The too-frequent equation of scientific medicine with laboratory science has tended to make historians regard resistance to bacteriology as resistance to science. Yet that was not necessarily the case. For nineteenth-century American physicians committed to the ideals of the Paris clinical school, for example, the pivot upon which a true science of medicine necessarily rotated was an ardent empiricism. Their vehement assaults upon rationalistic systems of medical practice inherited from the late eighteenth century and their efforts to reconstruct medicine on the foundation of clinical empiricism characterized American medicine during the first two thirds of the nineteenth century. Rationalism they denounced as an evil that had both misdirected medical practice and degraded the profession.

To practitioners of this mind, bacteriology, and the broader emergence after the Civil War of experimental medicine based upon laboratory science, constituted a grave threat to the clinical empiricism that defined scientific medicine. Just as the germ theory represented to the German physician Rudolph Virchow the dangerous delusions of the *naturphilosophische* medicine of his youth, so many Americans saw the rationalism that underlay extrapolations from experimental physiological and bacteriological findings as a revival of the speculative medicine their profession had fought so hard to dismantle. Such physicians were neither more nor less scientific than bacteriology's enthusiasts; they were scientific by other standards.

A PROGNOSIS

The historiographic reconstruction of the history of medicine has altered the questions asked most frequently about science and medicine but has not diminished the net attention to science. The stridency of earlier rhetoric contending

[45] Richard Harrison Shryock, "American Indifference to Basic Science during the Nineteenth Century," reprinted in *Medicine in America: Historical Essays* (Baltimore: Johns Hopkins Press, 1966), pp. 71–89; Nathan Reingold, "American Indifference to Basic Research: A Reappraisal," in *Nineteenth-Century American Science: A Reappraisal,* ed. George H. Daniels (Evanston, Ill.: Northwestern Univ. Press, 1972), pp. 38–62. See also Shryock, *American Medical Research: Past and Present* (New York: Commonwealth Fund, 1947). To be sure, more also remains to be said about the handful of American biomedical heroes from this period, as was demonstrated in Ronald L. Numbers, "William Beaumont and the Ethics of Human Experimentation," *J. Hist. Biol.,* 1979, *12*:113–135; and Numbers and William J. Orr, Jr., "William Beaumont's Reception at Home and Abroad," *Isis,* 1981, *72*:590–612.

that the history of medicine ignores social concerns in its iatrocentric pre-occupation with medical science is no longer intellectually or professionally conscionable; nor is the countering charge that the social history perspective neglects science. The extent to which studies from diverse orientations have illuminated different facets of the meaning of science in medicine reveals the field's eclecticism to be a strength. Insights will continue to come largely from studies that consider science as but one among the many elements that figured in health care. At the same time, explicit analyses stressing the multiplicity of meanings of science in American medicine certainly could reward the effort: there is still no attempt to give a synthetic account of those assorted meanings, their weights, and their interactions for orthodox American physicians at any given time, much less for the variety of healers and laypersons across time.

In the next decade, as in the last, most of the scholars who write the history of medicine are unlikely to regard their works chiefly as contributions to the history of science. Yet there is no doubt that science will remain a leading focus, especially as the emerging inclination to study twentieth-century medicine, so intermeshed with science and its culture, grows stronger. If the questions asked about science and medicine during the past decade are pressed further and a wider-ranging history of meanings developed, historians should be led back to a closer analysis of the content of medical science by their own research pro-gram. On the most fundamental level, consideration of what science meant com-mands attention to what it was. Beyond generating a fuller understanding of the place of science in a profession so squarely on the interface between the Amer-ican public and the authority, image, and products of the natural and social sci-ences, such attention should refine and discipline historical discourse. Historians who ought to know better continue to speak of "science" and "scientific med-icine" as if each term had one concrete, timeless, unambiguous meaning. More cautious, qualified use of language, reflecting the fact that science in medicine meant many things to different Americans at various times, will encourage the elucidation of such meanings instead of closing off inquiry through the use of "science" as a catchword.

Science and Religion

By Ronald L. Numbers*

MILITARY METAPHORS HAVE DOMINATED the historical literature
on science and religion since the last third of the nineteenth century,
when the Americans Andrew Dickson White and John William Draper published
their popular surveys of the supposed conflict between religion and science. Regardless of the scholarly merits of their works, they succeeded, as Donald
Fleming pointed out years ago, in setting "the terms of the debate" for future
generations of historians.[1] Even during the past few decades, which have witnessed the appearance of a number of excellent studies critical of the warfare
theme, few students of the subject have managed to avoid phrasing their discussions in martial language or addressing the question of whether scientists and
theologians have been enemies or allies. Because of the seminal influence of
White and Draper—and a lingering confusion about their views—it seems appropriate to begin our review with a brief look at their opinions before turning
our attention to the more narrowly focused works of their twentieth-century
successors.

"THE BATTLE-FIELDS OF SCIENCE"

The historiographical war between science and religion began in 1869, when
White, then president of Cornell University, lectured to a large audience at the
Cooper Union in New York City on "The Battle-Fields of Science." In this ad-
dress, carried the next day by the *New-York Daily Tribune,* he argued:

> In all modern history, interference with Science in the supposed interest of reli-
> gion—no matter how conscientious such interference may have been—has resulted
> in the direst evils both to Religion and Science, and *invariably*. And on the other
> hand all untrammeled scientific investigation, no matter how dangerous to religion
> some of its stages may have seemed, temporarily to be, has invariably resulted in
> the highest good of Religion and Science.

Although White drew most of his examples from the European theater, he admitted that America, too, had been the scene of conflict. For example, earlier
in the century the geologists Benjamin Silliman and Edward Hitchcock had suffered at the hands of some of "the pettiest and narrowest of men" for their views
on earth history, and in recent years his own university, Cornell, had come

* Department of the History of Medicine, University of Wisconsin, Madison, Wisconsin 53706.
 I wish to acknowledge support provided by the John Simon Guggenheim Memorial Foundation
and the National Science Foundation (Grant No. SES-8308523). An early version of this essay was
presented in 1984 at West Virginia University, which also provided support for the research.
 [1] Donald Fleming, *John William Draper and the Religion of Science* (Philadelphia: Univ. Pennsylvania Press, 1950), p. 131.

*Andrew Dickson
White in 1878.*

under attack for its nonsectarian policies and for daring to give "scientific studies
the same weight as classical studies."[2]

In the years following his Cooper Union address White fleshed out his account
of the relations between science and religion, adding detail to previously described conflicts and identifying numerous new battlefields, including many on
American soil. In 1876 he published a brief, 151-page survey entitled *The Warfare of Science*, and from time to time the *Popular Science Monthly* carried
"New Chapters in the Warfare of Science" by White. Finally, in 1896 he brought
out his *magnum opus*, an impressively documented, two-volume *History of the
Warfare of Science with Theology in Christendom*, which, deservedly or not,
has remained the standard account down to the present. As recently as 1965 the
historian Bruce Mazlish, in introducing an abridged paperback edition of White's
Warfare, noted that it still commanded "immense respect and continued
reading." In his opinion, White had established his thesis "beyond any reasonable doubt."[3]

In introducing the 1896 version of his *Warfare* White noted the existence of
Draper's earlier *History of the Conflict between Religion and Science*, published
in 1874. Despite the apparent similarity between his own book and Draper's,

---

[2] "First of the Course of Scientific Lectures—Prof. White on 'The Battle-Fields of Science,' "
New-York Daily Tribune, 18 Dec. 1869, p. 4.

[3] Bruce Mazlish, Preface to Andrew Dickson White, *A History of the Warfare of Science with
Theology in Christendom,* abridged ed. (New York: Free Press, 1965), p. 13. On White, see Glenn
C. Altschuler, *Andrew D. White—Educator, Historian, Diplomat* (Ithaca, N.Y.: Cornell Univ.
Press, 1979).

wrote White, there was a fundamental difference: Draper regarded "the struggle as one between Science and Religion," while White viewed it as "a struggle between Science and Dogmatic Theology." Draper and White did, indeed, offer strikingly different interpretations of the relationship between science and religion, but not in the way suggested by White. White's statement, sometimes accepted uncritically, implied that Draper viewed all religion, not just dogmatic theology, with suspicion—an interpretation supported by the respective titles of their books. In fact, both the titles and White's introduction are misleading.[4]

Draper's book would more accurately have been titled "History of the Conflict between Roman Catholicism and Science." The Catholic church, Draper argued, had thwarted science at every turn since it acquired temporal power. Protestants, on the other hand, had always welcomed science; their insistence on the right of private interpretation of the Scriptures made the Reformation the "twin-sister" of modern science. Besides, because Protestants were so divided, they rarely acquired the power to quash science, had they been inclined to do so. Because America was, at the time of Draper's writing, ovewhelmingly Protestant, it followed that relations between science and religion in America had been characterized more by peace than war. Indeed, Draper explicitly contrasted the negative attitude of Catholic Europe toward science with the positive attitude of Protestant America. Significantly, not one of Draper's "conflicts" took place in Protestant America.[5]

From the very beginning White insisted that Protestants were no less guilty than Catholics for the war on science. In one important respect, however, his thinking did change. In 1869 he repeatedly referred to relations between "Religion and Science" and expressed the hope that in the future the two would "stand together as allies, not against each other as enemies." In 1876 he concluded *The Warfare of Science* with a summary of "the long war between Ecclesiasticism and Science." Not until 1896 did he define the war strictly in terms of "a struggle between Science and Dogmatic Theology" and spell out the differences between religion and theology. The latter, according to White, made untestable statements about the natural world and regarded the Bible as a scientific text; religion consisted simply of recognizing "a Power in the universe" and living by the Golden Rule. Religion, so defined, often fostered science; theology smothered it. Although hints of this distinction appear in his earlier works, the focus on dogmatic theology seems to have been more of an afterthought— a misleading effort to distance his own views from Draper's—than an essential premise.[6]

SCIENCE AND RELIGION IN THE COLONIES

The first two centuries of European settlement in North America coincided with one of the greatest intellectual upheavals in history. The generation of the Pilgrims lived in a world designed by the ancient philosophers Aristotle and Ptolemy, in a stationary home near the center of a finite cosmos. God lived not

[4] Andrew Dickson White, *A History of the Warfare of Science with Theology in Christendom*, 2 vols. (New York: Appleton, 1896), Vol. I, p. ix.

[5] John William Draper, *History of the Conflict between Religion and Science* (New York: Appleton, 1874), pp. 182, 200, 286.

[6] "First of the Course of Scientific Lectures," p. 4; Andrew Dickson White, *The Warfare of Science* (New York: Appleton, 1876), p. 145; White *History of Warfare* (1896), Vol. I, p. xii.

very far away and often reminded the colonists of his power and presence—to
say nothing of his displeasure—by sending lightning and earthquakes, comets
and meteors, famine and disease. Frightened—and sometimes repentant—Americans fasted and prayed and clung to the Bible as the repository of spiritual and
temporal truth.

By the time of the American Revolution, Copernicus, Newton, and their
fellow revolutionaries had substantially remodeled the world. Earth's residents
now whirled through space in an infinite universe, the magnitude and complexity
of which telescopes and microscopes daily revealed. The world ran like a giant
machine, regulated by immutable natural laws that determined astronomical and
meteorological events. As this mechanical philosophy pushed God further and
further into the distance, says the English historian Keith Thomas, it "killed the
concept of miracles, weakened the belief in the physical efficacy of prayer, and
diminished faith in the possibility of direct divine inspiration."[7] Reason replaced
revelation as the surest guide to truth.

Although only a minority of Americans experienced the diminution of belief
to the extent described by Thomas, the events associated with the Scientific
Revolution have provided numerous historians with classic illustrations of the
alleged antagonism between science and theology. White, for example, in pas-
sages scattered throughout his large book, traced the progress of the colonial
conflict from the days of the seventeenth-century Puritan divine Increase
Mather, through the transition period represented by his son Cotton, and on to
the scientific victories of Benjamin Franklin and John Winthrop in the mid-eigh-
teenth century. For White, the elder Mather, who persecuted witches and saw
divine portents in virtually everything unusual, epitomized clerical superstition.
Cotton Mather, a minister with broad scientific interests, also persecuted
witches and defended the Scriptures, but at the same time took advanced positions on many scientific issues: "he favoured inoculation as a preventive of
smallpox when a multitude of clergymen and laymen opposed it; he accepted
the Newtonian astronomy despite the outcries against its 'atheistic tendency';
he took ground against the time-honoured dogma that comets are 'signs and
wonders.'" Still later, Professor Winthrop of Harvard bested the Reverend
Thomas Prince in their debate over the cause of earthquakes, and Franklin's
kite experiment in 1752 single-handedly brought down "the whole tremendous
fabric of theological meteorology reared by the fathers, the popes, the medieval
doctors, and the long line of great theologians, Catholic and Protestant."[8]

White's militaristic interpretation went virtually unchallenged until the 1930s,
when students of Puritanism in both the Old and New Worlds began to reassess
the relationship between the Puritans and science. The pioneer revisionists
among the Americanists were Samuel Eliot Morison, Theodore Hornberger, and
Perry Miller. In an influential essay in 1934 Morison argued that "the Puritan
clergy, instead of opposing the acceptance of the Copernican theory, were the
chief patrons and promoters of the new astronomy, and of other scientific discoveries, in New England"—a theme that reappeared two years later in his

---

[7] Keith Thomas, *Religion and the Decline of Magic*: *Studies in Popular Beliefs in Sixteenth and
Seventeenth Century England* (London: Weidenfeld & Nicolson, 1971), p. 643.
[8] White, *History of Warfare* (1896), Vol. I, pp. 173, 194–197, 207, 227, 335, 361, 364–366; Vol.
II, pp. 56–57, 127, 146–147.

books on Harvard College and on intellectual life in colonial New England. In a series of biographical studies published in the late 1930s Hornberger provided additional evidence of Puritan interest in scientific matters, at times correcting White's version of events and undermining Herbert W. Schneider's contention that science "was something entirely irrelevant to [the Puritans'] interests and problems."[9]

In contrast to Morison and Hornberger, who primarily documented and dated Puritan interest in science, Miller in his two seminal volumes on *The New England Mind* (1939, 1953) attempted to place Puritan attitudes toward science in cultural context and to analyze the ways in which they used science to further their own ends. In his first volume, on the seventeenth century, he argued that colonial Puritans paid little attention to the scientific revolution that so agitated their brethren abroad. However, given assurances "by the authoritative propagandists for the [Royal] Society, that the scientific innovations did not undermine, but rather supported the doctrines of piety, that the most important service of the new physics was its service to religion," they accepted the new science with equanimity, if not enthusiasm.[10]

In his second volume, in which he carried his analysis to about 1730, Miller again stressed the relative indifference, but lack of hostility, with which American Puritans greeted scientific developments. In a chapter devoted to the controversy over inoculation for smallpox in the early 1720s, he illustrated the error of reducing the episode to a simple conflict between science and religion. In contrast to White, who had praised the heroism of Cotton Mather and Zabdiel Boylston for advocating inoculation while being bombarded with scriptural objections, Miller emphasized Mather's insufferable arrogance and his recklessness in risking the lives of his neighbors merely to vindicate his own social position. The controversy, Miller insisted, involved a struggle for status and authority between Mather's ministerial party and Dr. William Douglass's anticlerical faction; only scholars ignorant of colonial history would interpret it as "one of the engagements in [the] warfare of science with theology."[11]

One might imagine that Miller's contemptuous dismissal of the warfare school of historians marked the end of the White tradition. For some historians, it did. Raymond P. Stearns, for example, in his magisterial survey of *Science in the British Colonies of America* (1970) said bluntly, without offering supporting arguments, that "there was no conflict between science and religion, nor were there any controversies of this nature either in colonial America or in the homeland." But many historians clung to the notion of warfare. Hornberger claimed that "an awareness of areas of conflict" began to appear in the years after 1690.

---

[9] Samuel Eliot Morison, "The Harvard School of Astronomy in the Seventeenth Century," *New England Quarterly*, 1934, 7:3–24; Morison, *The Intellectual Life of Colonial New England* (New York: New York Univ. Press, 1956), pp. 241–275 (first publ. 1936 as *The Puritan Pronaos*); Morison, *Harvard College in the Seventeenth Century* (Cambridge, Mass.: Harvard Univ. Press, 1936), pp. 208–251; Herbert Wallace Schneider, *The Puritan Mind* (1930; Ann Arbor: Univ. Michigan Press, 1958), pp. 42–47; Theodore Hornberger, "Puritanism and Science: The Relationship Revealed in the Writings of John Cotton," *New Engl. Quart.*, 1937, 10:503–515. For a list of Hornberger's publications, see William D. Andrews, "A Bibliography of the Writings of Theodore Hornberger," *Early American Literature*, 1973, 7:261–264.

[10] Perry Miller, *The New England Mind: The Seventeenth Century* (New York: Macmillan, 1939), pp. 207–235.

[11] Perry Miller, *The New England Mind: From Colony to Province* (Cambridge, Mass.: Harvard Univ. Press, 1953), pp. 345–366, 437–446.

In *Seeds of Liberty* (1949) Max Savelle described the relationship between science and religion in colonial America in language that even White would have admired. In 1967 Michael G. Hall wrote of a "covert war between science and theology" that broke out in the late seventeenth century as secular ideas began to clash with religious beliefs. "Not the removal of the earth from the center of the universe, but the removal of God from the earth brought about the war between science and theology," he asserted. However, the only evidence of conflict he offered involved astrology, not science. In 1980 Rose Lockwood sought to identify the scene of Hall's covert action. "The arena of this controversy was not the sermons and essays of the learned puritan establishment," she wrote; "it was the humble literary companion of ordinary New Englanders—the almanac." Having said this, she hastened to explain why the Puritans accepted the new astronomy "with such apparent lack of discomfort."[12] Such discussions lead to a suspicion that the war between science and theology in colonial America has existed primarily in the cliché-bound minds of historians.

In recent years historians have shied away from broad studies of science and religion in the colonies, choosing to focus instead on particular individuals or events. Not surprisingly, seventeenth-century scholars have tended to concentrate on the reasons why the Puritans of New England suffered so little trauma coming to terms with Copernicanism. (Hall is one of the few historians to question the consensus about the easy assimilation of the new astronomy. "Judging from their repeated polemics" in defense of the Copernican doctrine, he argues, "the younger generation had a hard time persuading its elders.") The most important contribution to this topic is a 1964 essay by Donald Fleming, which, because it appeared in an obscure European work, is rarely cited. Fleming took issue with Miller's assertions that New England Puritans passively accepted the new astronomy and that they decided between old and new cosmologies on the basis of "the verification of experience." On the contrary, wrote Fleming, the New Englanders greeted Copernicanism with "a compulsive embrace"—not because of the empirical evidence in its favor (of which there was little), but because the Copernican theory liberated them from the confinement of sensory experience and "drew deep and welcome drafts upon their faith in reason."[13]

Eighteenth-century historians of science and religion have displayed a fondness for studying Cotton Mather[14] and responses to the great earthquakes of

[12] Raymond Phineas Stearns, *Science in the British Colonies of America* (Urbana: Univ. Illinois Press, 1970), p. 160; Theodore Hornberger, *Scientific Thought in the American Colleges, 1638–1800* (Austin: Univ. Texas Press, 1945), p. 82; Max Savelle, *Seeds of Liberty: The Genesis of the American Mind* (New York: Knopf, 1949), pp. 84–150; Michael G. Hall, "Renaissance Science in Puritan New England," in *Aspects of the Renaissance: A Symposium,* ed. Archibald R. Lewis (Austin: Univ. Texas Press, 1967), pp. 123–136; Rose Lockwood, "The Scientific Revolution in Seventeenth-Century New England," *New Engl. Quart.,* 1980, *53*:76–95.

[13] Hall, "Renaissance Science," p. 128; Donald Fleming, "The Judgment upon Copernicus in Puritan New England," in *Mélanges Alexandre Koyré,* 2 vols. (Paris: Hermann, 1964), Vol. II, pp. 160–175.

[14] On Mather, see, e.g., Robert Middlekauff, *The Mathers: Three Generations of Puritan Intellectuals, 1596–1728* (New York: Oxford Univ. Press, 1971); Charles W. Bodemer, "Natural Religion and Generation Theory in Colonial America," *Clio Medica,* 1976, *11*:233–243; Margaret Humphreys Warner, "Vindicating the Minister's Medical Role: Cotton Mather's Concept of the *Nishmath-Chajim* and the Spiritualization of Medicine," *Journal of the History of Medicine and Allied Sciences,* 1981, *36*:278–295; as well as the earlier work by Otho T. Beall, Jr., and Richard H. Shryock, *Cotton Mather: First Significant Figure in American Medicine* (Baltimore: Johns Hopkins Press, 1954).

1727 and 1755, especially the exchange between Winthrop and Prince following the latter event, which White, overlooking the professor's religious orthodoxy, viewed as a skirmish in the colonial warfare between science and theology. Much of the earthquake literature has contributed little to our understanding of science and religion: in 1936 Hornberger tried (and failed) to set White straight, in 1940 Eleanor M. Tilton corrected Hornberger, and so forth. Perhaps the most radical revision of the old historiography that celebrated the victories of science was John E. Van de Wetering's description in 1965 of Prince's efforts to fuse science and religion as a "tragedy"—not because of the harm they did to science, but because of the danger they posed to theology. According to Van de Wetering, Prince erred in viewing "the new science from a scientist's frame of reference rather than from a cleric's point of view."[15]

Almost the entire body of literature on science and religion in the colonies has focused on the Puritans of New England; consequently, we know virtually nothing about relations to the south. Whether this situation has resulted from historical neglect or from the paucity of pertinent documents—or both factors— remains uncertain. Richard Beal Davis, in his prize-winning three-volume *Intellectual Life in the Colonial South, 1585–1763* (1978), devotes substantial sections to both science and religion but virtually ignores possible interaction. Readers might conclude that the Scientific Revolution failed to influence southern thinking were it not for Davis's description of the second half of the colonial period in the South as "an age of rationalism and Newtonian science" and his statement that "Newtonian science permeated . . . religion."[16] If true, the story remains to be told—for the entire region south of Connecticut.

Even for New England, our knowledge is based largely on the views of a small group of ministers, teachers, and almanac makers. Nevertheless, we glibly generalize about the "New England mind," ignoring the differences that may have existed between the educated elite and the unschooled masses. This tendency to homogenize colonial society has undoubtedly led us into error. Henry F. May, who has written extensively on the American Enlightenment, has recently questioned the common opinion that the eighteenth century witnessed "a decline of religion" as the diffusion of science steadily undermined the biblical idea of Providence. Very few Americans, he claims, became skeptics, or deists, and if religious orthodoxy declined, as countless revivalists alleged, it probably resulted more from urban prosperity and complacency and from "frontier ignorance and sectarian bickering" than from the secularizing influence of science.

[15] White, *History of Warfare* (1896), Vol. I, pp. 364–366; Theodore Hornberger, "The Science of Thomas Prince," *New Engl. Quart.*, 1936, 9:26–42; Eleanor M. Tilton, "Lightning-rods and the Earthquake of 1755," *New Engl. Quart.*, 1940, 13:85–97; John E. Van de Wetering, "God, Science, and the Puritan Dilemma," *ibid.*, 1965, 38:494–507. See also I. Bernard Cohen, "Prejudice against the Introduction of Lightning Rods," *Journal of the Franklin Institute*, 1952, 253:393–440; Charles Edwin Clark, "Science, Reason and an Angry God: The Literature of an Earthquake," *New Engl. Quart.*, 1965, 38:340–360; William D. Andrews, "The Literature of the 1727 New England Earthquake," *Early Amer. Lit.*, 1973, 7:281–294; Maxine Van de Wetering, "Moralizing in Puritan Natural Science: Mysteriousness in Earthquake Sermons," *Journal of the History of Ideas*, 1982, 43: 417–438.
[16] Richard Beale Davis, *Intellectual Life in the Colonial South, 1585–1763*, 3 vols. (Knoxville: Univ. Tennessee Press, 1978), Vol. II, p. 697–698. Exceptions to the general absence of non-Puritan studies of science and religion include Alfred Owen Aldridge, *Benjamin Franklin and Nature's God* (Durham, N.C.: Duke Univ. Press, 1967); and Daniel J. Boorstin, *The Lost World of Thomas Jefferson* (Boston: Beacon, 1960).

At any rate, advises May, in the future if "one says that Newtonian ideas spread, or scepticism increased, or religion declined, one must say when, where, and among whom."[17]

SCIENCE AND SCRIPTURE IN THE EARLY REPUBLIC

The years between the Revolution and Civil War, especially those after the turn of the century, witnessed the rapid growth of both evangelical Christianity and scientific inquiry in America. At the very time Bible-thumping revivalists were filling the pews of the nation's churches, many scientists were laboring to substitute a dynamic, naturalistic history of the world for the static, supernatural view found in the Scriptures. Such activity threatened the harmonious relationship between science and religion that had characterized the colonial period—and provided White with fresh examples of alleged conflict. According to the Cornell historian, the nebular hypothesis regarding the origin of the solar system ignited a fierce war that raged "throughout the theological world," discoveries in historical geology provoked skirmishes between American theologians and geologists, and suggestions that the human race had descended from several pairs of ancestors (not just Adam and Eve) brought the anthropologists under attack. Throughout the antebellum period, White suggested, the progress of science was retarded by its "subordination to theology and ecclesiasticism" in church-run colleges. White identified the passage of the Morrill Act in 1862 as the turning point in the struggle to free scientific instruction from theological controls—in part because this legislation, which established the land-grant colleges, had led to his own involvement in the warfare between science and theology.[18]

In contrast to the relatively early attention paid to science and religion in the colonies, subsequent developments attracted little notice before the 1960s. One of the few exceptions to this generalization was Conrad Wright's 1941 essay "The Religion of Geology," which focused on the United States and, in theme if not in substance, perpetuated the notion of science at war. For three decades after the appearance of Charles Lyell's *Principles of Geology* (1830–1833), claimed Wright, "the conflict between the geologists' account and the Mosaic story of Creation raged furiously, both in England and America"; yet most of his evidence indicated that even the theological community responded positively to attempts by Benjamin Silliman, Edward Hitchcock, and James Dwight Dana to reconcile Genesis and geology. In fact, the sharpest criticism of Silliman came from a fellow professor of geology, the free-thinking Thomas Cooper of South

---

[17] Henry F. May, "The Decline of Providence?" in *Ideas, Faiths, and Feelings: Essays on American Intellectual and Religious History, 1952–1982* (New York: Oxford Univ. Press, 1983), pp. 130–146. In *The Enlightenment in America* (New York: Oxford Univ. Press, 1976) May says surprisingly little about the influence of scientific ideas. Herbert Leventhal, *In the Shadow of the Enlightenment: Occultism and Renaissance Science in Eighteenth-Century America* (New York: New York Univ. Press, 1976), stresses the limits of the Enlightenment. In "Beyond Reason and Revelation: Perspectives on the Puritan Enlightenment," *Studies in Eighteenth-Century Culture*, 1981, *10*:165–179, Bruce Tucker laments the lack of adequate studies of the relationship between Puritans and the Enlightenment; as I have suggested, the problem is much worse for other religious groups, outside New England.

[18] White, *Warfare* (1876), pp. 134–135; White, *History of Warfare* (1896), Vol. I, pp. 17–18, 120, 224–225, 271, 412–414.

Carolina, who accused the Yale professor of an "absolute unconditional surrender of his common sense to clerical orthodoxy."[19]

Despite such early attention, the history of geology and theology in antebellum America remains, by and large, to be written. We still have no American equivalent of Charles C. Gillispie's British classic *Genesis and Geology* (1951); no biographical accounts of Silliman, Hitchcock, and Dana—or Cooper—equal to Nicolaas A. Rupke's portrait of the English geologist-cleric William Buckland; and not a single significant article on American responses to the subsidence (in geological literature) of Noah's flood.[20] Americanists have, however, contributed a number of useful articles on topics pertaining to geology and religion, including several on the bitter exchange in the 1850s between Dana and Tayler Lewis over the meaning of Genesis 1.[21]

In an important 1969 article, which prophesied the direction studies of science and religion would soon be taking, Morgan B. Sherwood challenged attempts to read the Dana-Lewis debate as another example of theology at war with science. The disagreement, Sherwood argued convincingly, had little to do with geological empiricism versus biblical literalism. Rather, the debate arose over questions of intellectual jurisdiction: whether philologists, trained in biblical languages, or geologists, who studied rocks, were better qualified to interpret the Scriptures. Recently James R. Moore has likewise stressed the territorial dimensions of nineteenth-century debates over Genesis and geology, arguing that professional geologists and professional biblical scholars formed an alliance against amateur exegetes and scriptural geologists.[22]

Of course, it is not always a simple matter, even for experts in the history of science, to sort out the tangled threads of science and religion that appear in nineteenth-century works on geology. For example, in two studies of Hitchcock that appeared simultaneously in 1972, Philip J. Lawrence emphasized the cleric-geologist's reluctance "to yield up his religion to his science," while Stanley M.

[19] Conrad Wright, "The Religion of Geology," *New Engl. Quart.*, 1941, *14*:335–358. In the brief "Mosaic Geology in America," in *The Age of the World: Moses to Darwin* (Baltimore: Johns Hopkins Press, 1959), pp. 250–265, Francis C. Haber emphasizes the resistance of the religiously orthodox to developments in geology.

[20] Charles Coulston Gillispie, *Genesis and Geology: A Study in the Relations of Scientific Thought, Natural Theology, and Social Opinion in Great Britain, 1790–1850* (Cambridge, Mass.: Harvard Univ. Press, 1951); Nicolaas A. Rupke, *The Great Chain of History: William Buckland and the English School of Geology (1814–1849)* (Oxford: Clarendon, 1983).

[21] Bert James Loewenberg, in "The Reaction of American Scientists to Darwinism," *American Historical Review*, 1933, *38*:698–699, contrasts Lewis's alleged biblical literalism and "unreasoned dogmatism" with Dana's humble, scientific spirit. Regarding an earlier debate, see John H. Giltner, "Genesis and Geology: The Stuart-Silliman-Hitchcock Debate," *Journal of Religious Thought*, 1966–1967, *23*:3–13; Giltner describes this episode as the beginning of "the American phase of the long, sometimes obscure, and usually heated controversy between biblical literalists and geologists"—but he then emphasizes "accommodation" more than "conflict."

[22] Morgan B. Sherwood, "Genesis, Evolution, and Geology in America before Darwin: The Dana-Lewis Controversy, 1856–1857," in *Toward a History of Geology*, ed. Cecil J. Schneer (Cambridge, Mass.: MIT Press, 1969), pp. 305–316; James R. Moore, "Geologists and Interpreters of Genesis in the Nineteenth Century," in *God and Nature: Historical Essays on the Encounter between Christianity and Science*, ed. David C. Lindberg and Ronald L. Numbers (Berkeley/Los Angeles: Univ. California Press, in press). On the Dana-Lewis debate, see also John C. Greene, "Science and Religion," in *The Rise of Adventism: Religion and Society in Mid-Nineteenth-Century America*, ed. Edwin S. Gaustad (New York: Harper & Row, 1974), pp. 50–69; and Ronald L. Numbers, *Creation by Natural Law: Laplace's Nebular Hypothesis in American Thought* (Seattle: Univ. Washington Press, 1977), pp. 95–100.

Guralnick, citing virtually the same evidence, concluded that Hitchcock was "scrupulously scientific."[23] The truth remains to be seen.

John C. Greene has argued that "the sciences of geology and paleontology" presented "the main challenge" to traditional religious beliefs before 1859, and it is true that geological concerns elicited the greatest comment from contemporaries. However, Richard H. Popkin has maintained that "the most fundamental challenge" came from anthropology—from notions of pre-Adamism and polygenism that "threatened to destroy the traditional picture of the nature and destiny of man, based on the Biblical account." Regardless of the merits of this claim, we do know more about antebellum anthropology than geology—thanks in large part to William Stanton's study of the so-called American school of anthropology, *The Leopard's Spots* (1960). In this much-cited work, Stanton argued that theologically orthodox Southerners passed up a unique opportunity to defend slavery scientifically (according to polygenist theory, blacks and whites were unrelated) in favor of the biblical story of the curse on Noah's son Ham, an interpretation with which Popkin concurred. Neither Stanton nor Popkin, however, adequately studied the critics of polygenism; Stanton, for example, dismissed the clergyman-naturalist John Bachman as "half theologian, half scientist, lost and confused between the hemispheres of his own personality." Bachman and like-minded Southerners deserve better, and Thomas Virgil Peterson has taken a first step toward a deeper understanding of them in his monograph *Ham and Japheth* (1978), which shows the social and religious functions served by the traditional story of Ham.[24]

The nebular hypothesis, the third front of White's antebellum war between science and theology, has received extensive treatment in my own book entitled *Creation by Natural Law* (1977), in which I attempt to show "how open warfare between the forces of science and religion was averted by reinterpreting natural and Biblical theology to accommodate the nebular hypothesis." Although this study refutes White's claim of fierce warfare, it fails, as J. H. Brooke has pointed out, to explain adequately the various factors, particularly of a theological nature, that allowed Americans to harmonize their religious beliefs with the nebular cosmogony so painlessly—a failing all too common among historians of science who write about religious matters.[25]

To explain the prevailing harmony between science and religion in the antebellum period, seemingly countless historians in recent years have appealed to the Baconianism (as taught by the Scottish school of common-sense philosophy) so pervasive among nineteenth-century Americans. The historian of science

[23] Philip J. Lawrence, "Edward Hitchcock: The Christian Geologist," *Proceedings of the American Philosophical Society,* 1972, *116*:21–34; Stanley M. Guralnick, "Geology and Religion before Darwin: The Case of Edward Hitchcock, Theologian and Geologist (1793–1864)," *Isis,* 1972, *63*: 529–543.

[24] Greene, "Science and Religion," p. 56; Richard H. Popkin, "Pre-Adamism in 19th Century American Thought: 'Speculative Biology' and 'Racism,' " *Philosophia,* 1978–1979, 8:205–239; William Stanton, *The Leopard's Spots: Scientific Attitudes toward Race in America, 1815–59* (Chicago: Univ. Chicago Press, 1960); Thomas Virgil Peterson, *Ham and Japheth: The Mythic World of Whites in the Antebellum South* (ATLA Monograph Series, 12) (Metuchen, N.J.: Scarecrow, 1978). See also Edward Lurie, "Louis Agassiz and the Races of Man," *Isis,* 1954, *45*:227–242; and H. Shelton Smith, *In His Image, But . . . : Racism in Southern Religion, 1780–1910* (Durham, N.C.: Duke Univ. Press, 1972), pp. 152–165.

[25] Numbers, *Creation by Natural Law* (cit. n. 22), p. vii; J. H. Brooke, "Nebular Contraction and the Expansion of Naturalism," *British Journal for the History of Science,* 1979, *12*:200–211.

George H. Daniels pointed the way in 1968, arguing in *American Science in the Age of Jackson* that the Baconian emphasis on factual, nontheoretical science fostered a sense of "perfect harmony" with religion, as one contemporary phrased it, while at the same time contributing to what Daniels saw as "the virtual lack of achievement" among American scientists during the years 1815 to 1845. A decade later the church historian Theodore Dwight Bozeman published an influential case study of the attitudes of Old School Presbyterians toward science, in which he argued that the Baconian philosophy not only eliminated "any apparent discord between scientific and religious views of the world," but even determined acceptable methods of biblical interpretation.[26]

The influence of Baconianism and the Scottish philosophy on matters pertaining to science and religion has been shown for groups as diverse as the professors of Princeton Theological Seminary, the preachers of the Old South, and the practitioners of spiritualism.[27] In the most ambitious survey of antebellum science and religion yet attempted, *Science and Religion in America, 1800–1860* (1978), Herbert Hovenkamp accorded Baconianism the central role in the period's "broad experiment in the unification of knowledge and belief." In fact, he hyperbolically claimed that "the 'Baconian method' determined practically everything the orthodox theologian thought was important." Unfortunately, Hovenkamp's whiggish orientation, confusing categories, and tendency to generalize from one or two examples reduce the value of what could have been a landmark contribution to our understanding of science and religion.[28] Carefully written syntheses remain one of our greatest needs.

A second element that, along with Baconianism, fostered harmony was what Bozeman has called "doxological science," that is, science pursued for the glorification of God, often to demonstrate his wisdom and power. Despite the immense popularity and historical significance of such "natural theology" in nineteenth-century America, historians have tended to discuss it only in passing. A major exception is Neal C. Gillespie's recently published analysis of the ways in which Anglo-American conchologists used natural theology in the years preceding 1859.[29]

[26] George H. Daniels, *American Science in the Age of Jackson* (New York: Columbia Univ. Press, 1968); Theodore Dwight Bozeman, *Protestants in an Age of Science: The Baconian Ideal and Antebellum American Religious Thought* (Chapel Hill: Univ. North Carolina Press, 1977).

[27] See, e.g., Mark A. Noll, ed., *The Princeton Theology, 1812–1921: Scripture, Science, and Theological Method from Archibald Alexander to Benjamin Breckinridge Warfield* (Grand Rapids, Mich.: Baker Book House, 1983); E. Brooks Holifield, "Science and Religion in the Old South: Scientists and Theologians," in *Science and Medicine in the Old South*, ed. Ronald L. Numbers and Todd L. Savitt (forthcoming); Holifield, *The Gentlemen Theologians: American Theology in Southern Culture, 1795–1860* (Durham, N.C.: Duke Univ. Press, 1978); R. Laurence Moore, *In Search of White Crows: Spiritualism, Parapsychology, and American Culture* (New York: Oxford Univ. Press, 1977); George M. Marsden, "Everyone One's Own Interpreter? The Bible, Science, and Authority in Mid-Nineteenth-Century America," in *The Bible in America: Essays in Cultural History*, ed. Nathan O. Hatch and Mark A. Noll (New York: Oxford Univ. Press, 1982), pp. 79–100.

[28] Herbert Hovenkamp, *Science and Religion in America, 1800–1860* (Philadelphia: Univ. Pennsylvania Press, 1978), pp. x, 43. For critical appraisals of Hovenkamp, see, e.g., Stanley M. Guralnick, review of *Science and Religion in America, 1800–1860*, in *Isis*, 1979, 70:627–628; and Theodore Dwight Bozeman, review of *Science and Religion in America, 1800–1860*, in *Theology Today*, 1979, 36:301–304. For a laudable example of basing generalizations on precisely defined samples, see Holifield, "Science and Religion in the Old South."

[29] Neal C. Gillespie, "Preparing for Darwin: Conchology and Natural Theology in Anglo-American Natural History," *Studies in History of Biology*, 1983, 7:93–145. For other discussions of natural

The role of religion—natural or otherwise—in motivating scientific investigation is an elusive but potentially rewarding area for future research. Is it significant, for example, that over half (50.9 percent) of America's scientific leaders on the eve of the Civil War belonged to the Episcopal church—at a time when Episcopalians ranked no higher than fifth or sixth among American religious bodies? Or are these figures mere statistical curiosities?[30] Charles E. Rosenberg has argued that some Americans sought scientific careers because "science, like religion, offered an ideal of selflessness, of truth, of the possibility of spiritual dedication." But his own studies of nineteenth-century agricultural chemists and public-health reformers are among the few attempts to pursue this link between religion and science.[31]

We are also indebted to Rosenberg for providing in *The Cholera Years* (1962) a model case study of the secularization of theories of disease in the nineteenth century—a development that, because of its intimate connection with the most basic of human feelings, may well have exerted a greater effect on the behavior and beliefs of Americans than did the more intellectually remote theories of geology, anthropology, and astronomy. In this vein, James Whorton and others have shown how moralistic health reformers fashioned a "Christian physiology" in preparation for the millennium.[32]

THE DARWINIAN DEBATES

No aspect of the interaction between science and religion in America has attracted more scholarly attention—with proportionally fewer positive results—than the debates over organic evolution in the late nineteenth century. White, a participant-observer, viewed the controversy generated by Darwin's theories as merely the latest battle in the warfare of science with theology, a conflict in which "older theologians, who since their youth have learned nothing and forgotten nothing, sundry professors who do not wish to rewrite their lectures, and a mass of unthinking ecclesiastical persons of little or no importance save in making up a retrograde majority in an ecclesiastical tribunal" tried, in predict-

---

theology, see, e.g., Daniel Walker Howe, *The Unitarian Conscience: Harvard Moral Philosophy, 1805–1861* (Cambridge, Mass.: Harvard Univ. Press, 1970), pp. 69–92; and Numbers, *Creation by Natural Law,* (cit. n. 22), pp. 77–87.

[30] Sally Gregory Kohlstedt, *The Formation of the American Scientific Community: The American Association for the Advancement of Science, 1848–60* (Urbana: Univ. Illinois Press, 1976), pp. 219–220; Edwin Scott Gaustad, *Historical Atlas of Religion in America*, rev. ed. (New York: Harper & Row, 1976), pp. 166–167, 176.

[31] Charles E. Rosenberg, *No Other Gods: On Science and American Social Thought* (Baltimore: Johns Hopkins Univ. Press, 1976), pp. 3, 109–122, 135–152. See also John C. Greene, "Protestantism, Science, and American Enterprise: Benjamin Silliman's Moral Universe," in *Benjamin Silliman and His Circle: Studies on the Influence of Benjamin Silliman on Science in America*, ed. Leonard G. Wilson (New York: Science History, 1979), pp. 11–27.

[32] Charles E. Rosenberg, *The Cholera Years: The United States in 1832, 1849, and 1866* (Chicago: Univ. Chicago Press, 1962); James C. Whorton, *Crusaders for Fitness: The History of American Health Reformers* (Princeton, N.J.: Princeton Univ. Press, 1982). On religion and medicine, see also Ronald L. Numbers, *Prophetess of Health: A Study of Ellen G. White* (New York: Harper & Row, 1976); Norman Dain, *Concepts of Insanity in the United States, 1789–1865* (New Brunswick, N.J.: Rutgers Univ. Press, 1964), pp. 183–193; William Gribben, "Divine Providence or Miasma? The Yellow Fever Epidemic of 1822," *New York History,* 1972, *53*:283–298; and Ronald L. Numbers and Janet S. Numbers, "Millerism and Madness: A Study of 'Religious Insanity' in Nineteenth-Century America," *Bulletin of the Menninger Clinic,* 1985, *49,* in press.

able fashion, to stifle "the thinking, open-minded, devoted men who have lis-
tened to the revelation of their own time as well as of times past, and who are
evidently thinking the future thought of the world."[33]

Writing in the aftermath of the antievolution crusade of the 1920s, Bert James
Loewenberg and Windsor Hall Roberts, authors of the first dissertations on the
subject, also stressed the military nature of the encounter and charged the religious party with aggression. Loewenberg, whose three articles on Darwinism
in America achieved near-classic status, argued that *The Origin of Species* (1859)
"opened another battle in the perennial warfare between science and theology
which was destined to rage until almost the close of the century," while Roberts
claimed that "some of the ablest American theologians and . . . laymen writing
in defense of theology" launched "a furious assault" upon Darwin's work.[34]

In 1944 Richard Hofstadter suggested a significant modification of the warfare
thesis in his book on *Social Darwinism in American Thought*. Although he persisted in using military rhetoric, he portrayed the evolutionists as the aggressors
and emphasized "not the strength of the resistance but the rapidity with which
the new ideas won their way in the better colleges and universities." In a subsequent work, *The Development of Academic Freedom in the United States*
(1955), he and Walter P. Metzger insisted that "the attack of the evolutionists
upon religious authority was not a 'war between science and religion' ":

> there was not *a* war, but many particular wars; a war between two kinds of men of
> knowledge—the clerical and the scientific; between two sorts of educational con-
> trol—the sectarian and the secular; between two fundamental ways of knowing—the
> authoritarian and the empiricist; between two basic approaches to instruction—the
> doctrinal and the natural. We can summarize these conflicts by saying that science
> and education joined forces to attack two major objectives—the authority of the
> clergy and the principles of doctrinal moralism—and that one of the effects of this
> coalition was the hastening of academic reform.[35]

In a dissertation on Christian Darwinism that, unfortunately, has remained
unpublished, Michael McGiffert in 1958 continued the reassessment of the Darwinian debates by focusing on "reconciliation" instead of "controversy" and by
underlining, like Hofstadter, "the remarkably rapid adjustment of substantial
sections of Protestant thought to evolution." Two years later, Joseph A. Borome
pushed the revisionist position to its limits, arguing that "religion, far from being
antagonistic to science [i.e., evolution], was welcoming it as a bulwark of su-
pernaturalism." By 1979, when James R. Moore published his extensive analysis
of the "post-Darwinian controversies," he could claim to be writing in support

[33] White, *History of Warfare* (1896), Vol. I, pp. 68–86, 313–319.

[34] Bert James Loewenberg, "The Controversy over Evolution in New England, 1859–1873," *New
Engl. Quart.*, 1935, 8:232–257, on p. 232; Windsor Hall Roberts, *The Reaction of American Protestant Churches to the Darwinian Philosophy, 1860–1900* (Chicago: Univ. Chicago Libraries, 1938),
p. 1. Loewenberg's other writings on the subject include "The Impact of the Doctrine of Evolution
on American Thought" (Ph.D. diss., Harvard Univ., 1934); "The Reaction of American Scientists
to Darwinism," *Amer. Hist. Rev.*, 1933, 38:687–701; and "Darwinism Comes to America, 1859–
1900," *Mississippi Valley Historical Review*, 1941, 28:339–368.

[35] Richard Hofstadter, *Social Darwinism in American Thought* (1944), rev. ed. (Boston: Beacon,
1955), pp. 21, 24; Richard Hofstadter and Walter P. Metzger, *The Development of Academic
Freedom in the United States* (New York: Columbia Univ. Press, 1955), pp. 344–346. For another
account of the attempt to create scientific autonomy, see William E. Leverette, Jr., "E. L. You-
man's Crusade for Scientific Autonomy and Respectability," *American Quarterly*, 1965, 17:12–32.

of "the standard revisionist thesis that Christian theology has been congenial to the development of modern science."[36] Though far from the final word on the subject, Moore's book quickly established itself as the point of departure for any future studies of science and religion in the age of Darwin, and his devastating critique of the military metaphor made it virtually impossible for any self-respecting historian to continue defending the simplistic science-versus-theology paradigm popularized by White.

Despite repeated warnings—from Hofstadter to Moore—about the inadequacies of the military metaphor, the language of war lived on. However, in recent years it has survived more in the form of off-the-cuff remarks and attention-getting rhetorical devices than as an organizing principle. For example, Paul F. Boller, Jr., in surveying American thought in the late nineteenth century, devoted an entire chapter to "The Warfare of Science and Religion"—while arguing that, after the early 1870s, "reconciliation became the order of the day."[37] It is also not uncommon to find historians of antebellum science, including ones who minimize conflict between science and religion before 1859, almost reflexively adopting warlike figures of speech when they look ahead to the post-Darwinian period.[38]

Even the leading critics of the warfare school of historiography have stopped short of denying all conflict. Hofstadter and Metzger, as we have seen, described a multiplicity of wars, and Moore himself proposed a "nonviolent and humane" interpretation that focused on conflict within the minds of persons struggling to come to terms with the new science. Ironically, the very year Moore seemingly laid White's version of the military metaphor to rest, Neal C. Gillespie brought out a sophisticated study of the shift from creationism to positivism in the Anglo-American scientific community in which he sought to revitalize "the much abused concept of the conflict of science and religion." In response to the growing inclination among historians to deny any hostilities, Gillespie insisted that there was genuine conflict, not between scientists and clerics, but within science, between advocates of an older theologically grounded science and proponents of a new positivistic science that ruled out the supernatural.[39]

[36] Michael McGiffert, "Christian Darwinism: The Partnership of Asa Gray and George Frederick Wright, 1874–1881" (Ph.D. diss., Yale Univ., 1958), pp. 3–4; Joseph A. Borome, "The Evolution Controversy," in *Essays in American Historiography: Papers Presented in Honor of Allan Nevins,* ed. Donald Sheehan and Harold C. Syrett (New York: Columbia Univ. Press, 1960), p. 170; James R. Moore, *The Post-Darwinian Controversies: A Study of the Protestant Struggle to Come to Terms with Darwin in Great Britain and America, 1870–1900* (Cambridge: Cambridge Univ. Press, 1979), p. ix.

[37] Paul F. Boller, Jr., *American Thought in Transition: The Impact of Evolutionary Naturalism, 1865–1900* (Chicago: Rand McNally, 1969), pp. 22–46. For similar examples, see, e.g., Edward A. White, *Science and Religion in American Thought: The Impact of Naturalism* (Stanford: Stanford Univ. Press, 1952), p. 110; John C. Greene, *Darwin and the Modern World View* (Baton Rouge: Louisiana State Univ. Press, 1961), p. 4; and Edwin S. Gaustad, ed., *A Documentary History of Religion in America: Since 1865* (Grand Rapids, Mich.: William B. Eerdmans, 1983), p. 327.

[38] See, e.g., Kohlstedt, *Formation of the American Scientific Community* (cit. n. 30), p. 114; Bozeman, *Protestants in an Age of Science,* (cit. n. 26), p. xiv; and Hovenkamp, *Science and Religion in America,* (cit. n. 28), pp. 48–49, 210.

[39] Moore, *Post-Darwinian Controversies,* pp. 13, 100; Neal C. Gillespie, *Charles Darwin and the Problem of Creation* (Chicago: Univ. Chicago Press, 1979), pp. 53, 152. For a critical evaluation of Gillespie, see James R. Moore, "Creation and the Problem of Charles Darwin," *Brit. J. Hist. Sci.,* 1981, *14*:189–200. See also Frank M. Turner, "The Victorian Conflict between Science and Religion: A Professional Dimension," *Isis,* 1978, *69*:356–376, which, though British in orientation, offers an analytical model applicable to America.

If historians remained divided over the very nature of the relationship between evolution and religion, it should not be surprising that they also failed to agree on such issues as the course, effects, and focus of the Darwinian debates. Loewenberg, for example, in 1935 identified the years 1859–1873 as "the period of absolute rejection," characterized by "the wholesale denunciation of frightened theologians"; six years later, without explanation, he extended the "stage of probation" to 1880. Frank Hugh Foster in 1939 offered a markedly different chronology, which a number of historians have adopted. According to Foster, American theologians largely ignored evolution before 1874, when "the grand battle was joined" following the publication of Charles Hodge's *What Is Darwinism?* Thus, as Stow Persons pointed out, Loewenberg's "controversy" between 1859 and 1874 never existed. More recently, Cynthia Eagle Russett has written in her text *Darwin in America* (1976) that "the main body of religious thought in this country gradually embraced some scheme of reconciliation with science in the years after 1870," while Moore has tentatively marked 1886, when Hodge's son Archibald published an introduction to Joseph S. Van Dyke's *Theism and Evolution,* as "a turning point for the acceptance of evolution among American Protestants."[40] Such confusion suggests the need for a simple plotting of the course of conversion to evolution among various segments of the American population.

The question of *who* embraced organic evolution remains as uncertain as the timing. Two anthologies on the Darwinian debates in America that appeared almost simultaneously in the late 1960s illustrate the problem. In one George Daniels claimed that "most religious thinkers quickly accommodated themselves to the theory," while in the other R. J. Wilson (more correctly, I suspect) pointed out that "a great majority of working clergymen and religious laymen" rejected Darwinism. Russett provides an even more striking example of guesses passing as scholarship. In her book she moves from the questionable claim that "the majority of American churchmen . . . satisfied themselves that the challenge of Darwinism could be met and weathered without any serious remodeling of the orthodox theological edifice" to the incredible assertion that "for most Christians in America, the reconciliation of Darwinism and religion left the old beliefs in all essentials intact."[41] Although we may know at present what a few hundred Americans thought about the religious implications of Darwinism (most studies generalize on the basis of the same dozen or so writers), we can only surmise how the majority of American Christians felt about Darwinism—or any other theory of evolution.

The focus of religious concern aroused by Darwin's work also remains in dispute. Did the appearance of *The Descent of Man* (1871) create "a great stir" as

[40] Loewenberg, "Controversy over Evolution," pp. 233–234; Loewenberg, "Darwinism Comes to America," p. 340; Frank Hugh Foster, *The Modern Movement in American Theology: Sketches in the History of American Protestant Thought from the Civil War to the World War* (New York: Revell, 1939), p. 43; Stow Persons, *Free Religion: An American Faith* (New Haven, Conn.: Yale Univ. Press, 1947), p. 110; Cynthia Eagle Russett, *Darwin in America: The Intellectual Response, 1865–1912* (San Francisco: Freeman, 1976), pp. 27–28; Moore, *Post-Darwinian Controversies,* pp. 241–242. Foster (p. 54) picked 1892 as the date by which "evolution was fully recognized by leaders in the liberal movement."

[41] George Daniels, ed., *Darwinism Comes to America* (Waltham, Mass.: Blaisdell, 1968), p. 95; R. J. Wilson, ed., *Darwinism and the American Intellectual: A Book of Readings* (Homewood, Ill.: Dorsey, 1967), p. 40; Russett, *Darwin in America,* pp. 29, 43.

White and most historians have maintained, or did Americans give the book a "calm reception," as Edward J. Pfeifer claims?[42] Did religious opposition arise primarily from a commitment to the inspiration of Scripture (Loewenberg and Frederick Gregory), from a desire to preserve natural theology (Persons, Wilson, and Pfeifer), or from an allegiance to particular scientific and philosophic principles (Moore)?[43] Given the present state of our knowledge, we can still offer only tentative answers.

We should not, however, allow the cloud of uncertainty that hovers over much of Darwinian scholarship to obscure some real gains that have been made in our understanding of the issues. Loewenberg, writing before much was known about science and religion in the pre-Darwinian period, believed that "evolution was so alien an idea and so at variance with the accepted formulas in science, philosophy and theology that an immediate acceptance could not be expected." I can imagine no historian defending this position today—though one might quarrel over the manner in which familiarity with evolution biased Americans during the Darwinian debates. Pfeifer, for instance, has argued that the furor created by the evolutionary *Vestiges of the Natural History of Creation* (1844) prejudiced Americans against Darwinism, while I have tried to show that widespread acceptance of planetary evolution before 1859 helped to prepare the way for organic evolution.[44]

Some of the most insightful studies of the relationship between evolution and theology have been biographical: the portrait of Asa Gray by A. Hunter Dupree, Louis Agassiz by Edward Lurie, John Fiske by Milton Berman, William Dawson by Charles F. O'Brien, and Joseph LeConte by Lester D. Stephens.[45] The attraction of such works is the detailed knowledge they provide of the ways in which individual minds have wrestled with the respective claims of science and

[42] White, *History of Warfare* (1896), Vol. I, p. 74; Edward J. Pfeifer, "United States," in *The Comparative Reception of Darwinism,* ed. Thomas F. Glick (Austin: Univ. Texas Press, 1974), p. 189. See also McGiffert, "Christian Darwinism," pp. 109–110; and Boromé, "Evolution Controversy," p. 173. For opinions similar to White's, see Loewenberg, "Controversy over Evolution," p. 238; Hofstadter, *Social Darwinism,* p. 25; Boller, *American Thought in Transition,* pp. 26–27; and Russett, *Darwin in America,* p. 31.

[43] Loewenberg, "Controversy over Evolution," p. 254; Frederick Gregory, "The Impact of Darwinian Evolution on Protestant Theology in the Nineteenth Century," in *God and Nature,* ed. Lindberg and Numbers (cit. n. 22); Stow Persons, "Evolution and Theology in America," in *Evolutionary Thought in America,* ed. Stow Persons (New Haven, Conn.: Yale Univ. Press, 1950), p. 425; Wilson, *Darwinism and the American Intellectual,* p. 39; Pfeifer, "United States," p. 182; Moore, *Post-Darwinian Controversies,* pp. 205–206, 214.

[44] Loewenberg, "Impact of the Doctrine of Evolution," p. 271; Pfeifer, "United States," p. 169; Numbers, *Creation by Natural Law,* pp. 105–118.

[45] A. Hunter Dupree, *Asa Gray: 1810–1888* (Cambridge, Mass.: Harvard Univ. Press, 1959); Edward Lurie, *Louis Agassiz: A Life in Science* (Chicago: Univ. Chicago Press, 1960); Milton Berman, *John Fiske: The Evolution of a Popularizer* (Cambridge, Mass.: Harvard Univ. Press, 1961); Charles F. O'Brien, *Sir William Dawson: A Life in Science and Religion* (Philadelphia: American Philosophical Society, 1971); Lester D. Stephens, *Joseph LeConte: Gentle Prophet of Evolution* (Baton Rouge: Louisiana State Univ. Press, 1982). Biographers of other key figures in the Darwinian debates often devote a chapter or section to science and religion; see, e.g., Ira V. Brown, *Lyman Abbott: Christian Evolutionist: A Study in Religious Liberalism* (Cambridge, Mass.: Harvard Univ. Press, 1953); Ralph E. Weber, *Notre Dame's John Zahm: American Catholic Apologist and Educator* (Notre Dame, Ind.: Univ. Notre Dame Press, 1961); William G. McLoughlin, *The Meaning of Henry Ward Beecher: An Essay on the Shifting Values of Mid-Victorian America, 1840–1870* (New York: Knopf, 1970); Clifford E. Clark, Jr., *Henry Ward Beecher: Spokesman for a Middle-Class America* (Urbana: Univ. Illinois Press, 1978); and J. David Hoeveler, Jr., *James McCosh and the Scottish Intellectual Tradition: From Glasgow to Princeton* (Princeton, N.J.: Princeton Univ. Press, 1981).

religion. Unfortunately, biographers rarely pay sufficient attention to the influence of their subjects, which they are inclined to assume or to overemphasize. Among the biographers named above, Berman provides the most detailed analysis of influence; Dupree, the most candid assessment. Gray's scheme for reconciling theism and Darwinism, he admits, "lacked the sharpness to sway the minds which were more genuinely disturbed"—an accurate but suprising conclusion in view of the frequency with which Gray appears in other studies as representative of the manner in which devout Americans came to terms with evolution.[46]

To avoid the bias of the single biography, Moore in *The Post-Darwinian Controversies* surveyed the ways in which twenty-eight British and American authors dealt with the implications of Darwinism for their faith. On the basis of this sample he concluded that "Christian anti-Darwinians" rejected Darwin primarily for scientific and philosophical rather than theological reasons; that "Christian Darwinists," nearly all of whom were theological liberals, preferred non-Darwinian theories of evolution; and that "Christian Darwinians"—and *only* such persons—successfully reconciled orthodox Darwinism with orthodox Christianity.[47] In view of Moore's herculean labors, it seems petty to fault him for not including more subjects. Nevertheless, before yielding to the temptation to extrapolate from his authors to all Christians—a temptation Moore himself resisted—we should bear in mind that his conclusion about the affinity between orthodoxy and Darwinism is derived from the opinions of only four men (including two Americans) and that for every Asa Gray and George Frederick Wright who espoused a modified version of Darwinism, there was an equally orthodox Charles Hodge or William Dawson opposing it.

Characterizing the attitudes of particular denominations has also proved hazardous. "The consensus," writes Moore, "seems to be that Unitarians in America were the most receptive to evolution, Congregationalists the most influential in interpreting and propagating it, Presbyterians alternately very hostile or quite accommodating, and Methodists, Baptists, and Lutherans reluctant but generally uninvolved." Yet even this supposed consensus is far from unanimous. Regarding the Unitarians, for example, Persons has argued that they "remained hostile for some years," and Sidney Ahlstrom has written that "the Unitarian friends of Agassiz, having invested largely in the goodness and dignity of man, would join hands with the most strenuous conservatives."[48] The problem seems to derive from the fact that most of the so-called mainline denominations tolerated a broad range of opinions on the question of evolution; and even such

[46] Dupree, *Asa Gray*, pp. 381–382. On Gray, see also Gillespie, *Charles Darwin*, pp. 111–117. On the failure of theistic evolution to attract scientific support, see Peter J. Bowler, *The Eclipse of Darwinism: Anti-Darwinian Evolution Theories in the Decades around 1900* (Baltimore: Johns Hopkins Univ. Press, 1983), pp. 15, 44–57.

[47] Moore, *Post-Darwinian Controversies*, esp. p. 303.

[48] *Ibid.*, p. 11; Persons, *Free Religion*, p. 111; Sydney E. Ahlstrom, *A Religious History of the American People* (New Haven, Conn.: Yale Univ. Press, 1972), p. 769. The lack of consensus regarding Presbyterians is reflected in the contrasting statements of Windsor Hall Roberts and Herbert W. Schneider. According to the former, the Presbyterians "put up the most stubborn, the most persistent and no doubt the ablest fight against evolution"; according to the latter, "Presbyterians found Darwinism congenial." Roberts, *Reaction of American Protestant Churches* (cit. n. 34), p. 40; Herbert W. Schneider, "The Influence of Darwin and Spencer on American Philosophical Theology," *J. Hist. Ideas*, 1945, 4:5.

distinctive and authoritarian groups as the Catholics, Mormons, and Seventh-day Adventists rarely agreed on all points.[49] We thus remain a long way from being able to rank with any confidence the relative importance of denominational affiliation, profession, education, or, as Charles Rosenberg has suggested, region and class in determining attitudes toward evolution.[50]

Darwin's contribution to the secularization of American thought—long assumed by historians to have been substantial if not seminal—also deserves a fresh look. In recent years the British church historian Owen Chadwick has downplayed the influence of science, concluding: "So far as can at present be discerned, Darwin and Darwinianism had no direct influence whatever in the secularization of the British working-man, and probably not much in that of any other worker of the nineteenth century." And inasmuch as a number of the most prominent scientific participants in the Darwinian debates maintained the same theological stance before and after 1859, Dupree has lately proposed a "non-revolutionary view of the effect of Darwin's ideas on scientists' philosophical and religious views."[51] To what extent Darwin's ideas were nonrevolutionary for other Americans remains to be shown.

A decade ago Michele L. Aldrich took American historians to task for concentrating almost exclusively on "the acceptance rather than the rejection of natural selection." Despite the contributions of Moore and others, her criticism remains valid today. Few historians would go as far as Loewenberg in dismissing anti-Darwinian views as "rationalizations unworthy of a sophomoric intelligence, arguments conceived in utter defiance of logic, superstitions more reasonable in an African tribesman than a nineteenth-century American," but they continue to slight those Americans—undoubtedly the majority—who resisted Darwin's theories.[52] A systematic and sympathetic study of such persons is long overdue.

Both historically and historiographically, the Darwinian debates have tended to eclipse other facets of the late nineteenth-century interaction between science and religion. Valuable studies have been written on such disparate topics as the rise of Christian Science, the influence of scientific values on the writing of church history, and the ways in which American missionaries enlisted science to help Christianize the world; but many subjects remain unexplored.[53] For ex-

[49] John L. Morrison, "A History of American Catholic Opinion on the Theory of Evolution, 1859–1950" (Ph.D. diss., Univ. Missouri, 1951); John Rickards Betts, "Darwinism, Evolution, and American Catholic Thought, 1860–1900," *Catholic Historical Review*, 1959–60, 45:161–185; Duane E. Jeffrey, "Seers, Savants and Evolution: The Uncomfortable Interface," *Dialogue*, 1973, 8:41–75; Richard Sherlock, "A Turbulent Spectrum: Mormon Reactions to the Darwinist Legacy," *Journal of Mormon History*, 1978, 5:33–59; Ronald L. Numbers, " 'Sciences of Satanic Origin': A History of Adventist Attitudes toward Evolutionary Biology and Geology," *Spectrum*, 1979, 9(4):17–30.

[50] Rosenberg, *No Other Gods*, pp. 2–4. Two recent studies have addressed the response of Canadians to evolution: A. B. McKillop, *A Disciplined Intelligence: Critical Inquiry and Canadian Thought in the Victorian Era* (Montreal: McGill-Queen's Univ. Press, 1979); and Carl Berger, *Science, God, and Nature in Victorian Canada* (Toronto: Univ. Toronto Press, 1983).

[51] Owen Chadwick, *The Secularization of the European Mind in the Nineteenth Century* (Cambridge: Cambridge Univ. Press, 1975), p. 106; A. Hunter Dupree, "Christianity and the Scientific Community in the Age of Darwin," in *God and Nature,* ed. Lindberg and Numbers (cit. n. 22). In contrast to Frank Hugh Foster in *The Modern Movement in American Theology,* William R. Hutchison in *The Modernist Impulse in American Protestantism* (Cambridge, Mass.: Harvard Univ. Press, 1976), p. 47, minimizes the contribution of science to the development of theological liberalism.

[52] Michele L. Aldrich, "United States: Bibliographical Essay," in *The Comparative Reception of Darwinism,* ed. Glick (cit. n. 42), p. 210; Loewenberg, "Controversy over Evolution" (cit. n. 34), p. 257.

[53] Stephen Gottschalk, *The Emergence of Christian Science in American Religious Life* (Berkeley:

ample, American historians have virtually ignored the relationship between the physical sciences and religion, despite the opinion of some contemporaries that the principle of conservation of energy posed a greater threat to belief than the principle of natural selection.[54]

SCIENCE AND RELIGION IN MODERN AMERICA

By the late nineteenth century the age-old conflict between science and religion appeared to be drawing to a close—or so it seemed to Andrew Dickson White as he penned the last version of his military history. As a result of the successful evolutionist offensive, he announced confidently, "the old theory of direct creation is gone forever."[55] Little could he have known at the time that the bitterest debates over evolution were yet to come, or that revolutionary developments in the behavioral sciences would soon open a new chapter in the relations between science and religion.

The fundamentalist crusade against evolution, launched in earnest in the early 1920s, has long been the object of scholarly derision. The earliest histories, such as Maynard Shipley's *The War on Modern Science* (1927), aimed more at condemning than at comprehending the antievolutionists. Shipley himself portrayed them as anti-intellectual boobs "bitterly opposed not only to the scientific method as applied to study of the Hebrew Scriptures, but to the method of science in general."[56] Although some creationists undoubtedly fit this description, the negative image, so long dominant, seems unlikely to survive the current reevaluation of fundamentalism, typified by George M. Marsden's standard-setting *Fundamentalism and American Culture* (1980). The fundamentalists, Marsden argues convincingly, harbored no ill will toward science as such—so long as it conformed to their outdated Baconian principles. "It is a mistake, then," he concludes, "to regard the fundamentalist controversies as at bottom a conflict between science and religion."[57]

Historians of the 1920s have produced a veritable library of studies—local, biographical, and denominational—describing the fundamentalist campaign to

---

Univ. California Press, 1973); Henry Warner Bowden, *Church History in the Age of Science: Historiographical Patterns in the United States, 1876–1918* (Chapel Hill: Univ. North Carolina Press, 1971); Peter Buck, *American Science and Modern China, 1876–1936* (Cambridge: Cambridge Univ. Press, 1980).

[54] Erwin N. Hiebert, "Modern Physics and Christian Faith," in *God and Nature*, ed. Lindberg and Numbers.

[55] White, *History of Warfare* (1896), Vol. I, p. 86.

[56] Maynard Shipley, *The War on Modern Science: A Short History of the Fundamentalist Attacks on Evolution and Modernism* (New York: Knopf, 1927), p. xii. This negative image persisted for decades; see, e.g., Norman F. Furniss, *The Fundamentalist Controversy, 1918–1931* (New Haven, Conn.: Yale Univ. Press, 1954); and Gail Kennedy, ed., *Evolution and Religion: The Conflict between Science and Theology in Modern America* (Boston: Heath, 1957).

[57] George M. Marsden, *Fundamentalism and American Culture: The Shaping of Twentieth-Century Evangelicalism, 1870–1925* (New York: Oxford Univ. Press, 1980), p. 214. Ernest R. Sandeen, who initiated the reappraisal of fundamentalism with his book *The Roots of Fundamentalism: British and American Millenarianism, 1800–1930* (Chicago: Univ. Chicago Press, 1970), relegated creationism to the periphery of the fundamentalist movement. The emergence of antievolutionism as a major concern of the fundamentalists is well documented in Ferenc M. Szasz, *The Divided Mind of Protestant America, 1880–1930* (University: Univ. Alabama Press, 1982). See also William E. Ellis, "Evolution, Fundamentalism, and the Historians: An Historiographical Review," *The Historian*, 1981, 44:15–35; and Willard B. Gatewood, Jr., ed., *Controversy in the Twenties: Fundamentalism, Modernism, and Evolution* (Nashville, Tenn.: Vanderbilt Univ. Press, 1969).

outlaw the teaching of evolution in public schools.[58] Not surprisingly, much of their attention has focused on the dramatic Scopes trial in Tennessee in 1925, when a high-school teacher was convicted of breaking a state law against teaching human evolution—a technical defeat for evolutionists, but, according to many accounts, a moral and public-relations victory. However, in an important reassessment of the results of the trial, published in 1974, Judith V. Grabiner and Peter D. Miller showed that despite the evolutionists' victory "in the forum of public opinion," they lost ground in their effort to keep evolution in the high-school curriculum. More recently, Paul M. Waggoner has challenged the notion that contemporaries viewed the trial as a creationist debacle and that 1925 represented a turning point in the antievolution movement. Rather, he argues, the crusade continued to flourish for two or three years after the trial, which, until the 1930s, was almost universally viewed as a setback to the cause of evolution.[59]

The current creation-evolution controversy, which erupted unexpectedly in the 1960s, has already prompted the writing of dozens of books, most of them nonhistorical, some of them unabashedly polemical. Among the few attempts to provide a historical perspective, Dorothy Nelkin has sketched a useful, if sometimes inaccurate, account of the textbook controversies; and I have traced the intellectual roots of "scientific creationism" back to the Seventh-day Adventist geologist George McCready Price, emphasizing the shift from a biblical to a scientific defense of special creation.[60]

Because of the historian's fascination with conflict and controversy, we now know considerably more about fundamentalist attitudes toward science than about the views of more liberal religious thinkers—the opposite of the situation for the late nineteenth century. James Ward Smith has claimed that, in contrast

[58] The best local studies of the antievolution movement are Kenneth K. Bailey, "The Enactment of Tennessee's Antievolution Law," *Journal of Southern History*, 1950, *16*:472–510; Willard B. Gatewood, Jr., *Preachers, Pedagogues & Politicians: The Evolution Controversy in North Carolina, 1920–1927* (Chapel Hill: Univ. North Carolina Press, 1966); and Virginia Gray, "Anti-Evolution Sentiment and Behavior: The Case of Arkansas," *Journal of American History*, 1970, *57*:352–366, an innovative analysis of voting patterns. Lawrence W. Levine, *Defender of the Faith: William Jennings Bryan: The Last Decade, 1915–1925* (New York: Oxford Univ. Press, 1965), sets a standard unmatched by other biographers of fundamentalists. Among the better denominational studies are John L. Morrison, "American Catholics and the Crusade against Evolution," *Records of the American Catholic Society of Philadelphia*, 1953, *64*:59–71; and James J. Thompson, Jr., "Southern Baptists and the Antievolution Controversy of the 1920's," *Mississippi Quarterly*, 1975–76, *29*: 65–81.

[59] Judith V. Grabiner and Peter D. Miller, "Effects of the Scopes Trial," *Science*, 1974, *185*:832–837; Paul M. Waggoner, "The Historiography of the Scopes Trial: A Critical Re-evaluation," *Trinity Journal*, 1984, *5* (in press). See also Ray Ginger, *Six Days or Forever? Tennessee v. John Thomas Scopes* (Boston: Beacon, 1958); and Ferenc M. Szasz, "The Scopes Trial in Perspective," *Tennessee Historical Quarterly*, 1971, *30*:288–298.

[60] Dorothy Nelkin, *Science Textbook Controversies and the Politics of Equal Time* (Cambridge, Mass.: MIT Press, 1977), revised and reprinted as *The Creation Controversy: Science or Scripture in the Schools* (New York: Norton, 1982); Ronald L. Numbers, "The Creationists," in *God and Nature*, ed. Lindberg and Numbers, an abridged version of which appeared as "Creationism in 20th-Century America," *Science*, 1982, *218*:538–544. See also Richard P. Aulie, "Evolution and Special Creation: Historical Aspects of the Controversy," *Proc. Amer. Phil. Soc.*, 1983, *127*:418–462; James R. Moore, "Interpreting the New Creationism," *Michigan Quarterly Review*, 1983, *22*:321–334; George M. Marsden, "Creation versus Evolution: No Middle Way," *Nature*, 1983, *305*:571–574; Marsden, "Understanding Fundamentalist Views of Science," in *Science and Creationism*, ed. Ashley Montagu (New York: Oxford Univ. Press, 1984), pp. 95–116; and Edward J. Larson, *Trial and Error: The American Legal Controversy over Creation and Evolution* (New York: Oxford Univ. Press, forthcoming).

to the "persistently superficial" accommodation of theology to science before 1900, "the philosophical [i.e., theological] revolution of the twentieth century is predicated upon taking science seriously." Granted the accuracy of this observation, it is lamentable that, so far as I am aware, the course of interactions between theology and modern science remains uncharted, and the changing objects of theological concern—whether in the physical, biological, or social sciences—remain largely unidentified.[61]

In the 1960s an American psychologist noted, with slight conceit, that "psychology and the social sciences are today occupying the center of the stage in the conflict between science and religious faith—a position occupied by astronomy in the 17th century, physics in the 18th, and biology and physical anthropology in the 19th." But despite the rise to prominence of the social and behavioral sciences, historians have said little about the interplay between religious belief and such developments as psychoanalysis and behaviorism. The best introduction to the subject—in fact, virtually the only historical account worth reading—is John C. Burnham's survey "The Encounter of Christian Theology with Deterministic Psychology and Psychoanalysis."[62] The psychology of religion, a related field pioneered by such Americans as G. Stanley Hall and William James, also remains underdeveloped historically, although a promising start has been made.[63]

The relationship between modern medicine and religious faith, altered but not severed by naturalistic explanations of disease, is also beginning to receive some of the scholarly attention it deserves. For example, in a collaborative project sponsored by Lutheran General Hospital in Chicago, nearly two dozen medical historians and church historians are attempting to determine the extent to which religious beliefs have influenced attitudes toward health and healing.[64] In other studies suggestive of future research, John Ettling has identified one way in which evangelical Christianity continued to influence public-health reform in the twentieth century, and David Edwin Harrell, Jr., has described in captivating detail the ambivalent feelings of millions of pentecostal Americans toward the miracles of modern medicine.[65]

---

[61] James Ward Smith, "Religion and Science in American Philosophy," in *The Shaping of American Religion,* ed. James Ward Smith and A. Leland Jemison (Princeton, N.J.: Princeton Univ. Press, 1961), devotes a lengthy section (pp. 1043–1109) to "Theology and Modern Science." Among the best introductions to this subject are Ian G. Barbour, *Issues in Science and Religion* (Englewood Cliffs, N.J.: Prentice-Hall, 1966): and Keith E. Yandell, "Protestant Theology and Natural Science in the Twentieth Century," in *God and Nature,* ed. Lindberg and Numbers.

[62] John C. Burnham, "The Encounter of Christian Theology with Deterministic Psychology and Psychoanalysis," *Bull. Menninger Clin.,* 1985, *49,* in press, from which is taken the quotation by the psychologist Joseph Havens. Among the few other studies on the subject, Frances Arick Kolb, "The Reaction of American Protestants to Psychoanalysis, 1900–1951" (Ph.D. diss., Washington Univ., 1972), is far preferable to Ann Elizabeth Rosenberg, *Freudian Theory and American Religious Journals, 1900–1965* (Ann Arbor, Mich.: UMI Research, 1980).

[63] Orlo Strunk, Jr., ed., *Readings in the Psychology of Religion* (Nashville, Tenn.: Abingdon, 1959); Dorothy Ross, *G. Stanley Hall: The Psychologist as Prophet* (Chicago: Univ. Chicago Press, 1972); Henry S. Levinson, *The Religious Investigations of William James* (Chapel Hill: Univ. North Carolina Press, 1981); Benjamin Beit-Hallahmi, "Psychology of Religion, 1880–1930: The Rise and Fall of a Psychological Movement," *Journal of the History of the Behavioral Sciences,* 1974, *10:* 84–90.

[64] To be published in a volume tentatively titled *Caring and Curing: Historical Essays on Health, Medicine, and the Faith Traditions,* ed. Darrel J. Amundsen and Ronald L. Numbers (New York: Free Press, forthcoming).

[65] John Ettling, *The Germ of Laziness: Rockefeller Philanthropy and Public Health in the New*

CONCLUSION

This historiographical survey of writings on science and religion from White to the present provides, I hope, not only a convenient evaluation of past contributions but also a modest guide for future research. Throughout this essay I have tried to indicate why I and so many other historians have come to regard the polemically attractive warfare thesis as historically bankrupt. In the form proposed by White and Draper and adopted by countless others, it assumes the existence of two static entities, "science" and "religion," thus ignoring the fact that many of the debates focused on the questions of what should be considered "science" and "religion" and who should be allowed to define them; it distorts a complex relationship that rarely, if ever, found scientists and theologians in simple opposition; it celebrates the triumphs of science in whiggish fashion; and, all too often, it fails to treat religious ideas and institutions with the respect accorded to the realm of science. I do not wish to suggest, however, that Christianity has generally fostered science, as some would have us believe, or that conflict never arose—only that we carefully define the nature of the interaction and clearly identify the participants.[66]

South (Cambridge, Mass.: Harvard Univ. Press, 1981); David Edwin Harrell, Jr., *All Things are Possible*: *The Healing & Charismatic Revivals in Modern America* (Bloomington: Indiana Univ. Press, 1975).

[66] For some excellent advice about writing on science and religion, see Martin Rudwick, "Senses of the Natural World and Senses of God: Another Look at the Historical Relation of Science and Religion," in *The Sciences and Theology in the Twentieth Century,* ed. A. R. Peacocke (Notre Dame, Ind.: Univ. Notre Dame Press, 1981), pp. 241–261.

Science in the Federal Government

By Sharon Gibbs Thibodeau*

ALMOST THIRTY YEARS HAVE PASSED since A. Hunter Dupree published *Science in the Federal Government,* a comprehensive analysis of the topic extending to the eve of World War II.[1] Time has neither diminished the importance of Dupree's classic work nor dimmed the hope that a sequel may someday be prepared. Although no new, improved *Science in the Federal Government* has appeared, the topic has hardly languished unexplored. A surprising number of studies of science in a federal context have been published since 1957. This essay reviews several of them, concentrating on those covering the recent period—the foreground of Dupree's portrait—because much of the latest work does so. The studies are characterized by considerable diversity, which may only partly be attributed to the diverse nature of government's connection to science.

Dupree observed in closing *Science in the Federal Government* that "the year 1940 marked the beginning of a new era in the relations of the federal government and science."[2] Experience has shown that this "new era" is characterized by a major increase in the federal investment in scientific activity. In 1940, federal monies accounted for only nineteen percent of the total national expenditure for research and development. By 1953, the federal portion of national support for science had grown to 54 percent.[3] Before 1940, most of the scientists receiving federal funds were civil servants employed by federal departments or agencies. Since 1940, the majority of scientists receiving federal funds do so indirectly as beneficiaries of federal contracts and grants. The nature of the science funded by federal money in recent years has changed along with the nature of the funding itself. While not completely abandoning agency work on geology, oceanography, meteorology, and agricultural chemistry, the federal government has shifted the majority of its support to the "new" disciplines of nuclear physics, space science, radio astronomy, computer science, and microbiology. Both science and the federal government have changed over the last forty years, confronting historians with a variety of complex subjects for study and analysis.

* Scientific, Economic and Natural Resources Branch, National Archives, Washington, D.C. 20408.
[1] A. Hunter Dupree, *Science in the Federal Government: A History of Policies and Activities to 1940* (Cambridge, Mass.: Belknap Press of Harvard Univ. Press, 1957). In his article "A Historian's View of Advice to the President on Science: Retrospect and Prescription," *Technology and Society,* 1980, 2:175–190, p. 175, Dupree characterizes *Science in the Federal Government* as the mechanism by which the National Science Foundation elected to carry out its statutorily mandated responsibility for developing national science policy.

[2] Dupree, *Science,* p. 369.

[3] For the 1940 figure, see John R. Steelman, Chairman, *Science and Public Policy,* (Report of the President's Scientific Research Board) (Washington, D.C.: Government Printing Office, 1947). For recent years, see *National Patterns of Science and Technology Resources, 1981* (National Science Foundation Publication 81-311) (Washington, D.C.: GPO, 1981).

FEDERAL SCIENCE BEFORE 1940

A survey of the work of historians of American science published in the last thirty years reveals that, far from being considered the last word on the topic, Dupree's *Science in the Federal Government* has stimulated a number of valuable contributions to a continuing discussion of science and government prior to World War II. Some of the most recent work—John Greene's *American Science in the Age of Jefferson,* for example—has concentrated on the earliest period of government's interest in scientific activity. The growing pains of the American patent system alluded to by Dupree have been carefully examined through the vehicle of biography by Robert Post in *Physics, Patents, and Politics: A Biography of Charles Grafton Page.* Joining Jefferson and Page as the focus of book-length studies that amplify Dupree are Spencer Fullerton Baird, Harvey W. Wiley, and Robert Millikan. The collective biographies of communities of American scientists recently published by Curtis Hinsley, Daniel Kevles, and Margaret Rossiter effectively reveal the federal government's impact on the shaping of these communities. The scientists' perspectives on government that can be drawn from these works nicely complement the government's perspective on science as portrayed by Dupree.[4]

The federal fascination with exploration recognized by Dupree has received the full consideration it deserves in two works by William Goetzmann: *Army Exploration in the American West* and *Exploration and Empire.* Particularly in the latter study, Goetzmann demonstrates that "even when the agents of exploration were not federal servants, their constant referent was nevertheless the national government and the aid and protection it might be expected to provide." In *The Great United States Exploring Expedition of 1838–1842,* William Stanton has provided a wealth of detail about "the most dramatic single enterprise" of Dupree's age of the common man. Kenneth Bertrand's excellent book, little known among historians of science, considers the Wilkes Expedition in the context of a long tradition of federal support of scientific activity in the vicinity of the South Pole. Bertrand's comprehensive discussion of this important topic is a welcome addition to the two or three paragraphs that Dupree devotes to polar exploration.[5]

[4] John C. Greene, *American Science in the Age of Jefferson* (Ames: Iowa State Univ. Press, 1983) (Greene justifies his choice of topic in "A Research Account," in *News and Views: History of Science in America,* 1984, 3:1–2); Charles Robert Post, *Physics, Patents, and Politics: A Biography of Charles Grafton Page* (New York: Science History, 1976); Dean C. Allard, Jr., *Spencer Fullerton Baird and the U.S. Fish Commission* (New York: Arno, 1978); Oscar E. Anderson, Jr., *The Health of the Nation: Harvey W. Wiley and the Fight for Pure Food* (Chicago: Univ. Chicago Press, 1958); Robert H. Kargon, *The Rise of Robert Millikan: Portrait of a Life in American Science* (Ithaca, N.Y.: Cornell Univ. Press, 1982); Curtis M. Hinsley, Jr., *Savages and Scientists: The Smithsonian Institution and the Development of American Anthropology, 1846–1910* (Washington, D.C.: Smithsonian Institution Press, 1981); Daniel J. Kevles, *The Physicists: The History of a Scientific Community in Modern America* (New York: Knopf, 1978); Margaret W. Rossiter, *Women Scientists in America: Struggles and Strategies to 1940* (Baltimore: The Johns Hopkins Univ. Press, 1982), esp. Ch. 8, pp. 218–247.

[5] William H. Goetzmann, *Army Exploration in the American West, 1803–1863* (New Haven, Conn: Yale Univ. Press, 1957); Goetzmann, *Exploration and Empire: The Explorer and the Scientist in the Winning of the American West* (New York: Knopf, 1966), quoting p. xii; William Stanton, *The Great United States Exploring Expedition of 1838–1842* (Berkeley/Los Angeles: Univ. California Press, 1975) (see Dupree, *Science,* p. 63); Kenneth J. Bertrand, *Americans in Antarctica, 1775–1948* (New York: American Geographical Society, 1971). A rare historical consideration of one of the sciences that drew Americans to Antarctica is C. Stewart Gillmor, "Early History of

According to Dupree, one of the important lessons taught by the Wilkes Expedition was the need to establish permanent federal agencies within which scientific activity might be conducted.[6] Of the many institutions eventually established in response to what proved to be a continuing need, one—the Geological Survey—has been the subject of a number of post-Dupree publications. Most of these were occasioned by celebration of the centennial of the establishment of the USGS in 1879. None, not even Mary Rabbitt's impressive multivolume history, *Minerals, Lands and Geology for the Common Defense and General Welfare,* supersede the careful critical analysis of the Survey's early years published by Thomas Manning in 1967.[7] The USGS is not the only federal scientific agency for which a significant anniversary has inspired the writing of a comprehensive administrative history. Other examples in the recent literature include the National Bureau of Standards, the Bureau of Reclamation, and the Forest Service.[8] Conceived as they are in a celebratory vein, such books can be extremely informative, but often are disappointingly uncritical.

Although the recent biographies, theme studies, and institutional histories have added a wealth of detail to Dupree's portrait of federal science before 1940, some remarkable gaps still remain in the literature. We await a comprehensive study of the oldest federal scientific institution, the Coast Survey. Inexplicably, historians have neglected the Coast Survey's most effective scientist-administrator, Alexander Dallas Bache. The large amount of manuscript material available to document the scientific activities of employees of the Coast Survey has gone unexploited, even for the best known of these employees, George Davidson and Charles Sanders Peirce.[9] (The extensive body of literature on Peirce neglects his work on gravity, pursued as an assistant in the Coast Survey.) Similarly neglected until very recently has been the second oldest federal scientific institution, the Naval Observatory, although a few articles have begun to fill that gap. Writing in *Isis* in 1955, Nathan Reingold introduced historians of science

---

Upper Atmospheric Physics Research in Antarctica," in *Upper Atmospheric Research in Antarctica,* ed. L. J. Lanzerotti and C. G. Park (Antarctic Research Series, 29) (Washington, D.C.: American Geophysical Union, 1978), 236–262.

[6] Dupree, *Science,* p. 61.

[7] Thomas Manning, *Science in Government: The U.S. Geological Survey, 1867–1874* (Lexington: Univ. Kentucky Press, 1967); the centennial volumes include Mary C. Rabbitt, *Minerals, Lands, and Geology for the Common Defense and General Welfare,* Vol. I: *Before 1879,* Vol. II: *1879–1904* (Washington, D.C.: GPO, 1979, 1980); Rabbitt, *A Brief History of the U.S. Geological Survey* (Washington, D.C.: U.S. Geological Survey, 1979); E. L. Yochelson and C. M. Nelson, *Images of the U.S. Geological Survey* (Washington, D.C.: U.S. Geological Survey, 1979); and Morris M. Thompson, *Maps for America: Cartographic Products of the U.S. Geological Survey and Others* (Washington, D.C.: GPO, 1982).

[8] Rexmond Cochrane, *Measures for Progress* (Washington, D.C.: GPO, 1966); Michael C. Robinson, *Water for the West: The Bureau of Reclamation, 1902–1977* (Chicago: Public Works Historical Society, 1979); Harold K. Steen, *The U.S. Forest Service: A History* (Seattle: Univ. Washington Press, 1976). At least a portion of the critical analysis missing from Cochrane's history of NBS may be found in Carroll W. Pursell, Jr., "A Preface to Government Support of Research and Development: Research Legislation and the National Bureau of Standards," *Technology and Culture,* 1968, 9:145–164.

[9] For a brief examination of Bache's role see Nathan Reingold, "A. D. Bache: Science and Technology in the American Idiom," *Technol. Cult.,* 1970, 11:167–177. Reingold describes the series of letters received by the Coast Survey's superintendent, along with other equally rich sources in National Archives Record Group 23, in "Research Possibilities in the U.S. Coast and Geodetic Survey Records," *Archives Internationales d'Histoire des Sciences,* 1958, 45:337–346 (see n. 31 below on Record Group numbers).

to the excellent documentation available in the National Archives for these
and other federal scientific agencies established before World War II. Most of
the opportunities for original research outlined by Reingold thirty years ago have
yet to be pursued.[10]

FEDERAL SCIENCE SINCE 1940

Historians who have elected to fill in the foreground of Dupree's portrait rather
than to add detail to his delineation of the prewar era have tended to focus either
on institutions or on specific projects or developments within disciplines. Early,
and in some sense classic, examples of both focuses for the wartime period can
be found in the series Science in World War II, published by Little, Brown
between 1947 and 1948. The one institutional history in the series, Irwin Stewart's *Organizing Scientific Research for War,* still provides a valuable entrée to
the administrative history of the Office of Scientific Research and Development
(OSRD). The other titles in the series focus on projects: *Advances in Military
Medicine; Applied Physics: Electronics, Optics, Metallurgy; Chemistry; Combat
Scientists; New Weapons for Air Warfare;* and *Rockets, Guns, and Targets.* All
of the volumes constituting the series Science in World War II are essentially
factual accounts prepared by participants in the research sponsored by the
OSRD. For historical interpretation of these facts, readers must turn to *Scientists Against Time,* a separately published "official history" of the OSRD prepared by James Phinney Baxter III. Baxter's conclusions about the effectiveness
of federal scientific activity as it was conducted by the OSRD deserve to be
fully reconsidered by modern historians, although few have accepted the challenge. Among them is Carroll Pursell, who in two recent articles has gone beyond the OSRD's impact on the war to consider its impact on science policy in
the United States.[11]

The institutional history as a vehicle for examining federal scientific activity
is represented by a number of works published since Stewart's, among them
several on postwar agencies with their roots in OSRD. Richard Hewlett, working
first with Oscar Anderson and later with Francis Duncan, has published two

[10] Nathan Reingold, "The National Archives and the History of Science in America," *Isis,* 1955,
46:22–28. On the Naval Observatory see Steven Dick, "How the U.S. Naval Observatory Began,
1830–65," *Sky and Telescope,* 1980, *60*:466–471; Jan Herman, *A Hilltop in Foggy Bottom* (rpt. from
U.S. Navy Medicine) (Washington, D.C.: Naval Medical Command, 1984); and Howard Plotkin,
"Astronomers Versus the Navy: The Revolt of American Astronomers over the Management of the
United States Naval Observatory, 1877–1902," *Proceedings of the American Philosophical Society,*
1978, *122*:385–399.

[11] For the series, all published in Boston by Little, Brown, see Irvin Stewart, *Organizing Scientific
Research for War* (1948); E. C. Andrus et al., eds., *Advances in Military Medicine* (1948; on OSRD's
Committee on Medical Research); C. G. Suits, George R. Harrison, and Louis Jordan, eds., *Applied
Physics: Electronics, Optics, Metallurgy* (1948; on Divisions 13, 15, 16, 17, and 18 of OSRD); W. A.
Noyes, Jr., ed., *Chemistry* (1948; on Divisions 8, 9, 10, 11 and 19); Lincoln R. Thiesmeyer and
John E. Burchard, *Combat Scientists,* ed. Alan T. Waterman (1947; on the Office of Field Service
of OSRD); John E. Burchard, ed., *Rockets, Guns and Targets* (1948; on Divisions 1, 2, and 3); and
Joseph C. Boyce, ed., *New Weapons for Air Warfare* (1947; on Divisions 4, 5, and 7). See also
James Phinney Baxter III, *Scientists Against Time* (Boston: Little, Brown, 1946); Carroll Pursell,
"Alternative American Science Policies During World War II," in *World War II: An Account of Its
Documents,* ed. James E. O'Neill and Robert W. Krauskopf (Washington, D.C.: GPO, 1976), pp.
151–162; and Pursell, "Science Agencies in World War II: The OSRD and Its Challengers," in *The
Sciences in the American Context: New Perspectives,* ed. Nathan Reingold (Washington, D.C.:
Smithsonian Institution Press, 1979), pp. 359–378.

volumes of a history of the Atomic Energy Commission which combines facts and analysis in a highly readable account. A third volume, now in progress, will continue the history of the AEC through the Eisenhower administration. Two recently published accounts of another of OSRD's descendants, the National Science Foundation, offer readers an opportunity to compare informal and formal institutional history. Predictably, the informal approach, adopted by Milton Lomask in 1976, was greeted with reviews that praised its readability and criticized its failure to address significant themes—in particular, NSF's impact on science and science education. A formal, extensively footnoted history of the birth and early development of the NSF, completed by J. Merton England in 1982, comes no closer to evaluating the agency's impact but succeeds in characterizing the impact of politics on bureaucracy.[12]

Although this aspect of postwar science in government has been examined thoroughly,[13] the phenomenon of federally funded research and development centers (FFRDCs) has yet to be fully analyzed by historians. Fortunately, Clayton R. Koppes has provided an excellent model study of a single such center, the Jet Propulsion Laboratory. In Koppes's hands, JPL's history effectively illustrates the degree to which federal policies (both military and civilian) can act at a distance to mold a scientific institution. The FFRDCs funded by the Department of Energy and those funded by the National Science Foundation constitute areas that might be explored following Koppes's example. Koppes's study is complemented by two project-oriented studies. One, R. Cargill Hall's history of Project Ranger, examines the working relationship between JPL and the headquarters of the National Aeronautics and Space Administration. The differences between research and development in a privately managed FFRDC and in a federally managed facility emerge when the accounts of Koppes and Hall are compared with David Allison's study of the origins of radar.[14]

Since the deterioration of Wilkes's astronomical observatory in 1842,[15] little, if any, federal scientific activity has actually taken place on Washington's Capitol Hill. Activity on Capitol Hill, however, accounts for, shapes, and assures

[12] Richard G. Hewlett and Oscar E. Anderson, Jr., *The New World, 1939–1946* (University Park: Pennsylvania State Univ. Press, 1962); Hewlett and Francis Duncan, *Atomic Shield, 1947–1952* (University Park: Pennsylvania State Univ. Press, 1969); Milton Lomask, *A Minor Miracle: An Informal History of the National Science Foundation* (NSF 76-18) (Washington, D.C.: NSF, 1976); and J. Merton England, *A Patron for Pure Science: The National Science Foundation's Formative Years, 1945–57* (NSF 82-24) (Washington, D.C.: NSF, 1982). For an analysis of NSF's peer review system, see Stephen Cole, Leonard Rubin, and Jonathan Cole, *Peer Review in the National Science Foundation: Phase One of a Study* (Washington, D.C.: National Academy of Sciences, 1978).

[13] For additional references to recently published histories of federal agencies engaged in scientific research, see Donald R. Whitnah, ed., *Government Agencies* (Westport, Conn.: Greenwood, 1983).

[14] Clayton R. Koppes, *JPL and the American Space Program* (New Haven, Conn.: Yale Univ. Press, 1982); R. Cargill Hall, *Lunar Impact: A History of Project Ranger* (Washington, D.C.: GPO, 1977); and David Allison, *New Eye for the Navy: The Origin of Radar at the Naval Research Laboratory* (Washington, D.C.: GPO, 1981). The Department of Energy's R & D centers include some administered by industrial firms, such as Oak Ridge National Laboratory (Martin-Marietta Corp.) and Sandia Laboratory (Western Electric Co., Inc.–Sandia Corp.); and others administered by universities and colleges, such as Brookhaven National Laboratory (Associated Universities, Inc.), Fermilab (Universities Research Association, Inc.), and Los Alamos Scientific Laboratory (University of California). All NSF centers are administered by colleges and universities; they include the National Center for Atmospheric Research, Boulder, Colorado, and several astronomical observatories, such as Kit Peak, Tucson, Arizona, and Cerro Tololo Inter-American Observatory, La Serena, Chile.

[15] Dick, "How the Naval Observatory Began," p. 468.

the conduct of science elsewhere. Curiously, historians of American science have neglected this crucial congressional role. They may have been daunted by the relative inaccessibility of documentation, for the records of Congress are not subject to the laws that discourage the destruction and encourage the availability of other historically valuable federal records.[16] A survey of the recent literature suggests that this obstacle has hindered historical study of congressional committees concerned with scientific activity, with one notable exception, a lengthy (1,074 pages) historical account of the House Committee on Science and Technology.[17] The work was prepared at the committee's request by Ken Hechler, a respected political scientist who served on the committee as representative of West Virginia's Fourth District until he lost reelection. One hopes that Hechler's thorough descriptive efforts will tempt others more inclined to explanation and analysis, and that they will be as successful as he in gaining access to original materials.

Numerous as they are, the various institutional histories published in recent years are far outnumbered by histories of research and development projects undertaken at federal expense. That the majority of these project histories emphasize America's space program has much to do with NASA's generous support of historical activity. To date NASA's history program has sponsored the publication of some thirty-five histories and reference works. NASA supports historical studies of its activities in part to carry out its legislated mandate to inform the public. Contributions to the NASA History Series are completed under contract and include both histories and reference works. The largely technical histories to date cover projects Vanguard, Mercury, Gemini, Ranger, and Apollo (in several volumes) and consider such subjects as space medicine, liquid hydrogen as a propulsion fuel, the early years of space science, the deep space network, and sounding rockets.[18]

In addition to its support of numerous retrospective studies, NASA may be

---

[16] One exception to this general neglect is Daniel J. Kevles, "The National Science Foundation and the Debate over Postwar Research Policy, 1942–1945," *Isis,* 1977, *68*:5–26. Federal law, codified in Title 44 of the U.S. Code, provides that records of executive agencies of the federal government may not be destroyed until the Archivist of the United States has had an opportunity to determine whether they have "sufficient administrative, legal, research or other value to warrant their continued preservation by the government" (44 U.S.C., Sec. 3303a). In language not nearly as strong, Title 44 states that the Secretary of the Senate and the Clerk of the House of Representatives "shall gather up all non-current records at the close of each Congress and transfer them to the National Archives, subject to the orders of the Senate and the House" (44 U.S.C., Sec. 2118). Many, but not all, Senate and House Committee records have been retired to the National Archives in response to this latter provision. Congressional records, whether transferred to the National Archives or not, are exempt from the provisions of the Freedom of Information Act (5 U.S.C., Sec. 552), which requires federal agencies to make all records publicly available unless they fall into a few specific exemption categories. Fortunately, the Senate recently has voted to open to researchers all of its unclassified records over twenty years old. Until the House takes similar action, access to all of its unpublished records is limited to researchers who have obtained the permission of the Clerk of the House.

[17] Ken Hechler, *Toward the Endless Frontier* (Washington, D.C.: U.S. House of Representatives, 1980).

[18] The most recent is Edward Clinton Ezell and Linda Neuman Ezell, *On Mars: Exploration of the Red Planet* (Washington, D.C.: GPO, 1984). Not all contributions to the History Series are project specific. In *Beyond the Atmosphere* (Washington, D.C.: GPO, 1980), Homer Newell provides an excellent overview of the impact of NASA on the development of space science. A helpful entrée into the workings of the NASA History Program is provided by Alex Roland in a *Guide to Research in NASA History* (Washington, D.C.: NASA Headquarters History Office, 1980).

distinguished from other civilian federal agencies by its willingness to include historical elements in ongoing projects. Working alongside scientists at the Space Telescope Science Institute are participants in the Space Telescope History Project.[19] The latter involves historians from the National Air and Space Museum, the Smithsonian Institution, and the Department of History of Science at Johns Hopkins University, who are monitoring research and development as it happens. Staff members of the History Project intend eventually to produce a history of the development and use of the Space Telescope. They are also working to ensure the preservation of documentation that will inform histories to supplement theirs.

Federal funding has proven essential to a wide variety of disciplines other than space science: cancer research, climatology, computer science, energy research, remote sensing, seismology—to name the most visible. Historians have considered developments in some of these areas and thereby assessed the impact of a federal connection. Two recent works enable readers to come to a better understanding of the convergence of private enterprise, public funding, and academic research in the development of the electronic digital computer. Kent Redmond and Thomas Smith cover public funding and academic research in their history of Whirlwind; Nancy Stern stresses private enterprise and public funding in her appraisal of the Eckert-Mauchly computers. In *Fusion,* Joan Bromberg has considered an as yet uncompleted project in the area of energy research. Her account shows how expensive hardware can take on a life of its own in government research and development. A few years ago the complicated politics of the war on cancer were examined by Richard Rettig, but the literature still lacks an analysis of the effects of the political maneuvering on the research it sought to control.[20] Work is in progress on studies of the federal role in radio research, remote sensing, and geophysics; but much of national science (and technology, for that matter) remains unexplored by American historians.

The federal government's relationship to scientific and technological research and development does not always involve extensive financial support. Government may also regulate these activities. The most recently published exploration of this connection is *Controlling the Atom: The Beginnings of Nuclear Regulation, 1946–1962,* the first of a planned multivolume history of the Nuclear Regulatory Commission. Its authors, George Mazuzan and J. Samuel Walker, view the regulatory phenomenon from Washington; views from outside are provided by Steven Del Sesto's overview of civilian nuclear power and Donald Stever's case study of the licensing of Seabrook. Various book-length studies of regulatory government's reaction to a technological crisis, one prepared by the Department of Energy, appeared in the aftermath of the nuclear accident at Three Mile Island, Pennsylvania, on 28 March 1979. Regulatory government's reaction

---

[19] Historians also "participated in" the Apollo-Soyuz Test Project. See Edward C. Ezell and Linda N. Ezell, *The Partnership: A History of the Apollo-Soyuz Test Project* (Washington, D.C.: GPO, 1978), esp. pp. 405–412, containing a discussion of the nature of the participatory history practiced.

[20] Kent C. Redmond and Thomas M. Smith, *Project Whirlwind: The History of a Pioneer Computer* (Bedford, Mass.: Digital, 1980); Nancy Stern, *From ENIAC to UNIVAC: An Appraisal of the Eckert-Mauchly Computers* (Bedford, Mass.: Digital, 1981); Joan Bromberg, *Fusion: Science, Politics, and the Invention of a New Energy Source* (Cambridge, Mass.: MIT Press, 1982); Richard Rettig, *Cancer Crusade: The Story of the National Cancer Act of 1971* (Princeton, N.J.: Princeton Univ. Press, 1977). Rettig's work builds on Stephen Strickland, *Politics, Science, and Dread Disease* (Cambridge, Mass.: Harvard Univ. Press, 1972).

to chronic problems resulting from technological innovation is carefully examined in Thomas Dunlap's *DDT: Scientists, Citizens, and Public Policy*.[21] As these examples indicate, studies of the impact of federal regulation for the most part emphasize its effects on technological development. By publishing standards for research involving DNA, the federal government has revised its traditional regulatory role to include scientific research, thus creating new possibilities for historical studies.

Although federal policymakers—in particular, those in the Executive Office of the president—have received advice, invited or unsolicited, from scientists since the beginning of the republic, it was not until after World War II that a highly visible formal advisory apparatus was created. The appointment of James R. Killian as Special Assistant for Science and Chairman of a President's Science Advisory Committee in 1957 linked scientific issues to presidential politics in an important new way. Historians have been slow to analyze this aspect of science in the federal government, but its significance has not been lost on political scientists and journalists. Unfortunately, only a few representatives of the latter groups draw on the evidential base provided by history.[22]

PERSPECTIVES

Differences in focus account for much of the variation in density of historical treatment of wartime and postwar science in the federal government. Stylistic variations in the recent secondary literature on this topic may be attributed to differences in author perspective. Consideration of the various perspectives reveals three basic types: that of the academic or professional historian, that of the reflective participant, and that of the critical observer-advocate.

The perspective of the academic historian has been described as the standard of objectivity and thoroughness against which other perspectives might be judged. However, while we may think we can recognize works that originated in academe, they are becoming more and more difficult to distinguish from works produced by historians under contract to or in the employ of the government under study. Those managing the extensive NASA history program discussed above are apparently quite conscious that the reading public might question the objectivity of its contract historians. A disclaimer appearing in one recent contribution to the History Series addresses this concern:

---

[21] George Mazuzan and J. Samuel Walker, *Controlling the Atom: The Beginnings of Nuclear Regulation, 1946–1962* (Berkeley/Los Angeles: Univ. California Press, 1984); Steven L. Del Sesto, *Science, Politics, and Controversy: Civilian Nuclear Power in the United States, 1946–1974* (Boulder, Colo.: Westview, 1979); Donald W. Stever, *Seabrook and the Nuclear Regulatory Commission: The Licensing of a Nuclear Power Plant* (Hanover, N.H.: Univ. Press of New England, 1980); Phil Cantelon and Robert C. Williams, *Crisis Contained* (Washington, D.C.: U.S. Department of Energy, 1980); Mark Stephens, *Three Mile Island* (New York: Random House, 1980); Philip Starr and William Pearman, *Three Mile Island Sourcebook: Annotation of a Disaster* (New York: Garland, 1983); and Thomas R. Dunlap, *DDT: Scientists, Citizens and Public Policy* (Princeton, N.J.: Princeton Univ. Press, 1981).

[22] See, e.g., Don K. Price, *Government and Science* (New York: New York Univ. Press, 1954); Daniel Greenberg, *The Politics of Pure Science* (New York: New American Library, 1971); and James E. Katz, *Presidential Politics and Science Policy* (New York: Praeger, 1978). For a less than optimistic view of the status of science policy studies in the United States, see A. Hunter Dupree's review of Thomas J. Kuehn and Alan L. Porter, eds., *Science, Technology and National Policy* (Ithaca, N.Y.: Cornell Univ. Press, 1981), in *Isis*, 1981, 72:649.

During the research phase, the authors conducted numerous interviews. Subsequently they submitted parts of the manuscript to persons who had participated in or closely observed the events described. Readers were asked to point out errors of fact and questionable interpretations and to provide supporting evidence. The authors then made such changes as they believed justified. The opinions and conclusions set forth in this book are those of the authors; no official of the agency necessarily endorses those opinions or conclusions.[23]

Reviewers have on the whole agreed that the NASA contracting effort has resulted in good contemporary history. Richard Hewlett has consistently shown that a "court historian" (the "court" in this case being the Atomic Energy Commission) can apply generally recognized standards of scholarship. Similarly, historians at the Department of Agriculture, primarily through journal articles, have convinced critics that high-quality historical analysis can be produced at close range.

The recently revised *Directory of Federal Historical Programs* lists over 150 major historical projects and offices in all three branches of government.[24] These programs employ some 500 historians, 300 of whom work within the military establishment. Many of these professional historians have contributed to the secondary literature of interest to historians of science. In the past, there has been a tendency to stereotype their contributions as lacking in excitement. Larry Badash's recent review of Hewlett and Duncan's *Nuclear Navy* suggests that this stereotype is no more appropriate than any other. Badash notes that the inside-government authors "have here superbly captured the drama of the administrative aspects of technological innovation."[25]

A major subset of the secondary literature on science after 1940 describes it from the perspective of participants. The reflective participants—not nearly so numerous as the historians described above—seem to divide into two groups: scientists associated with the Office of Scientific Research and Development or the Manhattan District of the Corps of Engineers; and science advisors to the president. Vannevar Bush, Arthur Compton, and James B. Conant are the best-known representatives of the former category. Contrasting with these views from the top is Leona Marshall Libby's *Uranium People*, the account of a working scientist involved in activities at Chicago, Argonne, and Hanford. Ernest C. Pollard has published similarly personal reflections on the wartime operations of the MIT Radiation Laboratory. Various reminiscences of Los Alamos have been gathered in a recently published volume edited by Larry Badash, Joseph Hirschfelder, and Herbert Broida. To these have since been added Robert Oppenheimer's recollections, edited by Alice Kimball Smith and Charles Weiner.[26]

[23] W. David Compton and Charles D. Benson, *Living and Working in Space* (Washington, D.C.: GPO, 1983), p. iv.

[24] *Directory of Federal Historical Programs and Activities* (Washington, D.C.: Society for History in the Federal Government, 1984), available from the American Historical Association.

[25] Larry Badash, review of Richard Hewlett and Francis Duncan, *Nuclear Navy* (Chicago: Univ. Chicago Press, 1974), in *Isis,* 1976, 67:147.

[26] Vannevar Bush, *Pieces of the Action* (New York: Morrow, 1970); Arthur Compton, *Atomic Quest: A Personal Narrative* (Oxford: Oxford Univ. Press, 1956); James B. Conant, *My Several Lives: Memoirs of a Social Inventor* (New York: Harper & Row. 1970); Leona Marshall Libby, *Uranium People* (New York: Scribners, 1979); Ernest C. Pollard, *Radiation: One Story of the MIT Radiation Laboratory* (Durham, N.C.: Woodburn, 1982); Larry Badash, Joseph Hirschfelder, and

In a recent special issue of *Technology in Society,* guest editor William Goldman, who had recommended to Truman the idea of appointing a science advisor, introduces an excellent comprehensive collection of insiders' commentary on science advice to the president with the observation that "no President after Thomas Jefferson can be his own Science Advisor." In pointing out why this is the case, at least one contributor to the collection, James R. Killian, could draw upon lengthier comments already published as *Sputnik, Scientists and Eisenhower: A Memoir of the First Special Assistant to the President for Science and Technology.* Killian's straightforward historical account of events surrounding the development of the science advisory mechanism nicely complements a more personal running commentary on later events as they occurred, that recorded in George Kistiakowsky's diary and published in 1976 as *A Scientist at the White House.*[27]

Several other firsthand commentaries on the various ways in which advice on science matters reaches the president have appeared in recent years. In *The Advisors: Oppenheimer, Teller, and the Superbomb,* Herbert York, first director of Lawrence Livermore National Laboratory, reflects upon the workings of the General Advisory Committee of the Atomic Energy Commission. Glenn Seaborg, chairman of the AEC from 1969 to 1971, has provided yet another valuable glimpse of the lines of communication between scientists and government policymakers in *Kennedy, Krushchev, and The Test Ban.* In *The Politics of American Science,* editors James Penick, Carroll Pursell, Morgan Sherwood, and Donald C. Swain let the policymakers speak for themselves via prepared statements, hearings, articles, and—most interesting of all—memos and letters not originally intended for publication.[28]

Participatory views of American scientific research are seldom known for their objectivity. The participants' subjectivity, however, pales in comparison with that of a growing group of critical observers who tend to emphasize the controversial aspects of the government's involvement in research and development. Examples of their works include Howard Rosenberg's *Atomic Soldiers,* James Jones's *Bad Blood,* Arthur Silverstein's *Pure Politics and Impure Science,* Adeline Levine's *Love Canal,* and Daniel Ford's *Cult of the Atom.*[29] The head-

Herbert P. Broida, eds., *Reminiscences of Los Alamos* (Boston: Reidel, 1980); and Alice Kimball Smith and Charles Weiner, *Robert Oppenheimer: Letters and Recollections* (Cambridge, Mass.: Harvard Univ. Press, 1980).

[27] William T. Goldman, ed., *Science Advice to the President* (New York: Pergamon, 1980), p. 2; James R. Killian, Jr., *Sputnik, Scientists and Eisenhower: A Memoir of the First Special Assistant to the President for Science and Technology* (Cambridge, Mass.: MIT Press, 1977); and George B. Kistiakowsky, *A Scientist at the White House* (Cambridge, Mass.: Harvard Univ. Press, 1976). In 1962 the science advisory mechanism expanded beyond these special assistants to become the President's Office of Science and Technology. Edward J. Burger, Jr., a former staff member, reports on the realities of this advisory structure in *Science at the White House* (Baltimore: Johns Hopkins Univ. Press, 1980).

[28] Herbert York, *The Advisors: Oppenheimer, Teller and the Superbomb* (San Francisco: Freeman, 1976); Glenn Seaborg, *Kennedy, Krushchev and the Test Ban* (Berkeley/Los Angeles: Univ. California Press, 1981); James L. Penick, Jr., Carroll W. Pursell, Jr., Morgan Sherwood, and Donald C. Swain, eds., *The Politics of American Science: 1939 to the Present,* rev. ed. (Cambridge, Mass.: MIT Press, 1972).

[29] Howard Rosenberg, *Atomic Soldiers: American Victims of Nuclear Experiments* (Boston: Beacon, 1980); James Jones, *Bad Blood: The Tuskegee Syphilis Experiment* (New York: Macmillan, 1981); Arthur Silverstein, *Pure Politics and Impure Science: The Swine Flu Affair* (Baltimore: Johns

line quality of these titles suggests a journalistic content that in almost all cases is confirmed on reading. For this reason, they are more accessible to the general public than work written from the perspective of an academic or professional historian. The critical observers are completing the picture that Dupree began in *Science in the Federal Government* in their own style.

MATERIALS

A study of the bibliographic notes in many recent contributions to the literature on science in the federal government reveals that the authors generally rely on a combination of primary source materials: the official record (published or unpublished) maintained by the government or its contractors, personal papers and works of individual scientists associated with the federal government, and oral interviews with participants. The source notes prepared by contributors to the NASA History Series are particularly good indicators of the resourcefulness required of a researcher intent on preparing a history of contemporary science as practiced in a federal context.[30] In reconstructing the history of the Jet Propulsion Laboratory, Clayton Koppes consulted not only JPL's own files, but also those of the California Institute of Technology, the Army Ordnance Corps, NASA, and various U.S. presidents.

On the face of it, the "official record" sounds like the least revealing of all of the sources available to the historian of science in the federal government. But in fact official federal records may include comments as candid as any found in collections of personal papers. A brief survey of some of the records on post-1940 science available in the National Archives in Washington suggests the evidential possibilities.[31]

The records of the Office of Scientific Research and Development are a remarkably rich and largely untapped source of commentary on wartime scientific activity. Included are the office files of Roger Adams, Lyman Briggs (in his capacity as chairman of the Uranium Committee), Vannevar Bush, Karl Compton, James Conant, Frank Jewett (in relation to his supervision of research on direction finders), and Richard Tolman. Some of the most useful materials in this record group were collected in the course of preparing manuscripts for the Science in World War II series described earlier.[32]

Hopkins Univ. Press, 1981); Adeline Levine, *Love Canal: Science, Politics and People* (Lexington, Mass.: Lexington, 1982); Daniel Ford, *The Cult of the Atom: The Secret Papers of the Atomic Energy Commission* (New York: Simon & Schuster, 1982).

[30] See esp. "A Note on Sources," in Arnold S. Levine, *Managing NASA in the Apollo Era* (Washington, D.C.: GPO, 1982), pp. 301–303.

[31] The National Archives holdings are divided according to their provenance into "record groups," each corresponding to a federal agency; the number assigned reflects nothing more than the timing of a given group's creation. Details about the various record groups described in this essay may be obtained directly from the responsible custodial branches within the National Archives. These are the Scientific, Economic, and Natural Resources Branch for Rec. Groups 7, 27, 54, 167, 227, 307, 310, 326, and 359; the Judicial, Fiscal, and Social Branch for Rec. Groups 51, 90, and 443; the Legislative and Diplomatic Branch for Rec. Group 128; the General Branch for Rec. Groups 17 and 255; the Navy and Old Army Branch for Rec. Groups 52 and 298; and the Modern Military Field Branch for Rec. Groups 112, 165, and 330.

[32] National Archives, Rec. Group 227. One MS intended for the Science in World War II series (cit. n. 11), Henry Guerlac's history of the development of radar, was never published but is available with illustrations in this record group.

The official records of the President's Office of Science and Technology constitute a productive source of commentary on the federal government's involvement in science and technology in the first decade of the Cold War. The records consist primarily of background materials accumulated during preparation of various public reports issued by OST between 1962 and 1971; these provide numerous clues to prevailing views of scientific and technological research and development. Topics addressed include civilian technology, energy, life sciences, oceanography, scientific information, and space.[33]

Yet a third particularly rich and largely untapped official record is that accumulated by F. W. Reichelderfer during his tenure as chief of the Weather Bureau from 1939 to 1963. These materials, consisting of correspondence, reports, and memoranda, document Reichelderfer's association with the American Geophysical Union, the International Geophysical Year, and the World Meteorological Organization, as well as his activities as Bureau Chief. A large subsection of the materials document the work of the Weather Bureau's Overseas Operations Division in the Arctic, Antarctic, and Greenland.[34]

The National Archives Record Group for the National Bureau of Standards (NBS) contains the files of its director through 1969. These files include materials relating to the ADX2 battery additive controversy, from which some contend the NBS has yet to recover. Also included among NBS records are the office files of J. Howard Dellinger, noted radio researcher, who retired from the NBS in 1948. The files of the National Advisory Committee for Aeronautics (NACA) through 1958 relate to virtually every aspect of aeronautical research and development.[35]

Only a small portion of the files of two of the most important postwar scientific agencies, the Atomic Energy Commission and the National Science Foundation, have been deposited in the National Archives. These include the office files of David Lilienthal through 1950 and of Alan Waterman through 1963. Lilienthal's files concern, for example, the AEC's national laboratories, nuclear testing, nuclear energy, civil defense, and nuclear biology and medicine. Waterman's subject files include documentation of NSF's interest in the International Geophysical Year, the establishment of a radio astronomy observatory, and the President's Science Advisory Committee. In the same record group are over 200 feet of records accumulated by NSF's Office of Antarctic Programs (1954–1969) and Office (later, Division) of Polar Programs (1970–1982). These consist primarily of subject files, closed grant and contract files, reports by exchange scientists and observers, reports by station scientific leaders, photographs, and newsclippings.[36]

Federal involvement in the life sciences is represented by the records of the Bureau of Entomology and Plant Quarantine through 1956, the records of the Bureau of Animal Industry through 1953, the records of the Bureau of Plant Industry through 1954, and the records of the Agricultural Research Service

[33] National Archives, Rec. Group 359. For further information on their content, see the list of OST's public reports appended to David Beckler, "The Precarious Life of Science in the White House," *Daedalus,* 1974, *103*:115–134.

[34] National Archives, Rec. Group 27.

[35] National Archives, Rec. Group 167 (NBS); Rec. Group 255 (NACA).

[36] National Archives, Rec. Group 326 (Lilienthal); Rec. Group 307 (Waterman; NSF Antarctic and Polar programs).

The National Archives Building. From Donald R. McCoy, The National Archives: America's Ministry of Documents, 1934–1968 *(Chapel Hill, N.C.: University of North Carolina Press). Used with permission of the publisher.*

through 1972.[37] The central file of the Public Health Service and the general records of the National Institutes of Health are particularly revealing of federal interest in diseases, drugs, hygiene, and public health education through 1951.[38]

The National Archives also contains records on science in the military. Military research and development is represented by almost 200 feet of records relating to the Defense Department's Research and Development Board, charged with oversight of all military R & D from 1947 to 1953, and the subject files of the Navy's Coordinator of Research and Development (1941–1945). The reports of ALSOS, the special mission to determine German progress in nuclear physics, are available among other records of the Army's Director of Intelligence (G-2). Two record groups pertaining to military medicine, Records of the Bureau of Medicine and Surgery (Navy) and Records of the Office of the Surgeon General (Army), include extensive collections of photographs of medical facilities and methods of treatment utilized during World War II.[39]

The National Archives also houses legislative records and records of the federal budgetary process likely to interest historians of science. Within the Legislative category, the voluminous records of the Joint Committee on Atomic Energy are the most notable. They include documentation of all aspects of the

[37] National Archives, Rec. Group 7 (entomology); Rec. Group 17 (animal industry); Rec. Group 54 (plant industry); Rec. Group 310 (agricultural research).

[38] National Archives, Rec. Group 90 (Public Health Service); Rec. Group 443 (National Institutes of Health).

[39] National Archives, Rec. Group 330 (R & D Board); Rec. Group 298 (Navy Coordinator); Rec. Group 165 (ALSOS); Rec. Group 152 (Navy Medicine); Rec. Group 112 (Surgeon General). Although photographic material is noted here only for these last two record groups, photographs may exist among the official records of every federal agency. The holdings of the National Archives include photographs of federal scientists at work, federal laboratories under construction, and federal instruments in use.

committee's interest in atomic energy from Argonne National Laboratory to Westinghouse Electric Corporation. Documentation of the budgetary process as monitored by the Bureau of the Budget, now the Office of Management and Budget, through 1970 is extensive. It contains reports, correspondence, memoranda, exhibits, and statistical tables relating to the development of the budget of every federal agency, including those engaged in scientific activity.[40]

The foregoing inventory of official records of the federal connection to science emphasizes those records available in National Archives facilities in Washington. National Archives facilities outside of the Washington area also house records of interest to historians of science. Principal among these are the Regional Archives Branches in Waltham, Massachusetts, and Atlanta, Georgia. The former facility houses a large collection of record material documenting the wartime work of MIT's Radiation Laboratory. The latter facility recently assumed custody of approximately 600 feet of records of operations at Oak Ridge, Tennessee, through 1966. The nine presidential libraries (Hoover to Carter) established to date also form part of the field facilities of the National Archives. Two of them contain small but significant collections of records documenting presidential interest in scientific activity: the records of the President's Scientific Research Board in the Harry S. Truman Library in Independence, Missouri, and the records of the President's Science Advisory Committee in the Dwight D. Eisenhower Library in Abilene, Kansas.[41]

This brief survey not only suggests the research potential of the holdings of the National Archives but reveals the limits of their coverage. There is very little for the period after 1960. Where, for example, are the records of the Director of the National Science Foundation after 1963? What of the post-1950 records of the chairman of the Atomic Energy Commission? Where are the official records of NASA? Where are the official records, the project files, of working scientists associated with any agency?

For answers to such questions the researcher must turn to the federal agency that created the records. The more recent records of most agencies remain in their custody. The records may actually be housed in one of fifteen strategically located warehouses constituting the Federal Records Center System; but the agency at issue must be approached for permission to use them.[42] In most agencies, the records officer is the point of contact for information about noncurrent

---

[40] National Archives, Rec. Group 128 (Joint Committee on Atomic Energy); Rec. Group 51 (Office of Management and Budget).

[41] Nine additional Regional Archives Branches are located in Bayonne, N.J.; Philadelphia, Pa.; Chicago, Ill.; Kansas City, Mo.; Fort Worth, Tex.; Denver, Colo.; San Bruno, Calif.; Laguna Niguel, Calif.; and Seattle, Wash. Records of the field operations of many federal agencies, such as the Department of Energy and the Corps of Engineers, have been allocated to these branches. The remaining libraries in the Presidential Library System are the Herbert Hoover Library in West Branch, Iowa; the Franklin D. Roosevelt Library in Hyde Park, N.Y.; the John F. Kennedy Library in Cambridge, Mass.; the Lyndon B. Johnson Library in Austin, Tex.; the Nixon Presidential Materials Project in Washington, D.C.; the Gerald R. Ford Library in Ann Arbor, Mich.; and the Carter Presidential Materials Project in Atlanta, Ga.

[42] Because the National Archives and Records Administration operates the Federal Records Centers, agency officials and, sometimes, researchers assume that records in one of these centers, such as the Washington National Records Center in Suitland, Md., are "in the National Archives" and thus accessible without the agency's permission. This is not the case; but it is easy to see how the confusion has arisen, especially since many of the Regional Branches of the National Archives occupy the same buildings as the Records Centers.

records. Thus questions about the post-1963 records of the director of the National Science Foundation should be directed to the NSF records officer. In the Headquarters of the Department of Energy, however, the agency historian has been designated the custodian of noncurrent, nondisposable records not yet transferred to the National Archives. Formal, published findings aids are generally not available for the records held in agencies of interest to historians of science: for example, no such aid exists for the DOE records, just mentioned. Two notable exceptions, however, describe holdings in naval laboratories. And NASA provides a guide to some of the official records remaining in its custody and housed at or near the various NASA facilities. The NASA guide also identifies the officials who must be contacted for access to the widely dispersed records of the federal space program.[43]

The question about the availability of the records of working scientists employed in federal agencies or FFRDCs can be answered one way in theory and another in actual practice. In theory, if the federal records disposition system prescribed by statute works, historically significant or representative research notebooks and project files will be identified when active use ceases, held for varying lengths of time by the creating agency, and eventually transferred to a National Archives facility. (Certain records of the OSRD, the National Advisory Committee for Aeronautics, and NBS, as examples, reflect this pattern.) In practice, however, the system may easily break down when either the significance or the official character of the records goes unrecognized. The victims of such a breakdown, the notebooks and project files, may either be destroyed or lost to the government (but saved for historians) through donation to private archival facilities or manuscript collections. As more and more federally sponsored scientific activity takes place outside of Washington, the possibility of loss of important documentation has greatly increased.[44]

CONCLUSION

By exploiting available source materials, the authors discussed in this essay have filled in the foreground of a picture of science in the federal government after 1940 with faithful renderings (in varying stages of completion) of a number of major federal agencies, at least one federally funded research and development center (JPL) and one government laboratory (NRL), much of the hardware launched into space and some of the science retrieved from it, wartime research on atomic energy and radar, several early computers, various science-policy makers, a few working scientists, and some scientific controversies. A sequel to Dupree based on these studies could well convey a sense of the various ways in which the federal government has mandated, funded, managed, and regulated

---

[43] David Allison, *Records Systems of the Naval Research Laboratory: Central Records and Directives System Records* (NRL Memorandum Report 4464) (Washington, D.C.: Naval Research Laboratory, 1981); Carol Nowicke, *Records of the Director of Navy Laboratories: Historical Files, 1960–1980* (David Taylor Naval Ship Research and Development Center Report 84/CT03) (Washington, D.C.: DTNSRDC, 1984); and Roland, *Guide to Research in NASA History*, (cit. n. 18).

[44] Recent studies by the Center for History of Physics of the preservation of documents at FFRDCs operated under contract with the Department of Energy suggest that historians interested in them may have particular difficulty locating relevant materials. The studies, available from the American Institute of Physics, should be read by any historian contemplating research involving Brookhaven, Oak Ridge, Lawrence Berkeley, or Argonne National Laboratories.

scientific and technological research and development. Such a sequel could also convey a sense of the strategems adopted by the scientific community to influence these various government roles. But the sequel would not have the pertinent studies to build on when it came to conveying a sense of how responsible the federal government may have been for failed research and development, or what the nature is of the federal government's relationship to science education, international scientific activity, or the exchange of scientific information. Other subjects as yet not addressed for which documentation is available are the grant-giving and contracting activities of the federal government and its varying attempts to coordinate scientific research. Once studies of these aspects of government's relation to science have been carried out, we may yet see a comprehensive portrayal of science in the federal government since 1940.

History of Geology

By Mott T. Greene*

TERRA INCOGNITA

THE HISTORY OF GEOLOGY IS THE YOUNGEST of the major branches of the professional history of science, a discipline that may be defined as history written by historians for historians. This field came into being in the last decade and a half and is far younger than the history of physics, chemistry, biology, astronomy, or mathematics. Nevertheless, there is a well-established and vigorous tradition of the history of geology written by and for geologists. Indeed, were it not for the vigor with which geologists have pursued this activity during the last few years, it is unlikely that historians would ever have awakened to the possibilities in the field.

There is ample evidence that most historians of science even now think of the history of science principally as the history of physics (inorganic) and/or biology (organic). It is no accident that the three largest industries of scholarship center on Newton, Einstein, and Darwin (though historians of geology have taken up the slack with a Lyell industry). The reasons for this preoccupation with physics and biology are not relevant here, but their effects are. Theories of scientific change based on a few examples and periods in the history of physics and biology continue to dominate philosophical thinking on science as a whole. This persists even as the history of geology and other sciences demonstrate with increasing sophistication that these theories of "scientific change" are either theories of change in physics or change in biology, and have neither the scope nor precision to account for the growth and change of other sciences.

It is clear that it will be some time before these historians and philosophers awaken from their dogmatic slumbers and see the history of geology as an area with broad implications for the study of science generally. It is correspondingly clear that the history of geology written by geologists for geologists will probably dominate the field for the indefinite future. It is therefore appropriate that in looking at the prospects for the history of geology we pause first to examine the reasons why geologists write the history of their subject and examine the effects of these motivations on both content and concept of the history thus produced.

GEOLOGISTS AND THE HISTORY OF GEOLOGY

When a geologist takes pen in hand or sits down at a word processor to write the history of geology, he or she usually has one of three aims in mind: a

* 15495 Sunrise Drive NE, Bainbridge Island, Washington 98110.

celebration, a review, or an attack. It is important that the would-be historian learn to differentiate among these approaches, as the motive has a controlling influence on the reliability and value of the history.

Rarely have geologists undertaken the history of their discipline in order to understand science generally, and, for the most part, there is no reason why they should. Recently there have been two exceptions, both of which have grown out of the discussion and eventual success of the theory known as plate tectonics (or continental drift). As to the first, when plate tectonics was proposed and debated in the 1960s, much of the most significant evidence came from study of past variations in the earth's magnetic field, rather than from more traditional fields of geology. A significant aspect of the debate concerning the theory was a historical and philosophical examination of the relationship of physics and geology, as geologists tried to understand and make way for, or get rid of, new and unfamiliar kinds of evidence. This new evidence seemed not only to change the content of the ruling theory, but to shift permanently the center of attention and significance away from stratigraphy and structural geology of continental rocks toward geophysical study of ocean floors and the earth's interior.

The second exception also concerns the history of the acceptance of plate tectonics and its unique and at times bizarre connection with the theory of scientific revolutions propounded in the early 1960s by Thomas Kuhn, a physicist turned historian.[1] Kuhn's notion of the "paradigm shift," in which sciences change abruptly from one major theory to another in a short time, rather than gradually over a long time, appealed to many geologists as a rational and reassuring account of what had seemed a rather embarrassing and irrational about-face on the behavior of the continents. This was all the more embarrassing because the new theory of plate tectonics, in many respects, was like a theory of continental motion proposed in 1912 by Alfred Wegener, but long since dismissed by most American and European geologists.

Not only was there a certain balm in Kuhn's theory, but it was published in so timely a fashion that a number of geologists began to argue that plate tectonics must be true because it fit the criteria for a "scientific revolution." Consequently there has been a good deal of crowing and preening concerning the new-found maturity of geology, which, like other sciences, has finally had its own revolution. This astounding instance of the fallacy of misplaced concretion is a unique and fascinating episode in the history of geology, but even here the component activities fall easily into the categorical divisions of geologically motivated history of geology proposed above: celebration, review, and attack.

Of these three categories the most common by far is the celebration. The death or retirement of a colleague, the anniversary of an association, the history of a department, or the semicentennial, centennial, sesquicentennial, or bicentennial of anything calls forth a review of geology's history appropriate to the occasion. The themes are gratitude, reverence, progress, success, and concord. Criticism is minor, muted, oblique, and often may be entirely lacking. The author is almost always a friend, member, employee, or believer. Although this material is invaluable (it would be impossible to write any history without it), at best it is always only half the story, and at worst it hides what is most im-

---

[1] Thomas S. Kuhn, *The Structure of Scientific Revolutions* (Chicago: Univ. Chicago Press, 1962).

portant. Discord is always more revealing than concord from a historical stand-point. Yet who would write an obituary emphasizing that Professor X was a toper who published the research of his graduate students? What department would celebrate a half century of plodding service to an undistinguished university? What association would commission a centennial history of acrimony, factionalism, and defection? The virtual nonexistence of this kind of material should remind us that every conclusion must be cross-checked against other accounts, and that in geology, as elsewhere, most history is written by the victors.

The next most common form of historical writing by geologists is the review of a field or subject or theory. The very appearance of such a review at a given time is indication that the topic is in transition, or at some significant stage in its development; if it were not, there would be no point, scientifically, in a review. Review literature is therefore an extremely valuable source for the historian. Indeed, one could construct a schematic and provisional history of geology by assembling a roster of significant review articles. Not the least benefit of the review article is that publications included in the bibliography (and those excluded!) alert the historian to the "official version" of the history of the topic at the time the review was written.

There are two principles to remember when using this material. The first principle is to distrust any citations in the bibliography more than a few decades old. Almost invariably, they will have been assembled by trusting the accuracy of earlier review articles, rather than by a direct search of the original literature. Errors accumulated in this way over several generations of review articles can render the citations unreliable for the historian. Material this old is generally deemed scientifically worthless by the author, who would otherwise have consulted the originals, and it is included only to give an air of completeness to the whole. From this vantage, one may take the review article as the barest starting point for the history of a subject.

The second principle to remember is that usually review articles are only the summary of the major tradition or line of advance that led to the present, and unsystematically exclude the research traditions that did not lead in a substantial way to the state of affairs at the time of writing. This is both honest and appropriate. From the standpoint of a review article, history of geology is largely an imaginative aid in the formulation of new questions and answers for working scientists. It does not have as a principal aim the exact transcription of the movement of a field of study. After all, it is history as a scientific activity within geology, not history as a cultural activity reflecting upon it. A convenient rule of thumb is always to remember this function and to see even the best reviews as provisional summaries of where to begin to look for the alternate lines of advance and competing traditions of research.

The most difficult material to evaluate historically is the literature of attack and defense, as it is rarely advertised as such and usually wears the somber garb of the calm reconsideration, particularly in the opening chapters of a textbook. The historical introduction of Charles Lyell's *Principles of Geology* (1830–1833) is a good example of an apparently measured and reasonable account of the history of geology which turns out to be a propagandistic distortion of the views of the author's opponents.[2]

[2] Martin Rudwick, "The Strategy of Lyell's *Principles of Geology*," *Isis*, 1970, 60:5–33; see also

In the American context, the debate over the origin of the Appalachian Mountains carried on in the pages of the *American Journal of Science* in the 1860s and 1870s was conducted in elevated and earnest terms by geologists desiring only to "set the historical record straight." These avowals notwithstanding, all parties in the debate were guilty of gross distortions of one anothers work, much of it clearly deliberate.[3] A convenient rule here is that when you read a scientist writing that he has undertaken a historical examination of a topic in order to set the record straight, you may be certain that you have unearthed some fraction of a significant controversy; the next step is to find the "disinterested" contributions of the author's opponents.

A minor irony is that historical interest and significance are about in inverse proportion to accuracy. A celebration will provide the positive half of something, probably accurate in what it says (if distorting in what it omits) but this something may be utterly insignificant and uninteresting in terms of the general movement of geology. A review can alert one to a shift or alteration of course in a field, but some of the information it supplies will be inaccurate, and it will certainly be incomplete. Like a celebration, a review rarely deliberately distorts or alters the record, and what it gives up in accuracy and completeness, it gives back in historical importance. Finally, the attack, whether open or posing as a calm reconsideration, locates a topic worthy of the historian's attention: if the topic was worth fighting about, it is worth writing about. However, an attack is almost invariably misleading as a historical document—using it alone is like trying to decide a judicial case working only from the brief of the plaintiff. Again, no wrong is implied in the judgment of one-sidedness, but the historian must be alert that here cases are being argued, not arbitrated.

In summary, the historian must be aware that there are a variety of reasons why geologists write the history of geology, and that most often they write from a standpoint other than what a historian would recognize as "historical." Moreover, the rules of evidence employed in such work are considerably relaxed from the standards employed by the same geologists in their scientific work. They are also well below what a professional historian would deem adequate in a work of critical scholarship.

AMERICAN HISTORIANS AND THE HISTORY OF GEOLOGY

If geologists have their own approach to the history of geology, one somewhat different from that of the historian of science, there is yet another group to be heard from, which writes from still another perspective. These are the American historians, particularly intellectual and social historians, but also those historians specializing in the topic "The Westward Expansion"—the exploration and initial exploitation of the lands west of the original colonies. In recent years a new

---

Alexander Ospovat, "The Distortion of Werner in Lyell's *Principles of Geology*," *British Journal for the History of Science*, 1976, 9:190–198; and Roy Porter, "Charles Lyell and the Principles of the History of Geology," *Brit. J. Hist. Sci.*, 1976, 9:92–104.

[3] James D. Dana, "On the Origin of Mountains," *American Journal of Science*, 1873, 3rd ser., 5:347–350; and Robert Mallet, "Note on the History of Certain Recent Views in Dynamical Geology," *ibid.*, p. 302; see also Joseph Leconte, "Letter to the Editor," *ibid.*, p. 156; and T. Sterry "On Some Points in Dynamical Geology," *ibid.*, pp. 264–270.

subspecialist group—environmental historians—has also taken part in this effort, concentrating on alteration of natural systems by settlement and exploitation, creation and management of public lands, administration of the resource base, and allied topics.

While this work is almost all performed by trained historians, and therefore generally more reliable than the historical work of geologists in its narrative content, such work is almost without exception what historians of science consider "externalist" history. Even when it deals directly with geology and geologists, the concern is the politics and sociology of scientific organizations and careers and the establishment and conduct of geology's institutional life.[4]

Because geologists and paleontologists had a distinguished formative role in the history of American scientific institutions, public and private, this interest is well merited. Still, the aim of this history often is not to understand geology but to understand geology's contribution to the national history.

Even the great number of large volumes that treat in ever finer detail the history of exploring expeditions and surveys of the West—the most thoroughly written-over aspect of American geological history—rarely concentrate on scientific practice and scientific results at a depth that would interest either a historian of science or a geologist. They are more concerned with establishing the links between science, public policy, and private exploitation of resources, and with telling the very exciting tale of exploration and adventure. One often receives the impression that the science was what got done between bureaucratic squabbles and the writing of long letters to Washington begging for more money. Although this may, in fact, have been the case, such studies overlook the obvious point that these explorations had an important scientific content. The work of Grove Karl Gilbert, Clarence Dutton, John Wesley Powell and others on the Colorado Plateau caused major changes in world geological thought in the 1870s. Their analyses of a large area of uplift without deformation (relatively speaking) cast down two or three promising theories of the earth's dynamic history, yet not one of the better known histories of these expeditions contains a discussion of the matter.

The obvious conclusion is not that the Americanists are not doing "their job," but that their work and the work of historians of science and scientists must be integrated and synthesized. These three groups of scholars are writing from different perspectives, with different emphases, according to different principles.

HISTORY OF GEOLOGY IN THE 1980S

This leaves the history of geology in America in the 1980s in a curious position. At a time when other branches of history of science are beginning to move from the detailed study of scientific ideas into the social and institutional history of the science in question, the history of geology already enjoys a robust tradition

[4] See e.g., Mark Wyman, *Hard Rock Epic: Western Miners and the Industrial Revolution, 1860–1910* (Berkeley/Los Angeles: Univ. California Press, 1979); William H. Goetzmann and Kay Sloan, *Looking Far North: The Harriman Expedition to Alaska, 1899* (New York: Viking, 1982); Leonard G. Wilson, ed., *Benjamin Silliman and His Circle: Studies on the Influence of Benjamin Silliman on Science in America* (New York: Science History, 1979); Alfred Runte, *National Parks: The American Experience* (Lincoln: Univ. Nebraska Press, 1979); and Henry Savage, Jr., *Discovering Amerca, 1700–1875* (New York: Harper & Row, 1979).

of writing about its social impact and institutional development. What it lacks is writing on the history of geology as a *science,* and this aspect of the subject finds few practitioners—the same dozen or so working to cover the whole field. This is the odder since the preliminary bibliographic work essential to such detailed history has largely been accomplished, leaving only the actual selection of topics and construction of histories.

Since geologists have science training without historical training, and American historians have the reverse, one need not look far to see whose responsibility it is to produce the history of geological science in America. The burden falls squarely on graduate programs in the history of science, which must permit and encourage their graduate students to take up thesis topics in areas not already covered; that certainly includes the history of geology. It seems, for instance, to border on irresponsibility to allow a student to write the nth thesis on the reception of Darwinism somewhere, or the $n + 1$ thesis on Newtonian minutiae, when an entire branch of science (and, moreover, one with similarly engrossing philosophical implications) is all but unexplored. In such a field, a student would have the chance to produce original and valuable historical work, rather than toting bones from one graveyard to another.

One hears a number of excuses as to why this exclusionary pattern persists. The most commonly heard is that since the thesis is written with an advisor whose field of study is that of the thesis student, and since there are few historians of geology, there are few graduate students in the history of geology. Such an argument might be offered to explain why, for instance, not a single one of the ninety-five dissertations in the history of science at the University of Wisconsin since 1952 concerns itself with the history of geology. Indeed, earth sciences do not even figure as an area in the categorical summaries of Wisconsin dissertations in a recent analysis.[5] The situation is much the same at the other distinguished centers of research in the history of science, for which similar explanations might be offered. It is fair to point out that had this reasoning prevailed two generations ago we would have no history of science at all. By defining competence to direct a dissertation as knowledge of the subject area sufficient to criticize the details of historical evidence involved, one reduces the dissertation from a work of original scholarship to an apprentice piece leading to a journeyman's credential. This is a pernicious and retrograde concept of graduate education if there ever was one. A graduate advisor should do more than attempt to erect a series of idols in his or her own image, and the best way to avoid such narcissism is to encourage students to work everywhere that work needs to be done. An additional advantage of this more expansive view of the dissertation is that it can more rapidly and effectively broaden the empirical base on which theories of scientific change can reasonably be erected.

A more serious obstacle is that dissertation advisors who are not historians of geology, though willing to supervise dissertations in the earth sciences, are unable to provide bibliographic guidance or suggest topics for research. I shall do my best to render this reasonable claim baseless in the remainder of this article, by providing a bibliographic orientation to the most recent work and by suggesting research opportunities in the history of American geology.

L. Hilts, "History of Science at the University of Wisconsin," *Isis,* 1984, 75:63–94, on

SOURCES FOR THE HISTORY OF GEOLOGY

Bibliographies. The best, most extensive and most recent is William A. S. Sarjeant's *Geologists and the History of Geology.*[6] This five-volume work with almost 5,000 pages of bibliographic information and indexes contains 350 pages on secondary works in the history of geology, with an additional 100 pages on the history of geological institutions. It also contains a comprehensive section on the history of the petroleum industry. The second and third volumes are devoted to the biographies of several thousand geologists, with short career sketches preceding biographical source information. It is no longer necessary, therefore, to search for this information via the toilsome route of annual indexes to periodical series. The fourth volume indexes geologists by country of birth and country of activity. This feature is exceptionally useful for the history of geology in a national setting, as is the case in this essay. The final volume indexes authors, editors, and translators of works on the history of geology. Although it serves primarily to provide access to the other volumes, in itself it is also a fascinating study of intellectual linkages.

The compiler, a distinguished geologist and historian of geology, has noted those classes of documents omitted from his comprehensive search. These include the following journals, essential to the study of the history of American geology: *Science, Journal of the Washington Academy of Sciences, Bulletin of the Philosophical Society of Washington,* publications of the Smithsonian Institution, and official publications of state and national geological surveys. While there are many citations from these periodicals in the bibliography, the review is not complete for their full runs. Coverage in all other respects, and within the imposed limits of the design, is particularly complete for English-language, and especially for North American documents.

Sarjeant's work must be supplemented by Harold Pestana's *Bibliography of Congressional Geology,* which indexes only geological work sponsored and published by the government. In addition, Robert M. Hazen and Margaret Hindle Hazen's *American Geological Literature 1669–1850* supersedes all other bibliographies of American geology for this period. It contains more than 11,000 entries covering all aspects of the history of geology, including even the relations of geology and agriculture, geology and medicine, geology and religion, and other topics.[7] It would be foolhardy to attempt research on any aspect of American geology without consulting the above-listed works.

A topically arranged *Bibliography and Index of Geology* is published annually by the Geological Society of America. The series supersedes the *Bibliography of North American Geology,* which was published regularly by the United States Geological Survey between 1923 and 1970, though there are several earlier volumes.

For geology after 1850, some older bibliographies are still useful.[8] In addition

---

[6] William A. S. Sarjeant, *Geologists and the History of Geology: An International Bibliography from the Origins to 1978,* 5 vols. (New York: Arno, 1980).

[7] Harold Pestana, *Bibliography of Congressional Geology* (New York: Hafner, 1972); Robert M. Hazen and Margaret Hindle Hazen, *American Geological Literature, 1669–1850* (Stroudsburg, Pa.: Dowden, Hutchinson & Ross, 1980).

[8] J. M. Nickles, *Geologic Literature on North America, 1785–1918* (U.S. Geological Survey Bulletin 746–747) (Washington, D.C.: GPO, 1923–1924); N. H. Darton, *Catalogue and Index of Contributions to North American Geology, 1732–1891* (U.S. Geological Survey Bulletin 127) (Wash-

to these general compendia, there are many bibliographies organized by state, province, and region, most of which are listed on pages 12–16 of Hazen and Hazen (1980). Since that publication, two additional state survey bibliographies have appeared, on Illinois and Nebraska. The *Biographical Dictionary of Rocky Mountain Naturalists* by Joseph Ewan and Nesta Dunn Ewan falls into this category.[9]

Finally, a great wealth of bibliographical information occurs in the form of published catalogues of major scientific library collections. These include the libraries of the U.S. Geological Survey, the Museum of Comparative Zoology at Harvard, the American Museum of Natural History, the Naval Observatory, the Arctic Institute of North America, as well as the Atmospheric Sciences Collection and the National Agricultural Library. The published catalogues of general research libraries, and special bibliographies drawn from their collections, are also of value for the history of American geology. For instance, there is the New York Public Library's *History of the Americas Collection*. The history of science collection at the University of Oklahoma is very strong in the history of geosciences. The Library of Congress has a helpful bibliography to local histories in its holdings. Every historian of geology in America should realize the extent of bibliographic resources at his or her disposal.[10]

I have chosen to exemplify the major categories of useful works; an exhaustive survey would be much larger. However, I cannot fail to mention the *Isis Cumulative Bibliography*, which collates and indexes the annual critical bibliography of the History of Science Society from the beginning to 1975. Anyone who has had to undertake a research project by consulting the original bibliographies volume by volume can attest to the life-extending character of this wonderful work.[11]

Dictionaries, Encyclopedias, Handbooks. The *Dictionary of Scientific Biography* contains material on more than 800 geologists and paleontologists in many different specialties. The publication dates and origin of the secondary biographical

ington, D.C.: GPO, 1896); Max Meisel, *A Bibliography of American Natural History: The Pioneer Century, 1769–1865,* facs. rpt. (3 vols.: 1924, 1926, 1927; New York: Hafner, 1967).

[9] Robert G. Hays, *State Science in Illinois: The Scientific Surveys, 1850–1978* (Carbondale: Southern Illinois Univ. Press, 1980); John H. Sandy and Jay Fussell, *Bibliography of Nebraska Geology, 1843–1976* (Lincoln: Univ. Nebraska Conservation and Survey Division, 1983); Joseph Ewan and Nesta Dunn Ewan, *Bibliographical Dictionary of Rocky Mountain Naturalists: A Guide to the Writings and Collections of Botanists, Zoologists, Geologists, Artists and Photographers, 1682–1932* (The Hague/Boston: W. Junk [Kluwer Boston], 1981).

[10] New York Public Library, *History of the Americas Collection: Bibliographic Guide to North American History,* 28 vols., 9-vol. suppl. (Boston: G. K. Hall, 1972–present); *U.S. Local Histories in the Library of Congress: A Bibliography,* 5 vols. (Baltimore: Magna Carta, 1975); Duane H. P. Roller and Marcia M. Goodman, *The Catalogue of the History of Science Collection of the University of Oklahoma Libraries,* 2 vols. (London: Mansell, 1976); *Dictionary Catalog of the National Agricultural Library* (1862 and after) (New York: Rowman & Littlefield, 1967–present); and the following catalogues also issued by G. K. Hall, Boston: *Catalog of the United States Geological Survey Library,* 25 vols., 14-vol. suppl. (1964–present); National Oceanographic and Atmospheric Administration, *Catalog of the Atmospheric Sciences Collection,* 24 vols. (1978–present); Harvard University, *Catalogue of the Museum of Comparative Zoology* (1968–present); *Catalog of the Library of the American Museum of Natural History* (1978); *Catalogue of the Library: Arctic Institute of North America* (1968–present); *Catalog of the Naval Observatory Library* (1976–present).

[11] Magda Whitrow, ed. *Isis Cumulative Bibliography, 1913–1965* (London: Mansell, 1971–1982); John Neu, ed., *Isis Cumulative Bibliography, 1966–1975* (London: Mansell, 1980).

material listed for these entries shows that most major figures in the history of American geology lack a scholarly biography. The *Dictionary of American Biography,* in many volumes, includes the biographies of scientists and engineers; many of these are of importance for geology. The *Biographical Dictionary of American Science* is useful and up to date but, as its compiler points out, does not include engineers and explorers to any great extent. These categories are, however, very important for the history of geology. For example, the best survey of theories of mountain origin in the United States in the nineteenth century was written by a civil engineer.[12]

Recent interest in the earth sciences kindled by the plate tectonics "revolution" has spawned a number of encyclopedias of the earth sciences, which run the gamut from coffee-table picture books to serious reference works. The best series of encyclopedia volumes is that published under the editorship (and frequent authorship) of Rhodes W. Fairbridge: the Encyclopedia of Earth Sciences Series. Designed for the nonspecialist but serious audience, they are a superb aid to the historian who needs to learn some geology, or some more geology. Volumes of particular interest are *Encyclopedia of World Regional Geology, Part I: Western Hemisphere* (1975); *Encyclopedia of Oceanography* (1966, rev. ed. 1983); *Encyclopedia of Geochemistry and Environmental Sciences* (1972); *Encyclopedia of Sedimentology* (1978); *Encyclopedia of Geomorphology* (1968); and *Encyclopedia of Paleontology* (1968).[13]

The Benchmark Series is also worthy of note. This series, now over sixty volumes, collects and reprints, with annotations by the editor, significant papers from the history of a field or subfield. The selection of papers in these volumes is often as biased as that of a review article. Historical coverage is uneven, most volumes are shaded heavily to recent work, and errors are not infrequent. Nonetheless, these collections can be useful in understanding the direction of work in a field as perceived by the editor at the time of compilation. Titles include fields as diverse as paleogeography, glacial isostasy and philosophy of geohistory. At the very least one ought to see whether a volume exists in one's area of historical interest.[14]

Periodicals. The secondary periodical literature of interest to historians of American geology—science journals being treated as primary sources—falls into four categories: journals and newletters devoted to earth science history, general history of science journals, geological journals that include historical articles, and journals of American history and culture.

Earth Sciences History, the journal of the History of Earth Sciences Society, is the only periodical devoted entirely to earth science history, and in its short existence it has already published a number of papers on the history of geology

[12] George L. Vose, *Orographic Geology, or the Origin and Structure of Mountains: A Review* (Boston: Lee & Shepard, 1866); see also *Dictionary of Scientific Biography,* ed. Charles C. Gillispie, 16 vols. (New York: Scribners, 1970–1980); *Biographical Dictionary of American Scientists* (Westport, Conn.: Greenwood, 1979).

[13] The Encyclopedia of Earth Sciences series is published by Dowden, Hutchinson & Ross, Stroudsburg, Pa.

[14] The Benchmark Papers in Geology series is also edited by Rhodes W. Fairbridge; see, e.g., John T. Andrews, ed., *Glacial Isostasy* (Benchmark Papers in Geology, 10) (Stroudsburg, Pa.: Dowden, Hutchinson & Ross, 1974); and Claude C. Albritton, ed., *Philosophy of Geohistory, 1785–1970* (Benchmark Papers in Geology, 13) (Stroudsburg, Pa.: Dowden, Hutchinson & Ross, 1975).

in America. It also contains research notes and news of the history of geology. This function is also performed by the *Newsletter of the History of Geology Division* of the Geological Society of America and the *Newsletter* of the historical section of the American Geophysical Union. *Eos,* the journal of the latter organization, will be publishing more articles on the history of geophysics in the near future and signals the rapid growth of historical interest in this field. Indeed, the AGU has recently announced the inauguration of a new journal, *The History of Geophysics.*

Among the history of science journals it would seem that *Isis, Annals of Science, Janus,* and *History of Science* publish more history of earth science than do other journals. In the last few years, at least, most articles have been on the eighteenth century and before and rarely have concerned American geology. This reflects availability of suitable manuscripts rather than editorial preference, to be sure. Recently, *Historical Studies in the Physical Sciences* and *Archive for the History of Exact Sciences* have begun to publish articles on the history of geology and geophysics, so the publication pattern may be changing. Articles on the history of paleontology and paleogeography can be found in the *Journal of the History of Biology.*

The *Journal of Geological Education* publishes much of historical interest, as do *Geology, Geotimes,* and *Earth Science Reviews. Journal of Geology* occasionally publishes an article with historical interest, as do numerous subfield journals like *Techtonophysics* and *Paleoecology, Paleoclimatology.* These journals are all well indexed through 1978 in Sarjeant's bibliography.

Finally, large amounts of material, particularly on the institutional history of American geology, can be found in *Journal of the West* and in local and regional journals serving other portions of the country. Mainstream journals like *Journal of American History* and *American Historical Review* carry little or no history of science. Most material on the history of geology in America is remanded to book-length studies, or, more commonly, to symposium volumes. In any event, journals devoted to general American history are not a fruitful source.

Public and Official Histories. These histories are of three kinds: histories of government agencies, histories of scientific societies, and histories of university departments of geology and geophysics. All three are informative and especially useful in establishing chronology, membership, and direction (who held what office when), but all are definitely in the celebratory class and should be used with the caution suggested at the outset of the article. Here I shall consider the major additions to the genre since 1978.

In public histories, Mary C. Rabbitt's multivolume history of the United States Geological Survey, *Minerals, Lands, and Geology for the Common Defense and General Welfare,* is an extremely useful work that declares its intent in its title. The establishment of a Historical Division of the Army Corps of Engineers has produced a number of official histories of this controversial agency. Notable among them are Frank N. Schubert's *Vanguard of Expansion: Army Engineers in the Trans-Mississippi West, 1819–1878* and his *Explorer of the Northern Plains,* an edition of Lieutenant Warren's reports on the exploration of Nebraska and the Dakotas between 1855 and 1857.[15]

---

[15] Mary C. Rabbitt, *Minerals, Lands and Geology for the Common Defense and General Welfare,*

Among the learned societies, Edwin B. Eckel's official history of the Geological Society of America is the most significant recent publication.[16]

Sarjeant's bibliography includes about twenty-five histories of academic departments of earth science in North America. As these histories are often locally printed for private circulation, the full list is probably much longer. Notable additions since 1978 are *The History of Geology and Geophysics at the University of Wisconsin–Madison, 1848–1980,* edited by Bailey Sturges, and Robert Shrock's giant *Geology at M.I.T., 1865–1965.* Despite the distinctly victorious tone of these works, they are extremely useful in charting the development and spread of subfields. If these works are used in conjunction with current and past issues of the American Geological Institute's *Directory of Geoscience Departments,* some interesting conclusions can be drawn about the influence of the relatively few departments that granted a Ph.D. in geology in the United States before World War II. One may also learn a great deal about the nature and scope of geology as a discipline, and the relation between academic geologists, mineral and petroleum companies, and geological surveys.[17]

Compilations. Edited, multi-author works, usually the results of symposia, are the mainstay of publication in the history of American geology, for better or worse. They are uneven in quality and difficult to review: contributions of great value and intellectual depth appear side by side with antiquarian excesses of numbing triviality; rarely does the volume as a whole have any thematic coherence.

An example of most of the above pitfalls is afforded by *Two Hundred Years of Geology in America,* proceedings of a bicentennial conference on the history of geology at the University of New Hampshire.[18] At this conference, more than thirty participants, including most active historians of geology in the United States and some foreign visitors, met to discuss themes for the history of geology. It is notable that in spite of the thematic impetus provided by the convener, who sponsored a conference with similar intent in 1967, no such themes emerged. Indeed, the only coherence in the volume is that introduced by the editor, Cecil Schneer, in an introduction that can only be described as heroic. Therein Schneer gives a straightforward assessment of the problems inherent in a historical effort combining scientists, American historians, and historians and philosophers of science, all with different aims and different levels of attainment

Vol. I: *Before 1879* (Washington, D.C.: GPO, 1979); Vol. II: 1879–1904 (1980; 2 more planned); Frank N. Schubert, *Vanguard of Expansion: Army Engineers in the Trans-Mississippi West, 1819–1879* (Washington, D.C.: GPO, 1980); Frank N. Schubert, *Explorer of the Northern Plains* (Washington, D.C.: GPO, 1981).

[16] Edwin B. Eckel, *The Geological Society of America: Life History of a Learned Society* (Boulder, Colo.: GSA, 1982).

[17] Bailey Sturges, ed., *The History of Geology and Geophysics at the University of Wisconsin–Madison, 1848–1980* (Madison: Univ. Wisconsin Dept. of Geology and Geophysics, 1980); Robert Shrock, *Geology at M.I.T., 1865–1965: A History of the First Hundred Years of Geology at the Massachusetts Institute of Technology* (Cambridge, Mass.: MIT Press), Vol. I: *The Faculty and Supporting Staff* (1977); Vol. II: *Departmental Operations and Products* (1982); American Geological Institute, *Directory of Geoscience Departments* (Falls Church, Va.: AGI, 1952–present; biennial 1960–1972, annual 1973–present).

[18] Cecil J. Schneer, ed., *Two Hundred Years of Geology in America: Proceedings of the New Hampshire Bicentennial Conference on the History of Geology* (Hanover, N.H.: Univ. Press of New England for the Univ. New Hampshire, 1979).

in history. This is a gentle way of saying that the papers in the volume are united only in that they all concern geology and all concern the past. While neither uniformly reliable nor terribly useful as a historical source, the volume exhibits in high relief the structural problems of the history of geology, and it is as germane now as it was when it was published.

Similar problems, on an even larger scale, are exhibited by Mary Sears and Daniel Merriman's *Oceanography: The Past*.[19] This gigantic (812 pages) volume, united only by the conjunction of the nouns of its title, consumed a great deal of effort by many authors and then sank like a stone. The same problems emerge here: lack of focus, publication of all offerings regardless of merit, and conjunction of scientists, American historians, and historians and philosophers of science, assembled without any methodological unity or rules of procedure.

Examination of this symposium literature reveals that several common strategies can be employed to overcome the above obstacles, and most important of them is limitation of focus. Those volumes that center on a person, a concept, a delimited geological region, a short span of time—all have more chance of offering a return on effort than scattershot efforts that wander through the whole science, over centuries.

An excellent example of a productive effort in which the range of authors' concerns is shaped by the topic is *The Scientific Ideas of G. K. Gilbert*, edited by Ellis L. Yochelson. Regionally organized efforts seem to provide less coherence, but are better than no coherence at all. Some interesting themes emerge in both *History of Geology in the Northeast*, edited by William M. Jordan, and *The Geological Sciences in the Antebellum South*, edited by James X. Corgan. A symposium organized around a single kind of geological effort seems promising, as in *Frontiers of Geological Exploration of Western North America*, edited by Alan E. Leviton and others. At least one recent effort attempts to organize itself around the history of subfields in a limited period, Shelby Boardman's *Revolution in the Earth Sciences: Advances in the Past Half-Century*. Here the problem of celebration (revolution), "centenniality" (last fifty years), and historical discomfort of some of the authors—who claw their way out of the 1930s and 1940s into their own professional lifetimes with an unseemly haste—combine to blunt an admirable effort whose overall design has succeeded from time to time in the past.[20]

The outstanding problem with this literature is that few of the volumes give convincing evidence of having been edited in the sense that an academic journal or book is edited, in which the editor uses his discretion to exclude inferior material and oversees a process of review whereby most accepted contributions are brought up to a uniform standard. This is not a reflection on the conveners

[19] Mary Sears and Daniel Merriman, eds., *Oceanography: The Past* (Third International Congress on the History of Oceanography, Woods Hole, Mass., 1980) (New York: Springer, 1980).

[20] Ellis J. Yochelson, ed., *The Scientific Ideas of G. K. Gilbert: An Assessment on the Occasion of the Centennial of the United States Geological Survey (1879–1979)* (Geological Society of America Special Paper 183) (Boulder, Colo.: GSA, 1980); William M. Jordan, ed., *History of Geology in the Northeast: Proceedings of a Symposium*, special issue of *Northeastern Geology*, 1981, *3*(1):1–103; James X. Corgan, ed., *The Geological Sciences in the Antebellum South* (University, Ala.: Univ. Alabama Press, 1982); Alan E. Leviton et. al., eds., *Frontiers of Geological Exploration of Western North America* (San Francisco: Pacific Division of the AAAS, 1982); Shelby Boardman, ed., *Revolution in the Earth Sciences: Advances in the Past Half-Century* (Dubuque, Iowa: Kendall/Hunt, 1983).

or editors, but on the state of the history of geology itself, trapped between a recreational activity for working scientists and an unsure future as an academic subdiscipline of the history of science. If one wishes to assemble enough bodies to hold a conference, one must include groups with divergent aims. As this pattern is unlikely to change any time soon, these volumes must be approached cautiously when used as sources in historical work.

Books by One Author. It is lamentable how few books on any aspect of the history of geology have appeared in the last few years, other than those that flog the exploration of the West in a scholarly industry of major proportions. Even in the few works outside that genre that have appeared, no two seem to share either a methodology or an aim in writing history in the first place. Martha C. Bray's *Joseph Nicollet and His Map* is a readable and beautifully produced account of an early American cartographic effort. The context into which it might fit, however, is lacking as other than a solid contribution to the literature of geology, *tout court*. Stephen Pyne's *Grove Karl Gilbert: A Great Engine of Research* is less a complete scientific biography than a theory of the connections between Gilbert's ideas, employing what some reviewers have found to be an unsuccessful and idiosyncratic metaphor for unity and recurrence. Lester Stephens's *Joseph Leconte: Gentle Prophet of Evolution* is an interesting and approving life of an important figure in mid-nineteenth-century geology, but it is definitely a life rather than a "life and times." The principal utility of biography in the history of science is to illuminate an epoch through the participation of a single individual, suitably chosen (L. Pearce Williams's biography of Michael Faraday is a model in this regard). In this respect, all of these biographies fall somewhat short of the mark.[21]

The other possible focus, thematic history, has two recent examples. Mott T. Greene's *Geology in the Nineteenth Century* concerns American geology only in part. A chronicle of the emergence of tectonics, and especially global tectonics, as the ultimate level of correlation for nineteenth-century geologists, it is an "internal" history of geological theory, with few explicit ties to social and intellectual themes outside the science. William Glen's *The Road to Jaramillo: Critical Years of the Revolution in the Earth Sciences* is a history of magnetic-polarity–reversal time scales, particularly those produced at Berkeley and Menlo Park. It has been criticized for its slighting of some foreign contributions to this theme, and its richness of detail makes it difficult to read. Moreover, since it chronicles events up to the year of its publication, there is some question about the extent to which it is history and the extent to which it is science journalism. The distinction, not invidious, concerns the author's vantage point and ability to reflect on the material. But it is a valuable source of institutional history of modern earth science, and the author's tireless efforts to interview participants in recent and major theoretical shifts have produced an archive without equal in the history of geology. The book is a gold mine of possibilities for further research.[22]

[21] Martha C. Bray, *Joseph Nicollet and His Map* (Memoirs of the American Philosophical Society, 140) (Philadelphia: APS, 1980); Stephen Pyne, *Grove Karl Gilbert: A Great Engine of Research* (Austin: Univ. Texas Press, 1980); Lester Stephens, *Joseph Leconte: Gentle Prophet of Evolution* (Baton Rouge: Louisiana State Univ. Press, 1982).

[22] Mott T. Greene, *Geology in the Nineteenth Century: Changing Views of a Changing World*

SOME SUGGESTIONS

Certainly, it is true that each problem dictates a choice of methods, and the development of appropriate methods is part of the solution to any problem. Nonetheless, there are some general approaches worth mentioning for the history of American geology.

First, in the phrase "American geology," "America" is not synonymous with "United States" as in the case of "American history." It concerns the United States, Canada, and Mexico and can be reasonably extended to include Central and South America for certain topics. A historian who does not recognize this generalization will fall into serious error and distort the historical record. Of course, the reservation applies that a history of a government survey stops, perforce, at a national boundary. Moreover, the arrest of geological surveys at national borders does have an effect on theory and practice of geology which a historian ought to chronicle and understand but should not imitate.

Second, the phrase "history of geology" in all periods up to World War II should be understood to include a number of fields now treated as independent sciences. Geophysics, geography and geomorphology, and even oceanography and meteorology (though to a lesser extent) are all a part of mainstream geology from the historical standpoint. One might well add that much of the history of paleontology is the history of a geological and not a zoological subject, the shift beginning in the 1870s toward the latter. For a time, even archaeology was a geological subject; recently the emergence of geoarchaeology makes it so once again.

Third, American geologists have always had close contact with their European counterparts and have participated extensively in the mapping and exploration of the world outside the Western Hemisphere. Americans have regularly traveled to Europe for graduate study since Benjamin Silliman studied Wernerian mineralogy at Edinburgh at the beginning of the nineteenth century. Americans not only participated in the International Geological Congresses since their inception, but James Hall of New York was the first president of the international organization. Therefore, one cannot write the history of American geology without knowing the history of European geology. Most major figures in American geology have written and spoken languages other than English, and they have both used and published in European journals on topics of interest to Europeans. Moreover, there are numerous instances of important geological work in this hemisphere performed by Europeans, with a great impact on the interpretation of the native geology.

Fourth, in a field in which the history is all but unwritten, the historian's first responsibility is the study of printed sources. Unless this task is undertaken before more detailed study of archival materials, there will be no context into which the resulting work can fit. Even professional historians of science need to be reminded of this state of affairs. Their training and experience stresses that the real story is to be found "behind" the published science in notebooks, letters, private papers, drafts and fragments, and private library contents. But

---

(Ithaca, N.Y.: Cornell Univ. Press, 1982); William Glen, *The Road to Jaramillo: Critical Years of the Revolution in the Earth Sciences* (Stanford, Calif.: Stanford Univ. Press, 1982).

until the history of printed sources is assembled and reviewed, the notion of "behind" has no meaning. Since graduate training in the history of science covers principally the history of those fields for which a comprehensive outline has been available for generations, it is sometimes difficult for some to grasp the idea that the history of any science could be in such a primitive state that it is as yet too early to *go to the archives* in major areas of interest. This is, incredible as it may seem, the state of affairs in the history of geology today.

By the history of printed sources I refer to scientific journals both general and geological, to geology texts, to the publications and reports of geological surveys, and to the biographical memoirs of geologists, whether private "life and letters" or the official obituaries of science journals and academies. Only when this material is surveyed and reconnoitered does it make sense to isolate a problem for deeper study involving the extensive use of unpublished materials. The scale of the labor involved here is large, but it recently has been considerably moderated by the existence of high-quality bibliographies.

Moreover, there is a certain rationality in studying the history of geology by subfield and region rather than by century. One might survey the history of structural geology in the Rocky Mountains without feeling a responsibility for global problems, except when they turn up in the course of research, or any need to carry the story back to the Flood. Neither is there any reason why the results of the survey of printed sources need be done or published instead of archival research; the presence of papers in a nearby archive often dictates the project in the first place. My point is simply that one cannot assume that a context for detailed archival research already exists for most fields of American geology. It makes little sense, then, to pursue any project without first creating and stating a context, however rudimentary, into which it might fit. Moreover, the challenge and excitement of actually writing, rather than revising or rewriting a field of history, is unparalleled, and ought to be welcomed.

OPPORTUNITIES FOR RESEARCH

In a field all but untouched by the historian of science, the opportunities are endless. The following list of topics exemplifies classes of problems, with each standing for a long list of problems with a similar structure. I chose them because each would make a contribution not only to the history of geology but to the understanding of science generally.

In the history of geological *concepts,* few ideas have had a greater impact than that of the geosyncline. While a number of reviews of this topic have been written, no historical monograph exists to give the detailed history of the idea. It was invented in North America, founded on the structure of the North American continent, and exerted a controlling influence on all fields of geological research through the first half of the twentieth century. There are dozens of such influential concepts, each of which deserves a history of its own.

The history of geological *methods,* itself a subset of conceptual history, touches on issues that concern all the sciences. The history of experimental simulation of geological processes is extremely important in American geology and serves as an exemplary topic for research. Questions of the scaling and simulation of long-term processes, the nature of dynamic modeling, the connection with engineering and physics through the technical and intellectual development of the study of the strength of materials could all figure as the focus of studies.

The history of most *fields* of geology is unknown in any scope or detail. The history of petrology in North America is unsurveyed, though it includes such important steps as the Cross, Pirrson, Iddings, and Washington system of classification; the Bowen Reaction Series; the development of the ternary-phase-diagram method of representation; and the collaboration of R. A. Daly and P. W. Bridgman on high-temperature and high-pressure studies; these are among many other topics. All are major steps in the development of petrology and had great international impact. They have in common that none have been historically investigated in any depth.

The history of stratigraphy in North America has not had even a cursory overview since Raymond Moore wrote a retrospective for the fiftieth anniversary of the GSA in 1938.[23] The difficulty of establishing correlations, their vulnerability to changing scales of mapping and interpretation of fossils, the conservatism mandated by the enormous labor of remapping, all could be studied individually or as an ensemble. Stratigraphy is what most geologists have done through most of the history of geology, and it borders on the ridiculous that no history of the subject in America has ever been attempted.

The connections between fields and the creation of *new fields* also need to be studied. The history of gravimetry in North America, which grew jointly out of issues in theoretical physics and the practical needs of geodesists, led to the creation of an entire new field of geophysics. In the work of John Hayford and William Bowie on isostasy this led to several competing models of the nature and dynamics of the earth's interior. Technological development of gravity instruments, the connection with mineral exploration, and the practical and academic power struggles over the results of such research are absorbing stories waiting to be written.

More generally, the connection between exploration geophysics (whether gravimetric, seismic, electrical, or magnetic) and the development of major geological theories in the 1920s and 1930s is extremely important and worthy of investigation. The financial stakes involved made petroleum geologists a good deal more daring theoretically than their academic colleagues. The celebrated 1926 New York symposium on Wegener's theory of continental drift was sponsored by the American Association of Petroleum Geologists and organized by a Dutch geologist who was vice president of the Marland Oil Company of Tulsa, Oklahoma. The story of the organization of this conference would be a revealing intersection of many themes.[24]

Another topic with interesting implications and close ties to economic geology is the history of theories of sedimentation. Some provocative exploratory work on this topic by R. H. Dott, Jr., ought to be followed up.[25]

The study of paleogeography, the comprehensive reconstruction and repre-

[23] Raymond C. Moore, "Stratigraphy," in Geological Society of America, *Geology, 1888–1938: Fiftieth Anniversary Volume* (New York: GSA, 1941), pp. 179–220.

[24] W. A. J. M. van Waterschoot van der Gracht, ed., *Theory of Continental Drift* (Tulsa, Okla.: American Association of Petroleum Geologists, 1928).

[25] R. H. Dott, Jr., "Tectonics and Sedimentation a Century Later," *Earth Science Reviews*, 1978, *13*:1–34.

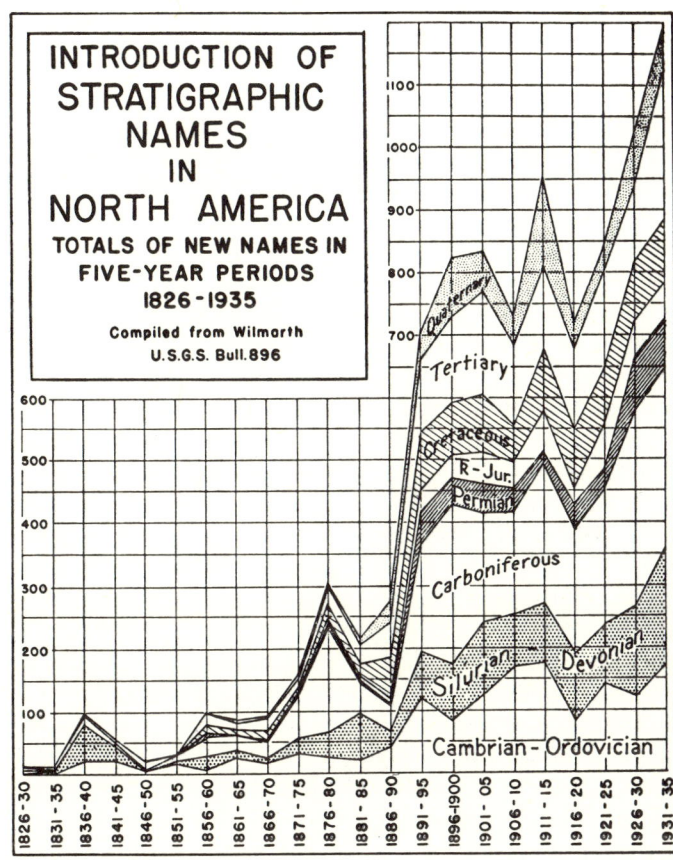

INTRODUCTION OF STRATIGRAPHIC NAMES IN NORTH AMERICA

TOTALS OF NEW NAMES IN FIVE-YEAR PERIODS 1826-1935

Compiled from Wilmarth U.S.G.S. Bull. 896

Graph from Raymond C. Moore, "Stratigraphy," in Geological Society of America, Geology, 1888–1938: Fiftieth Anniversary Volume (New York: GSA, 1941). Courtesy of the GSA.

sentation of the earth in the past, had a number of pioneers in North America. It is a subject that synthesizes the results of many subfields, and in which there is considerable tension on both a personal and disciplinary level. It has produced some agreeably ferocious controversies. Historical study of this field and its many practitioners—including Charles Schuchert, E. O. Ulrich, Bailey Willis, Thomas C. Chamberlin, and Walter Bucher—is a good way to tackle the relationship of European and North American geology in the first third of the twentieth century. It also highlights the changing relationship between geology and paleontology on the one hand and geology and geophysics on the other.

Continuing the intersection of general geological theory and economic geology, the history of theories of ore genesis unites the history of geochemistry with the practical problems of mineral prospecting and exploitation. Such a study, wending its way from assayers' methods to industrial chemistry texts, among other sources, might stretch all through the history of modern geology as a leading theme.

A noncommissioned, nonofficial history of the Geological Society of America might take up a number of issues not discussed in the official versions, particularly the secession, just after World War I, of a number of geologists of practical bent and nonacademic employment to found their own associations and journals. The creation of a three-tiered structure of geological publications—government, academic, and nonacademic professional—and its implications for

the development of the science could lead to at least one publishable dissertation.

In addition to concepts, fields, techniques, and other approaches, *biography*
is another wide-open field. While we are overwhelmed with short appreciative
essays, we lack substantial critical biographies of all but a few of even the most
important geologists. The recent return to respectability of the scientific biography, stigmatized by our more marxological colleagues as the epitome of bour-
geois individualism, seems to offer real possibilities. These prospects are
brightest when a life and career are used to illuminate a field or period.

Prime candidates for biographical treatment are James Hall, James Dana,
Bailey Willis, F. V. Hayden, Joseph Barrell, R. A. Daly, Charles Schuchert,
Walter Bucher, Marshall Kay, Raymond C. Moore, E. W. Berry, Henry Rogers,
John Hayford, Clarence Dutton, Frank Taylor, William Hobbs, T. Sterry Hunt,
Frank Adams, Beno Gutenberg, C. D. Walcott, William Maclure—there are
dozens of similarly obvious choices. However, virtually unknown figures could
also be attractive candidates. George L. Vose, a civil engineer from Boston,
wrote the best review of theories of mountain origin to appear in the United
States in the nineteenth century.[26] His work in this area, combined with his work
on structural engineering and mathematics, would seem to provide fertile ground
for an illuminating biography, especially since it would underline the importance
of engineers in advancing theoretical science in the period after the Civil War.

Beyond this roster of topics organized by field, intersection of fields, interplay
of theory and practice, and biography, there are surely other approaches. Certain geological phenomena (Mauna Loa, Meteor Crater, the Grand Canyon,
Crater Lake, the Front Range of the Rockies, the Piedmont, the Squantum tillite)
have attracted interpretation and reinterpretation in every period of American
geology. A monograph on the literature of any one of these phenomena, following the example of Stephen Pyne's work on the Grand Canyon, would be a
valuable excursion through the history of geological theory and practice and
allied social and political concerns. Other such topics would include a history
of geological cartography of the Appalachians from colonial times to the recent
deep seismic profiles of the Consortium for Continental Reflection Profiling (CO-
CORP), a history of the redefinitions of granite in American geological history,
or a history of the impact of any number of instruments on geology. These latter
might include the pendulum gravimeter, the petrographic microscope, electric
well-logging apparatus, or the seismometer. Each of these created whole new
fields of investigation for geology in America.

The reader can easily extend this list of topics further. A convenient and desirable way to do this is to consult a geologist. An easy entree to this method
is the *Directory of Geoscience Departments* of the American Geological Institute. The directory's listing of emeritus professors can put one in touch with
scores of geologists who took their degrees in the 1930s and who were taught
by geologists who took their degrees at the turn of the twentieth century. This
living archive, which, to my knowledge, is almost unused and has certainly not
been systematically approached, is certainly the best way to identify problems
worthy of investigation and to locate papers, letters, and other materials. Ge-

---

[26] Vose, *Orographic Geology* (cit. n. 12).

ology is a historical science and geologists are, as a rule and regardless of age, historically interested, and have compendious memories for scattered individual items of evidence as part of their professional armament. In my experience, they are also gracious and willing to help the historian along the way.

PROSPECTS FOR A HISTORY OF GEOLOGY

Of all the reasons why the history of geology is undeveloped (there are many: worthy and unworthy, rational and silly), the most significant is the outlandish but ubiquitous tendency to train all historians of science to accept physics, particularly theoretical physics, as a template science, against which all others are to be measured and to which all others can and will be reduced. This predisposition seduces most graduate students into the history of the physical sciences in the hope and expectation that somehow this gets them closer to the heart of the matter and the essence of science.

The more we learn of the history of the many sciences, the more tendentious and presumptuous such as orientation appears and the more untenable it becomes. Theories of scientific revolutions, of research programs and similar formalistic representations, while presented as comprehensive and highly general, are instead closely and unreflectively bound to the heritage of physical reductionism, and in particular to the style of conceptual reductionism associated with the defunct program known as logical positivism. Most history of science today is hostage to the belief that if one knows the history of physics, one understands how science is done and, in some particularly deep sense, what all science is. *This is simply not true.* Nothing demonstrates this more clearly than the history of geology, where methods of work, intellectual orientation, choice of techniques, and visions of the whole bear little historical relation to developments in physics—and often proceed viably, bravely, and even correctly in defiance of the "known laws of physics."

This being so, it is time to cease conducting the profession as if everyone who does not work in the history of Newtonian mechanism or quantum mechanics is somehow in a collateral or peripheral study, best left till the fundamental work gets done. We shall make no further progress toward an understanding of the sciences until we can force a recognition of their irreducible variety.

In concluding his *Mechanization of the World Picture,* E. J. Dijksterhuis, speaking of Newton, argued that he had not gone far enough in his reformulation of the whole, had left the structure too much as he found it in spite of his revolutionary additions.[27] This was not a criticism but an acknowledgment of a more general state of affairs. We always keep too many of the concepts we learned in school, we never proceed radically enough, we always keep much of the familiar as requiring no revision, and we grudgingly give ground where cherished ideals are concerned, especially those promising unity and synthesis in the near future.

A certain complacency and orthodoxy of the kind Dijksterhuis noted dominates the history of science to an astonishing degree today. Therefore, to any graduate student reading this, I offer the following advice. Be bold. Pick a

---

[27] E. J. Dijksterhuis, *The Mechanization of the World Picture,* trans. C. Dikshoorn (Oxford: Clarendon Press, 1961), p. 465.

dissertation topic, not in some burned-over district, but in something new. Take something ample on which almost nothing has been done, and begin your career by making a substantial and original contribution to the history of science. Remember that libraries of books on the history of geology can still be written that do not hinge on the discovery of new minutiae bearing on old topics hidden in distant archives requiring expensive travel. Stay away from all topics that form part of an established scholarly industry. Do not be afraid of the late nineteenth and early twentieth centuries, in which the interactions and issues seem more complex—they are rewarding in proportion. Remember that the history of science is very young, and that the view of the whole bequeathed you by your professors is inevitably provisional and fallible. It is within your power, and largely your responsibility, to change it and refine it. One of many roads to this worthy goal, and certainly the most open, lies through the history of the earth sciences.

History of Astronomy

By Marc Rothenberg*

THE FIELD OF HISTORY OF AMERICAN ASTRONOMY has formed at the confluence of three intellectual streams. Historians of American science, whose source lies in turn in American cultural history, are interested primarily in the social and institutional framework of astronomy. Those who approach American astronomy from the perspective of the history of astronomy are more likely to be concerned with the intellectual content of the field. Finally, astronomers have always had considerable interest in the history of their discipline, although this interest has manifested itself primarily in the form of chronologies, biographies, and anecdotes.

This diversity of interest, background, and methodology on the part of its practitioners gives the history of American astronomy intellectual breadth. Lacking shared assumptions and even agreement on what constitutes the most significant questions to be asked, these practitioners range far and wide, ever sharpening our understanding of the evolution of the American astronomical community. This article seeks to provide some sense of this diversity.

GENERAL STUDIES

The starting point for any excursion into the history of American astronomy should be two recent, selective, annotated bibliographies, although for neither work is American astronomy the primary focus. David DeVorkin is concerned with the history of modern astronomy and astrophysics, while Marc Rothenberg deals with American science and technology. However, between them, the bibliographies include references to almost all the significant literature published before 1981.[1] This article is both an effort to place the historical literature in a more interpretive framework than that provided by the two bibliographies, despite their historiographic introductions, and an attempt to update the bibliographies through mid 1984. The growth of the field over the past few years is threatening to make both volumes obsolete.

Both bibliographies have one major restriction: they list published material only. For manuscripts, one must turn to the International Catalog of Sources for the History of Physics and Astronomy, maintained by the Niels Bohr Library of the American Institute of Physics. This catalogue includes references to most major manuscript collections in history of American astronomy. The *Journal for the History of Astronomy* contains an early checklist of collections.[2]

* Joseph Henry Papers, Smithsonian Institution, Washington, D.C. 20560.

[1] David H. DeVorkin, ed., *The History of Modern Astronomy and Astrophysics: A Selected, Annotated Bibliography* (New York: Garland, 1982); Marc Rothenberg, ed., *The History of Science and Technology in the United States: A Critical and Selective Bibliography* (New York: Garland, 1982).

[2] Charles Weiner and Joan N. Warnow, "Source Materials for the Recent History of Astronomy

It is a good thing that neophytes have bibliographies and a source catalogue to guide them, because there is no adequate overview of American astronomy. Nevertheless, a few general surveys are useful if their limitations are kept in mind. Carroll Pursell's brief pamphlet gives a fine nontechnical summary of institutional development in astronomy, while the bicentennial issue of *Astronomy* supplies details on intellectual content and personalities. However, neither publication was written with a scholarly audience in mind. Still suggestive, even if dated, is Carl Rufus's effort at dividing American astronomy into periods.[3]

Another effort at an overview is Stephen Brush's bicentennial contribution to the history of American astronomy. Brush presents both a historiographic argument and a survey of American astronomy from 1800 to 1950.[4] He begins by rejecting A. Hunter Dupree's vision of all history of science as social history and Dupree's belief that the proper subject of study for the historian of science is the scientist. Brush views the history of science as the study of the search for knowledge and believes that this search is essentially international (or rather, he implies, supranational). The kind of studies advocated by Dupree belong, according to Brush, under the rubric of American history, not history of science. Having made his historiographic point, Brush then attempts to measure the increasing significance of American contributions to astronomy. As a survey, Brush's essay does not demonstrate full understanding of what went on in the United States. Because he concentrates on the most consequential events and individuals as defined by current astronomical practice, he ignores the truly significant intellectual and institutional developments during the United States' first century, which formed the foundation for later major events.

EARLY EFFORTS: THROUGH 1825

Although the literature discussing astronomy in the United States during the colonial and early national periods is not extensive, it is of high quality and diversity of theme. One should begin with Donald Yeomans's survey of the progressive astronomical activities of English colonials, then contrast it with Michael Hall's study of the persistence of Renaissance elements in Puritan science. Hall uses astronomical theory as evident in almanacs to argue that the older form of science persisted until 1700. Placing seventeenth-century colonial astronomy in its broadest context, emphasizing the ties to the Old World, is Raymond Stearns's magnificent survey of all of British-American colonial science.[5]

For the eighteenth century, Stearns is less significant because of Brooke Hindle's earlier study of science in Revolutionary America, complemented by his

and Astrophysics: A Checklist of Manuscript Collections in the United States," *Journal for the History of Astronomy,* 1971, 2:210–218.

[3] Carroll W. Pursell, Jr., *Astronomy in America* (Chicago: Rand McNally, 1967); *Astronomy,* 1976, *4* (7); and W. Carl Rufus, "Proposed Periods in the History of Astronomy in America," *Popular Astronomy,* 1921, 29:393–404, 468–475.

[4] Stephen G. Brush, "Looking Up: The Rise of Astronomy in America," *American Studies,* 1979, *20* (2):41–67.

[5] Donald K. Yeomans, "The Origin of North American Astronomy—Seventeenth Century," *Isis,* 1977, *68*:414–425; Michael G. Hall, "Renaissance Science in Puritan New England," in *Aspects of the Renaissance: A Symposium,* ed. Archibald R. Lewis (Austin: Univ. Texas Press, 1967), pp. 123–136; and Raymond Phineas Stearns, *Science in the British Colonies of America* (Urbana: Univ. Illinois Press, 1970).

analytical biography of David Rittenhouse, the leading American astronomer of that period. In reading both books, the historian of astronomy should turn first to the material dealing with the 1769 transit of Venus. However, what makes these books especially valuable are Hindle's expositions of the problems and limitations facing early American scientific practitioners, including slow communications and a lack of patronage, and the collective efforts made to overcome the limitations.[6]

To gain a wider perspective on the American astronomical scene, Hindle's work can be supplemented by Silvio Bedini's. His study of colonial and early national period mathematical practitioners—surveyors, instrument makers, and the like—focuses on a group of individuals usually on the periphery of Hindle's concern. Bedini's biography of Benjamin Banneker is of less interest as a biography than as an indicator of the state of American surveying practice and almanac making at the time. Finally, to give Hindle and Bedini an international perspective, one should turn to Susan Cannon's essay on Humboldtian science, which parenthetically makes the argument that the applied form of late colonial American astronomy was as sophisticated as its European counterparts.[7]

Where Hindle leaves off chronologically, John Greene picks up. His broad study of American science carries the story forward through the first decades of the nineteenth century. As in the case of Hindle, to savor fully Greene's chapter on astronomy, one should read it in the context of the larger work. Interestingly, Greene had earlier published an article discussing the astronomy of the late colonial and early national periods.[8] Over the intervening thirty years, Greene's judgment of the quality of American astronomy during the early national period has grown harsher. While his earlier publication was fairly laudatory, his later one concludes that the early nineteenth century saw "little advance" over the efforts of colonial astronomy.[9] This turnabout may be the result of viewing astronomy in the context of gains in other disciplines.

A common thread running through the work of Bedini and Greene is the prominent role of surveyors in American astronomy. Because of the utilitarian orientation of geodesy, there is a temptation to divorce it from other forms of astronomical observations. But in reality, the practitioners of geodesy and of the purer forms of observational astronomy frequently were one and the same well into the middle of the nineteenth century.

ESTABLISHING FOUNDATIONS: 1825–1875

The literature of this period is a mix of the products of historians of American science who touch on astronomy in more inclusive studies on the social history

---

[6] Brooke Hindle, *The Pursuit of Science in Revolutionary America, 1735–1789* (Chapel Hill: Univ. North Carolina Press, 1956; rpt. New York: Norton, 1974); Hindle, *David Rittenhouse* (Princeton, N.J.: Princeton Univ. Press, 1964; rpt. New York: Arno, 1980).

[7] Silvio A. Bedini, *Thinkers and Tinkers: Early American Men of Science* (New York: Scribners, 1975); Bedini, *The Life of Benjamin Banneker* (New York: Scribners, 1972); and Susan Faye Cannon, *Science in Culture: The Early Victorian Period* (New York: Dawson/Science History, 1978), pp. 73–110.

[8] John C. Greene, *American Science in the Age of Jefferson* (Ames: Iowa State Univ. Press, 1984); Greene, "Some Aspects of American Astronomy, 1750–1815," *Isis,* 1954, *45:*339–358.

[9] Greene, *American Science,* p. 156.

of American science and of historians of American astronomy. Since the age was one from which few American astronomical contributions were ultimately deemed important by modern astronomers, historians have paid relatively little attention to the intellectual content of American astronomy during these years. In particular, there are few temptations for general historians of astronomy. Rather, the institutional and social structure of American astronomy, including journals, forms of funding, and education, took root in this period, and these topics have attracted considerable interest.

To understand this period, one must begin with three very different publications. Nathan Reingold's documentary history of nineteenth-century American science contends that geography was the dominant science in the United States during the first two thirds of the nineteenth century and that astronomy was geography's handmaiden. Reingold explains the widespread appeal of astronomical activity by viewing support of and accomplishments in the area as part of a larger intellectual and institutional commitment to the geographical sciences. In contrast, Deborah Warner is interested in astronomy's changing content during the antebellum period, a feature that made the field so much more attractive to study. She credits John Herschel's *Treatise on Astronomy* with introducing Americans to the changes in content. Finally, there is Elias Loomis's contemporary survey of institutions and ideas. More than 130 years after its compilation, Loomis's work is still reliable and valuable.[10]

Turning to more specific topics, it is not surprising, given the identification of the dome with astronomical research, to find that observatory histories form a sizable fraction, by bulk, of the literature of the history of astronomy. The topic has attracted particular interest for this period because of the great growth in the number of American observatories between 1825 and 1875.[11]

Histories of American observatories have taken three quite different forms. Most popular has been the straightforward history of a particular observatory. Typically, such works are celebrations of the institution, valuable perhaps for their chronology of events or list of staff. Even the exemplar of this approach, Bessie Jones and Lyle Boyd's history of the Harvard College Observatory, with its wealth of detail, reliance upon primary sources, and objectivity, cannot overcome the inherent limitations of the genre: the narrowness of focus, the lack of awareness of parallel events in American science, and the sacrifice of thematic issues to the need to record all the names, dates, and events. It is the history of astronomy's version of local history. Jones and Boyd's history should be on one's shelf, but the reader comes away unsatisfied.[12]

[10] Nathan Reingold, ed., *Science in Nineteenth Century America: A Documentary History* (New York: Hill & Wang, 1964; rpt. New York: Octagon, 1979); Reingold, "Cleveland Abbe at Pulkowa: Theory and Practice in the Nineteenth Century Physical Sciences," *Archives internationales d'histoire des sciences*, 1964, *17*:133–147; Deborah Jean Warner, "Astronomy in Antebellum America," in *The Sciences in the American Context: New Perspectives*, ed. Reingold (Washington, D.C.: Smithsonian Institution Press, 1979), pp. 55–75; and Elias Loomis, *The Recent Progress of Astronomy, Especially in the United States* (New York: Harper, 1856; rpt. New York: Arno, 1980).

[11] D. B. Herrmann, "Zur Statistik von Sternwartengründungen im 19. Jahrhundert," *Sterne*, 1973, *49*:48–52, on p. 51.

[12] An example of this genre is Willis I. Milham, *The History of Astronomy in Williams College and the Founding of the Hopkins Observatory* (Williamstown, Mass.: Williams College Press, 1937); see also Bessie Zaban Jones and Lyle Gifford Boyd, *The Harvard College Observatory: The First Four Directorships, 1839–1919* (Cambridge, Mass.: Harvard Univ. Press, 1971).

A more interesting approach to observatory history uses that history to illuminate larger themes in the history of American science. Take, for example, the studies of the early years of the Dudley Observatory. Why should the early history of this observatory, whose contributions to astronomical science have been relatively minor, have prompted an influential scholarly article and, in response, a dissertation? Richard Olson views Dudley's history as an illustration of both the confrontation in the antebellum period between popular and professional visions of science and the clash over the question of deference to an elite in a democratic nation. These have been controversial and important themes in the history of American science, having been used as explanatory tools in analyzing the alleged backwardness of American science during the mid-nineteenth century. Mary James rejects Olson's contention, recasting the conflict in terms of a clash between two elites—scientists versus local cultural leaders—each seeking deference from the other.[13]

Integrated, thematic analysis is yet a third approach to observatory history. David Musto's brief study of the rapid increase in the number of American observatories after 1830 is an example of such an analysis. He contrasts these successes with the failures during the early decades of the century, concentrating on the diversity of financial arrangements used to support astronomical activity.[14]

The dependence of the astronomer upon instrumentation might suggest that the history of telescopes and auxiliary equipment would have attracted attention from historians of American astronomy. It has, but unfortunately generally in the limited format of local history—the story of a particular telescope, often in the context of the history of an observatory. Like histories of observatories, these studies suffer from narrowness of focus. They do not, for example, tell us anything about the mutually supportive roles of technician and scientist, nor do they prove to be sources of information about the designers and makers of the instruments.

There are some exceptions. Deborah Warner, one of the few historians of American astronomy to pay attention to "the producers of works rather than words," has demonstrated how scientific discoveries can inspire technical improvements, which in turn may lead to new discoveries. She has also published a catalogue of telescopes manufactured by Alvan Clark and Sons, which not only contains useful information about the instruments but also supplies the alert reader with indications of the state of astronomical research, levels of funding, and institutional commitments. There is also an important study of the pioneering generation of American telescope makers, and Albert Van Helden's recently published summation of the state of the art of telescope making in the second half of the nineteenth century provides an international context for American activities.[15]

[13] Richard G. Olson, "The Gould Controversy at Dudley Observatory: Public and Professional Values in Conflict," *Annals of Science*, 1971, *27*:265–276; Mary Ann James, "The Dudley Observatory Controversy" (Ph.D. diss., Rice Univ., 1980).

[14] David F. Musto, "A Survey of the American Observatory Movement, 1800–1850," *Vistas in Astronomy*, 1968, 9:87–92.

[15] Deborah Jean Warner, "Lewis M. Rutherfurd: Pioneer Astronomical Photographer and Spectroscopist," *Technology and Culture*, 1971, *12*:190–216, quoting p. 190; Warner, *Alvan Clark & Sons: Artists in Optics* (Washington, D.C.: Smithsonian Institution Press, 1968); Robert P. Multhauf,

Other topics in the social history of astronomy have attracted less attention than observatory history, but the resulting work includes representatives of high quality. To understand the position of astronomy in the liberal arts curriculum, one can turn to Stanley Guralnick's study of science in antebellum colleges. Guralnick demonstrates how science, including astronomy, permeated the curriculum of these institutions and provides information on faculty members and course content. In contrast, Rothenberg analyzes the much more specific issue of the training of astronomers and the correlation between their training, whether informal or formal, and the quality and direction of their subsequent research. Even more limited in scope is Howard Plotkin's discussion of the founding of the Ann Arbor School (referring to astronomers who studied at the University of Michigan), a major influence on the development of American astronomy during the second half of the century.[16]

There are two significant studies of astronomical journals: Russell Mc-Cormmach's history of the *Sidereal Messenger* and D. B. Herrmann's article on the founding of the *Astronomical Journal*. McCormmach argues that the *Sidereal Messenger*, a popular journal, represented an unsuccessful effort to meld a dedication to research with a popular appeal for the support of astronomy; that is, the use of popularization to support research. In contrast, Herrmann, one of the few continental Europeans to study American astronomy, is interested in how a German scientific institution—in this case a specialized journal—served as a model for an American institution.[17]

Aside from Musto's brief survey of the funding of observatories, there has been no attempt specifically to analyze financial support of astronomy during the antebellum era. However, Howard Miller's study of private sector support for American science includes considerable material on astronomy, as does Hunter Dupree's monographic treatment of science in the federal government.[18]

Because of the perception that American astronomers produced little of importance during this period, historians have paid almost no attention to individual scientists. The *Biographical Memoirs of the National Academy of Sciences* is still the most useful source on this select group. For others, there are nineteenth-century lives and letters. An example of the best of this genre is Edward Holden's memorial of William Cranch Bond and George Phillips Bond.

---

et al., "Holcomb, Fritz, and Peate: Three 19th-Century American Telescope Makers," *U.S. National Museum Bulletin,* 1962, *228*:156–183; and Albert Van Helden, "Telescope Building, 1850–1900," in *General History of Astronomy,* ed. Owen Gingerich, Vol. IV, Part A: *Astrophysics and Twentieth-Century Astronomy to 1950* (Cambridge: Cambridge Univ. Press, 1984), pp. 40–58.

[16] Stanley M. Guralnick, *Science and the Ante-bellum American College* (Philadelphia: American Philosophical Society, 1975); Marc Rothenberg, "The Educational and Intellectual Background of American Astronomers, 1825–1875" (Ph.D. diss., Bryn Mawr College, 1974); and Howard Plotkin, "Henry Tappan, Franz Brünnow, and the Founding of the Ann Arbor School of Astronomers, 1852–1863," *Ann. Sci.,* 1980, *37*:287–302.

[17] Russell McCormmach, "Ormsby MacKnight Mitchel's *Sidereal Messenger,* 1846–1848," *Proceedings of the American Philosophical Society,* 1966, *110*:35–47; D. B. Herrmann, "B. A. Gould and His *Astronomical Journal*," *J. Hist. Astron.,* 1971, *2*:98–108.

[18] Musto, "American Observatory Movement" (cit. n. 14); Howard S. Miller, *Dollars for Research: Science and Its Patrons in Nineteenth-Century America* (Seattle: Univ. Washington Press, 1970) (Miller contends that "astronomy [was] the most richly supported of the physical sciences in the United States" in the mid-nineteenth century, p. 26); and A. Hunter Dupree, *Science in the Federal Government: A History of Policies and Activities to 1940* (Cambridge, Mass.: Harvard Univ. Press, 1957; rpt. New York: Arno, 1980).

More recent contributions include the studies of Maria Mitchell, who, as the first American woman astronomer, has great symbolic importance; Arthur Norberg's work on the early career of Simon Newcomb (Newcomb's later career is considered in the discussion of the years 1875–1945); and a dissertation on the Yale astronomer Denison Olmsted.[19] We still have much to learn about the generation of American astronomers who established the foundations of the community.

But if we lack modern investigations of the astronomers, we have even less on their work. Whether it is the debate over the nature of meteors, the controversy over the position of Polaris, or the calculations on the orbit of the Great Comet of 1843, there is general ignorance on the part of historians. The only area that has attracted attention is cosmogony, in the form of Ronald Numbers's monograph on American attitudes toward Laplace's nebular hypothesis. Numbers reviews American contributions to the hypothesis, especially the work of Daniel Kirkwood, and analyzes the role the hypothesis played in the growing acceptance by Americans of explanation by natural law.[20]

Not fitting neatly into any single category are Norriss Hetherington's two case studies of mid-nineteenth-century American astronomy. Hetherington is attempting to derive, from narrow research bases, some rather broad conclusions about the professionalization of American science and the state of science in a developing country. Although the papers are suggestive, their conclusions are unsupported.[21]

THE TRANSITIONAL PERIOD: 1875–1945

During these years, the center of gravity of astronomical research moved from traditional positional astronomy to astrophysics. For a number of institutional and intellectual reasons, including the willingness of the private sector to support astronomical research and the lack of a particular dominant research tradition, such as astrometrics, the United States was able to become one of the astrophysical superpowers by the early years of the twentieth century. Given the predilection of historians of astronomy to focus on great intellectual achievements, it is not surprising that the increasing significance of American research to the world astronomical community sparked their interest in studies of the lives and ideas of American astronomers. This does not mean that interest in the social history of American astronomy for these years disappeared. Quite the

---

[19] Edward S. Holden, *Memorials of William Cranch Bond, Director of the Harvard College Observatory, 1840–1859, and of His Son, George Phillips Bond, Director of the Harvard College Observatory, 1859–1865* (New York: Lemcke & Buechner, 1897; rpt. New York: Arno, 1980); Helen Wright, *Sweeper in the Sky: The Life of Maria Mitchell, First Woman Astronomer in America* (New York: Macmillan, 1949); Sally Gregory Kohlstedt, "Maria Mitchell: The Advancement of Women in Science," *New England Quarterly*, 1978, *51*:39–63; Arthur L. Norberg, "Simon Newcomb's Early Astronomical Career," *Isis*, 1978, *69*:209–225; Gary Lee Schoepflin, "Denison Olmsted (1791–1859), Scientist, Teacher, Christian: A Biographical Study of the Connection of Science with Religion in Antebellum America" (Ph.D. diss., Oregon State Univ., 1977).

[20] Ronald L. Numbers, *Creation by Natural Law: Laplace's Nebular Hypothesis in American Thought* (Seattle: Univ. Washington Press, 1977).

[21] Norriss S. Hetherington, "Cleveland Abbe and a View of Science in Mid-Nineteenth-Century America," *Ann. Sci.*, 1976, *33*:31–49; Hetherington, "Mid-Nineteenth-Century American Astronomy: Science in a Developing Nation," *ibid.*, 1983, *40*:61–80.

contrary. However, social history of astronomy is no longer so obviously dominant. Moreover, reflecting the evolution of American astronomy, social historians have changed the relative emphasis they give to particular themes.

To gain an international perspective on American events, the *General History of Astronomy* should ultimately prove invaluable. So far, only half of one of the projected four volumes has been published, but it contains considerable material on American contributions to astrophysics, astrophotography, variable star classification, and the construction of large telescopes.[22] Unfortunately, the published segment reveals a defect of the entire project—the use of astronomers to write historical essays. Having technical expertise in a scientific field is not a credential for writing competent history, as the contrast with the essays in the volume by professional historians makes even clearer.

Another potential source of international context, Herrmann's superficial comparison between events in the United States and those in Germany, is ultimately disappointing. Actually, Herrmann is analyzing the astrophysical contributions to the *Astronomische Nachrichten* and the *Astronomical Journal* during the second half of the nineteenth century. Although he makes some interesting points about Germany, his conclusions about the American situation are particularly cursory.[23]

Turning to observatory history, we find examples both of local history[24] and of efforts to consider these institutions in more fruitful ways. DeVorkin's discussion of the Yerkes Observatory during World War II is a case study of how a scientific institution's director responds to war mobilization. In particular, he shows how the director deflects the perceived threat to pure research represented by that mobilization. William Hoyt offers a different approach to observatory history, making the research of the observatory the central focus and viewing the institution and the individuals in terms of that research. His two studies on the Lowell Observatory should be read both for their own sake and as possible models for structuring observatory histories.[25] I find Hoyt's approach much more interesting than organizing a history around the regimes of observatory directors.

The history of instrumentation is another area in which historians are producing significant insights. John Lankford's case studies on the adoption of large refractors to photographic work show that the astronomical community was rather conservative in adopting innovative technology. Technology, economics, and human interaction all played a role in deciding the success of an innovation. The interaction between technologist and researcher is demonstrated in Deborah Mills's study of George Ritchey's technological contributions. Through his improvements of optical, mechanical, and photographic elements, Ritchey enabled the large photographic reflector to become the basic tool of twentieth-century

---

[22] Gingerich, ed., *Astrophysics and Twentieth-Century Astronomy* (cit. n. 15).

[23] D. B. Herrmann, "Zur Frühentwicklung der Astrophysik in Deutschland und in den USA," *NTM: Zeitschrift für Geschichte der Naturwissenschaften, Technik und Medizin*, 1973, *10*:38–44.

[24] Bessie Zaban Jones, *Lighthouse of the Skies: The Smithsonian Observatory, Background and History, 1846–1955* (Washington, D.C.: Smithsonian Institution Press, 1965).

[25] David H. DeVorkin, "The Maintenance of a Scientific Institution: Otto Struve, the Yerkes Observatory, and Its Optical Bureau during the Second World War," *Minerva*, 1980, *18*:595–623; William Graves Hoyt, *Lowell and Mars* (Tucson: Univ. Arizona Press, 1976); and Hoyt, *Planets X and Pluto* (Tucson: Univ. Arizona Press, 1980).

The 200-inch Hale reflecting telescope on Mount Palomar, initiated by George Ellery Hale (1868–1938), funded about 1928 by the Rockefeller International Education Board and General Education Board and later by the Rockefeller Foundation, and now under the administration of the California Institute of Technology. Palomar Observatory photograph.

astronomy. Edward Pershey has studied the early telescope manufacturing of Warner and Swasey, a machine-tool company that approached the design and construction of telescopes as an engineering problem. Sometimes important insights come from unexpected sources. Although George Webb's biography of A. E. Douglass is not focused on the history of instrumentation, his description of the travails Douglass faced in attempting to secure a research telescope for the University of Arizona brings home the dependence of the astronomer upon the glass industry.[26]

Miller and Dupree continue to be the authorities on support for astronomical research, although their works should be supplemented by Plotkin's. Plotkin's study of the Henry Draper Memorial focuses on Edward C. Pickering, one of American astronomy's leading fundraisers. Another article by Plotkin discusses federal funding of astronomy in the context of a struggle over civilian control of a governmental bureau—the U.S. Naval Observatory.[27]

A theme that has attracted additional attention in this period is the role of women in American astronomy. Women, who were generally welcome as a source of inexpensive labor in American astronomy, at the same time faced restrictions in employment and research opportunities. Pamela Mack surveys individuals active during the period 1875–1925, while Warner carries forward to the contemporary period the description of the efforts of women to become an integral part of the American astronomical community. Placing the efforts of female astronomers in a wider context is Margaret Rossiter's account of women scientists in the United States.[28]

The professionalization of American astronomy, the interaction between amateurs and professionals, and the organization of the astronomical community have also attracted considerable attention. Lankford examines the issue from the perspective of the division of research problems and differing attitudes toward instrumentation, carefully looking at both sides of the transatlantic arena, while Rothenberg focuses on the organization of the American astronomical community during the period of professionalization. Also touching on the questions of organization and, incidentally, of professionalization of astronomers are Trudy Bell's study of the first American Astronomical Society and Richard Berendzen's of the second. Bell traces the history of a brief-lived, amateur-dominated society, which failed in the atmosphere of increasing professional domination of the discipline. Berendzen is concerned with the current American Astronomical Society. He focuses on the interaction between traditional astronomers and astrophysicists, generally ignoring the amateur-professional tension that attracted Rothenberg's attention.[29]

[26] John Lankford, "Photography and the Long-Focus Visual Refractor: Three American Case Studies, 1885–1914," *J. Hist. Astron.*, 1983, *14*:77–91; Deborah J. Mills, "George Willis Ritchey and the Development of Celestial Photography," *American Scientist*, 1966, *54*:64–94; Edward J. Pershey, "The Early Telescope Work of Warner and Swasey" (Ph.D. diss., Case Western Reserve Univ., 1982); and George Ernest Webb, *Tree Rings and Telescopes: The Scientific Career of A. E. Douglass* (Tucson: Univ. Arizona Press, 1983).

[27] Miller, *Dollars for Research* (cit. n. 18), esp. Ch. 5; Dupree, *Science in the Federal Government* (cit. n. 18); Howard Plotkin, "Edward C. Pickering, the Henry Draper Memorial, and the Beginnings of Astrophysics in America," *Ann. Sci.*, 1978, *35*:365–377; and Plotkin, "Astronomers versus the Navy: The Revolt of American Astronomers over the Management of the United States Naval Observatory, 1877–1902," *Proc. Amer. Phil. Soc.*, 1978, *122*:385–399.

[28] Pamela Mack, "Women Astronomers in the United States, 1875–1920" (B.A. thesis, Harvard Univ., 1977); Deborah Jean Warner, "Women Astronomers," *Natural History*, 1979, *58* (5):12–26; and Margaret W. Rossiter, *Women Scientists in America: Struggles and Strategies to 1940* (Baltimore: Johns Hopkins Univ. Press, 1982).

[29] John Lankford, "Amateur versus Professional: The Transatlantic Debate over the Measurement of Jovian Longitude," *Journal of the British Astronomical Association*, 1979, *89*:574–582; Lankford, "Amateurs versus Professionals: The Controversy over Telescope Size in Late Victorian Science," *Isis*, 1981, *72*:11–28; Lankford, "Amateurs and Astrophysics: A Neglected Aspect in the Development of a Scientific Speciality," *Social Studies of Science*, 1981, *11*:275–303; Marc Rothenberg,

Interest in the lives of American astronomers is evident in the growing list of recent scholarly biographies and studies of individuals, including Donald Obendorf's account of Samuel P. Langley's astronomical activities, Webb's scientific biography of A. E. Douglass, Norberg's study of Simon Newcomb, Plotkin's scientific biography of Henry Draper, Donald Osterbock's study of Edward S. Holden, and his eagerly awaited biography of James E. Keeler. Perhaps attracting the most attention has been George Ellery Hale, arguably the central figure in American astronomy during this era. He is the subject of Helen Wright's fine biography, a number of articles by Daniel Kevles examining his activities in the larger scientific arena, and a volume attempting to delineate his legacy.[30]

Published autobiographies are also available for this era, including those of such important figures as Newcomb and Harlow Shapley. Of course, the use of autobiographies by nonspecialist historians is problematic because of their uncertain reliability. Those who are familiar with the literature and manuscripts would not, for example, uncritically accept Shapley's account of his years as director of the Harvard College Observatory. But what of the neophyte? With this problem in mind, historians and publishers might look upon the recent treatment of Cecilia Payne-Gaposchkin's autobiography as a model. The volume includes a historical introduction by Peggy A. Kidwell, which provides perspective, balance, and an independent evaluation of the events.[31]

Research on American contributions to astronomical theories, observations, and discoveries during this period far exceeds that for earlier years, no doubt a reflection of the perceived significance and sophistication of these contributions. Historical research has centered on three subject areas—the solar system, stellar astronomy, and galactic structure—with each subject attracting its own group of historians.

A central theme in some of the studies on the solar system has been the marriage of geological ideas to astronomical subjects.[32] Given the traditional disdain

---

"Organization and Control: Professionals and Amateurs in American Astronomy, 1899–1918," *ibid.*, 1981, *11*:305–325; Trudy E. Bell, "Towers Reaching to the Skies: The First American Astronomical Society, *Griffith Observer*, 1978, *42*:2–11; and Richard Berendzen, "Origins of the American Astronomical Society," *Physics Today*, Dec. 1974, *27* (12):32–39.

[30] Donald Leroy Obendorf, "Samuel P. Langley: Solar Scientist, 1867–1891" (Ph.D. diss., Univ. California at Berkeley, 1969); Webb, *Tree Rings and Telescopes;* Arthur L. Norberg, "Simon Newcomb and Nineteenth Century Positional Astronomy" (Ph.D. diss., Univ. Wisconsin, 1974); Howard Plotkin, "Henry Draper, a Scientific Biography" (Ph.D. diss., Johns Hopkins Univ., 1971); Donald E. Osterbrock, "The Rise and Fall of Edward S. Holden," *J. Hist. Astron.*, 1984, *15*:81–127, 151–176; Osterbrock, *James E. Keeler, Pioneer American Astrophysicist and the Early Development of American Astrophysics* (Cambridge Univ. Press, 1984); Helen Wright, *Explorer of the Universe: A Biography of George Ellery Hale* (New York: Dutton, 1966); Daniel J. Kevles, " 'Into Hostile Camps': The Reorganization of International Science in World War I," *Isis*, 1971, *62*:47–60; Kevles, "George Ellery Hale, the First World War, and the Advancement of Science in America," *Isis*, 1968, *59*:427–437; and Helen Wright, Joan N. Warnow, and Charles Weiner, eds., *The Legacy of George Ellery Hale* (Cambridge, Mass.: MIT Press, 1972); see also Nathan Reingold and Ida H. Reingold, eds., *Science in America: A Documentary History, 1900–1939* (Chicago: Univ. Chicago Press, 1981).

[31] Simon Newcomb, *The Reminiscences of an Astronomer* (Boston: Houghton Mifflin, 1903); Harlow Shapley, *Through Rugged Ways to the Stars* (New York: Scribners, 1969); Katherine Haramundanis, ed., *Cecilia Payne-Gaposchkin: An Autobiography and Other Recollections* (Cambridge: Cambridge Univ. Press, 1984).

[32] Stephen G. Brush, "A Geologist among Astronomers: The Rise and Fall of the Chamberlin–

of astronomers for study of the composition, as opposed to the position, of the components of the solar system, some might argue that these works belong under the rubric of the history of geology, while the historian of astronomy should stick to the obviously astronomical issues of the existence of a planet beyond Neptune or observations of a transit of Venus.[33] Indeed, with the rise of new disciplines at the intersection of two or more of the older, traditional disciplines, it is unclear where historians will place a field like planetology in some future collection of bibliographic essays.

Unquestionably part of the history of astronomy, however, are studies of our increasing understanding of the stars and galaxies. The dominant figure in American stellar work is Henry Norris Russell. Not unexpectedly, the historical literature reflects that domination.[34]

Galactic structure is a speciality for which historians have acknowledged American domination.[35] It may also be the only topic in the history of American astronomy that is beginning to face anything like overcrowding. Most topics, whether social or intellectual, have attracted few investigators. Overlap, conflicting interpretations, or even dialogues in print are the exception, not the rule. But the history of galactic structure has been thoroughly documented by Michael Hoskin, his student Robert Smith, Hetherington, Berendzen, and his students Richard Hart and Daniel Seeley.[36] At present, we have a clear understanding of the facts.[37] Now there is disagreement only over interpretation. Hence, a milestone of sorts has been reached.

Moulton Cosmogony," *J. Hist. Astron.*, 1978, 9:1–41, 77–104; William Graves Hoyt, "G. K. Gilbert's Contribution to Selenology," *ibid.*, 1982, 13:155–167.

[33] William Graves Hoyt, "W. H. Pickering's Planetary Predictions and the Discovery of Pluto," *Isis*, 1976, 67:551–564; Hoyt, *Planets X and Pluto*; Morton Grosser, "The Search for a Planet beyond Neptune," *Isis*, 1964, 55:163–183; and P. M. Janiczek and Lee Houchins, "Transits of Venus and the American Expedition of 1874," *Sky and Telescope*, 1974, 48:366–371 (which actually goes beyond the 1875 cutoff).

[34] David H. DeVorkin, "The Origins of the Hertzsprung-Russell Diagram," in *In Memory of Henry Norris Russell*, ed. DeVorkin and A. G. Davis Philip (Dudley Observatory Reports, 13) (Albany, N.Y.: Dudley Observatory, 1977), pp. 61–77; DeVorkin and Ralph Kenat, "Quantum Physics and the Stars," *J. Hist. Astron.*, 1983, 14:102–132, 180–222; and Karl Hufbauer, "Astronomers Take Up the Stellar-Energy Problem, 1917–1920," *Historical Studies in the Physical Sciences*, 1981, 11:277–303. Some stellar research histories are not concerned with Russell, including DeVorkin, "Michelson and the Problem of Stellar Diameters," *J. Hist. Astron.*, 1975, 6:1–18; Howard Plotkin, "Henry Draper, the Discovery of Oxygen in the Sun, and the Dilemma of Interpreting the Solar Spectrum," *ibid.*, 1977, 8:44–51.

[35] Robert W. Smith, for example, argues that 1914 marked the "last important observational contributions to the island universe debate by an astronomer working outside the United States"; *The Expanding Universe: Astronomy's 'Great Debate,' 1900–1931* (Cambridge: Cambridge Univ. Press, 1982), p. 11.

[36] M. A. Hoskin, "Ritchey, Curtis, and the Discovery of Novae in Spiral Nebulae," *J. Hist. Astron.*, 1976, 7:47–53; Hoskin, "The 'Great Debate': What Really Happened," *ibid.*, pp. 169–182; Robert W. Smith, "The Origins of the Velocity-Distance Relation," *ibid.*, 1979, 10:133–165; Smith, *The Expanding Universe*; Norriss S. Hetherington, "Edwin Hubble's Examination of Internal Motions of Spiral Nebulae," *Quarterly Journal of the Royal Astronomical Society*, 1974, 15:392–418; Hetherington, "The Simultaneous 'Discovery' of Internal Motions in Spiral Nebulae," *J. Hist. Astron.*, 1975, 6:115–125; Richard Berendzen and Richard Hart, "Adriaan van Maanen's Influence on the Island Universe Theory," *ibid.*, 1973, 4:46–56, 73–98; and Berendzen, Hart, and Daniel Seeley, *Man Discovers the Galaxies* (New York: Science History, 1976).

[37] In his review of Smith's book, E. Robert Paul notes that other historians have previously dealt with the issues of the discovery of an expanding universe and the Shapley-Curtis "great debate," mildly accusing Smith of rehashing older material; *Isis*, 1984, 75:596.

Finally, one study does not fit the neat tripartite division referred to earlier—Jeffrey Crelinsten's examination of American astronomers' reception of Einstein's general theory of relativity. Perhaps most stimulating is Crelinsten's attempt to identify national characteristics of American astronomers active in the second and third decades of this century.[38]

THE SPACE AGE: 1945 TO YESTERDAY

The application of space technology—missiles and spacecraft—to astronomical research and the creation of new institutions—the National Aeronautics and Space Administration (NASA) and the Jet Propulsion Laboratory—to support that technology have generated considerable excitement among historians, resulting in studies that integrate analysis of funding, administration, organization, technology, and scientific advances. These include Clayton Koppes's history of the Jet Propulsion Laboratory, Cargill Hall's probe of Project Ranger, and Homer Newell's study of the early years of space science as perceived from the headquarters of NASA. But not all studies of space age astronomy are as concerned with institutions. Brush's discussion of Harold Urey and lunar theory focuses on ideas, and Richard Hirsh's study of X-ray astronomy is a good example of an integrated history of a discipline.[39]

One problem resulting from the enthusiasm for space astronomy has been the tendency to forget that other research has been going on during the past four decades. Ground-based astronomy also has flourished, sometimes, as Joseph Tatarewicz has shown in his study of NASA's program of research support for planetary astronomy, because space science has needed the ground-based researchers.[40] Nevertheless, historians have not investigated the more traditional forms of astronomical research, focused on American contributions to new ground-based techniques like radio and radar astronomy, or considered in any detail the social history of the contemporary American astronomical community.[41]

[38] Jeffrey Crelinsten, "The Reception of Einstein's General Theory of Relativity among American Astronomers, 1910–1930" (Ph.D. diss., Univ. Montreal, 1981), a portion of this work has been published as "William Wallace Campbell and the 'Einstein Problem': An Observational Astronomer Confronts the Problem of Relativity," *Hist. Stud. Phys. Sci.*, 1983, *14*:1–91.

[39] Clayton R. Koppes, *JPL and the American Space Program: A History of the Jet Propulsion Laboratory* (New Haven, Conn.: Yale Univ. Press, 1982); R. Cargill Hall, *Lunar Impact: A History of Project Ranger* (Washington, D.C.: National Aeronautics and Space Administration, 1977); Homer E. Newell, *Beyond the Atmosphere: Early Years of Space Science* (Washington, D.C.: National Aeronautics and Space Administration, 1980); Stephen G. Brush, "Nickel for Your Thoughts: Urey and the Origin of the Moon," *Science*, 1982, *217*:891–898; for a fuller version of this paper see Brush, "Harold Urey and the Origin of the Moon: The Interaction of Science and the Apollo Program" in *Spacelab, Space Platforms, and the Future*, ed. Peter M. Bainum and Dietrich E. Koelle (San Diego: American Astronautical Society, 1982), pp. 437–470; see also Richard F. Hirsh, *Glimpsing an Invisible Universe: The Emergence of X-Ray Astronomy* (Cambridge: Cambridge Univ. Press, 1983).

[40] Joseph N. Tatarewicz, " 'Where Are the People Who Know What They Are Doing?' Space Technology and Planetary Astronomy, 1958–1975" (Ph.D. diss., Indiana Univ., 1984).

[41] One exception is Richard E. Berendzen, "On the Career Development and Education of Astronomers in the United States" (Ph.D. diss., Harvard Univ., 1968); another, from the sociology of science, is Thomas F. Gieryn, "Patterns in the Selection of Problems for Scientific Research: American Astronomers, 1950–1975" (Ph.D. diss., Columbia Univ., 1979). General histories of radio astronomy that mention American contributions include J. S. Hey, *The Evolution of Radio Astronomy* (New York: Science History, 1973); and Woodruff T. Sullivan, "Early Radio Astronomy," in Gingerich, ed., *Astrophysics and Twentieth-Century Astronomy*, pp. 190–198.

Yet there are unique opportunities for investigating the post–World War II history of astronomy. The Center for History of Physics of the American Institute of Physics has conducted over four hundred hours of interviews with over one hundred astronomers.[42] Conducted by interviewers well aware of the pitfalls of the technique, these oral histories and those conducted by the National Air and Space Museum and NASA focusing on space science will provide students of postwar astronomy with sources that those of us interested in earlier epochs can only dream about.

In addition, this period is very rich in a form of publication that falls somewhere between primary and secondary sources, namely, participant's history. This is a summation of events in contemporary astronomical fields by an astronomer who took part in at least some of those events or is still actively conducting research in the particular speciality.[43] While ostensibly a secondary account and presented as history, such an essay more often than not is more useful as a primary document than as historical literature. These accounts provide insight into the opinions, biases, blind spots, and concerns of the working astronomer. The genre is not new; it was popular among nineteenth-century astronomers.[44] Used properly, with sensitivity for its limitations, it will serve the historian well. Unfortunately, many astronomers and even some historians of astronomy do not recognize the difference between participant's history and the more analytical product of the professional historian. Even more tragic, some astronomers view participant's history as the true form of historical publication. At best, the historian will face indifference; occasionally, there is outright hostility. To do contemporary history, historians often must educate the astronomers about the historical craft.

CONCLUSION

The community of historians of American astronomy is not large. Probably no more than a few dozen scholars view the topic as their primary research interest, while a somewhat larger number have come to it from the history of astronomy or American science, worked on a particular problem, and then gone on to other interests. There are advantages to this. Members of the community never feel overwhelmed by the quantity of the literature. Keeping posted is relatively easy. In addition, opportunities for research, no matter what the chronological interest, are still abundant. No graduate student seeking a dissertation topic need search very hard or far. Rich manuscript holdings, still relatively untapped, intellectual and social problems, still unprobed, and important figures and institutions, still unexamined, await the scholar. With few exceptions, there

[42] Spencer R. Weart and David H. DeVorkin, "Interviews as Sources for History of Modern Astrophysics," *Isis*, 1981, *72*:471–477; also see Weart and DeVorkin, "The Voice of Astronomical History," *Sky Telesc.*, 1982, *63*:124–127.

[43] See, e.g., Bart J. Bok, "Harlow Shapley and the Discovery of the Center of Our Galaxy," in *The Heritage of Copernicus: Theories "More Pleasing to the Mind,"* ed. Jerzy Neyman (Cambridge, Mass.: MIT Press, 1974), pp. 26–62; also the section "Astronomy and Geophysics in Space" in Paul A. Hanle and Von Del Chamberlain, eds., *Space Science Comes of Age: Perspectives in the History of the Space Sciences* (Washington, D.C.: Smithsonian Institution Press, 1981), pp. 1–75; and Richard Berendzen, ed., *Education in and History of Modern Astronomy* (Annals of the New York Academy of Sciences, 198) (New York: New York Academy of Sciences, 1972).

[44] Loomis, *Recent Progress of Astronomy* (cit. n. 10), is an example.

is little worry about bumping into someone else's project. A final advantage of smallness is the ability to gather together for fruitful interaction. Thanks to the advent of the Historical Astronomy Division of the American Astronomical Society, historians of American astronomy can annually meet with each other and, for those interested in more contemporary topics, with some of the subjects of their research.

There is, however, a major disadvantage. The field is underdeveloped. With so many of the pieces still not well known, putting together a synthetic view of events is extremely difficult. If the historian's goal is to build upon strength, then the period from about 1890 to 1940 should be attractive. More is known about what went on in American astronomy during those years than about any other period. Yet even here there is no survey, no attempt to treat the American astronomical community as a community. Traditionally, essays of this type end with a list of needs and opportunities, but such an exercise seems unnecessary when those needs and opportunities are almost limitless, as they are in American astronomy.

But what of the future? The field will face a major problem in keeping up its numbers unless it can attract converts by sheer intellectual excitement. A relatively large fraction of the active historians hold nonacademic positions, for example, at the Smithsonian Institution, while a number of the academic historians are at institutions without graduate programs. The problem is not insurmountable. Dissertations in the subject need not be supervised by historians of astronomy, let alone by historians of American astronomy (mine was written under a historian of American civilization). Perhaps more historians of astronomy, both in Europe and in the United States, will finally come to recognize the wealth of resources available. More likely, however, recruitment into the field will be slow, and the history of American astronomy will remain the intellectual property of a small group.

History of Chemistry

By John W. Servos*

SEVERAL YEARS AGO, in my graduate seminar on science in modern America, a bright student whose previous study had been confined to the Scientific Revolution commented that he was surprised by the weakness of the secondary literature. "There is just not much substance to it," he complained. In a sense I believe he was correct. There certainly is some good literature on the history of science in America, but by comparison with the profusion of exciting and intellectually challenging books and articles on science in the sixteenth and seventeenth centuries, or with the riches of scholarship on Darwin, that literature is very limited, both in breadth and depth of treatment. Moreover, if we divide the historical writings on science in America into categories determined by scientific discipline, then it seems fair to say that the category dealing with chemistry shows some of the greatest deficiencies. The development of chemistry in the American setting has simply not received the attention given the more glamorous disciplines of physics and biology. To judge by its treatment by historians, chemistry, far more than economics, deserves the title, "the dismal science."

There is more than a little irony in this situation, for the development of chemistry in America ought to be judged a strategic site for historical research. Chemists have constituted the largest scientific community in the United States since at least the mid-nineteenth century, when reliable figures first become available; by World War I, the American Chemical Society was the largest disciplinary society in the world. Among the natural sciences, chemistry traditionally has had the closest relationship with industry, and chemists were in the vanguard of those who first succeeded in developing science-based technology, both in Europe and the United States. When authors write of "scientific medicine" or "scientific agriculture," it is often the drugs and diagnostic tests, the fertilizers, insecticides, and soil analyses of the chemist that they have in mind. And just as chemists working for industry have produced drugs and insecticides, so too it has been chemists, working for agencies of the federal, state, and local governments, who have carried responsibilities for protecting consumers from risks sometimes associated with those products. The work of the physiologist and molecular biologist depends, in large measure, on chemistry, and chemistry has long played an important role in the work of the petrologist and geologist. Sixty years ago, Wilder Bancroft, professor of physical chemistry at Cornell, called chemistry "the central science," citing its many and diverse links with other disciplines and its intimate connections with social life. Although the physicist might challenge Bancroft's use of the definite article, few would dispute that chemistry has had a vital role in the evolution of science in modern America or that it has directly affected American industry, agriculture, and medicine.[1]

* Department of History, Amherst College, Amherst, Massachusetts 01002.

[1] Wilder D. Bancroft, "The Future in Chemistry," *Science,* 1908, *27*:979–980.

So, the picture is, at present, one of great opportunity for historians and limited achievement, and, in view of this asymmetry, the paper that follows will be as much a review of topics awaiting study as a review of the existing literature. My strategy will be to divide the possible range of topics into two broad and admittedly imperfect categories. In the first, the accent will be on chemistry as an evolving body of ideas, techniques, and institutions, and as a set of conceptual resources affecting and affected by the development of other sciences. For convenience, let us call this the disciplinary genre. Work in this vein must observe certain limits. The United States has, during the twentieth century, assumed leadership in many branches of chemistry, but there is no branch of the science in which Americans enjoy a monopoly. Hence, it is impossible to write a satisfying history of chemistry, or any one of its specialties, solely out of American sources. Nevertheless, there are important episodes in the history of chemistry in which action took place largely in the United States, and there are significant issues to be examined that relate both to the reception of ideas or methods from abroad and to the conditions favoring the emergence or elaboration of certain traditions of thought or practice in the United States.

In the second category, which we may label the social and economic genre, the emphasis will be on the relations between chemistry and the history of American social and economic institutions and practices. Instead of taking as its main aim the study of how American conditions have affected chemical science, work in this category focuses on how chemistry, both as a profession and as a body of expertise, has affected American conditions. The disciplinary literature has its most natural affinities with the study of science as an international enterprise; the literature on chemistry's impact on society is more closely related to themes and topics in American social and economic history.

Basically this classification depends upon the motives of the historian. Is he or she looking at American developments because they exemplify important trends in the history of chemistry or illuminate such general issues in the history of science as professionalization, the cross-national transmission of scientific traditions, and the evolution of national styles in science? Or is he or she primarily interested in the development of American institutions and the relationship between science and social values?

There are obvious difficulties in using these two categories. Motives of historians, no less than those of their subjects, are difficult to ascertain. And those motives, separate as they may be, do not and should not completely specify research strategies and sources. It is, for example, often necessary to discuss the conceptual development of the science in order to treat its impact on society, and it is often essential to understand social context in order to explain the timing and geographical locus of, and differences in the receptivity of scientists to, conceptual innovations. Nevertheless, unless this or some similarly arbitrary arrangement is imposed, it will be impossible to discern any order in the scholarship in a field so poorly developed as this, for, as is often the case in ill-tended gardens, growth has proceeded along diverse and seemingly random lines.

THE DISCIPLINARY GENRE

The history of chemistry has not found its Daniel Kevles; there is no comprehensive history of the discipline in America that approaches the breadth, detail, and grace of Kevles's *The Physicists*. Aaron Ihde's *The Development of Modern Chemistry* remains a magnificent resource, although its principal concern is not with chemistry in a particular national context. Chemistry in America has been the subject of brief synoptic accounts by Kenneth L. Taylor and Edward H. Beardsley. They succeed in packing a surprising amount of information into small packages, but neither's work may be considered an entirely satisfactory survey. Several years ago, a quartet of historians of chemistry at the University of Pennsylvania prepared an exhaustive quantitative study of chemistry in America over the century from 1876 to 1976. Their extensive statistical results form an indispensable reserve of information on everything from investment in chemical industry to the number of pages on chemistry in various editions of encyclopedia yearbooks. Although accompanied by a valuable interpretive essay, this study was designed as a tool for historians rather than as a definitive history.[2]

Instead of looking for comprehensive breadth, the tendency in recent years among historians of American chemistry has been to look at the specialties of which chemistry is composed, and at the research schools that, in modern times, have constituted the building blocks of those specialties.[3] By comparing American and European developments at this level, it becomes possible to specify with greater confidence the ways in which local circumstances affect the physiognomy and physiology, the external profile and internal workings, of science.

The most impressive and stimulating contribution to this literature is Robert E. Kohler's *From Medical Chemistry to Biochemistry*.[4] The book is novel in both method and content. Kohler's analysis rests largely on the use of analogies between the expansion of biochemistry and the growth of business enterprises into new markets or the expansion of biological species into new territories. His aim is to show that institutional structures and markets for services can exercise a dominant influence on specialties—on their boundaries, on their rate and direction of growth, and on the scientific style of their members. Scientific specialties, no less than biological species or business firms, evolve in ways that allow them to fill available niches.

Biochemistry, Kohler argues, expanded rapidly in turn-of-the-century America because such a niche existed in the medical colleges then undergoing curricular reform; once entrenched in medical schools, American biochemists tended to emphasize the clinical side of their science—this by contrast with some of their European colleagues whose institutions were not so deeply em-

---

[2] Aaron J. Ihde, *The Development of Modern Chemistry* (New York: Harper & Row, 1964); Kenneth L. Taylor, "Two Centuries of Chemistry," in *Issues and Ideas in America*, ed. Benjamin J. Taylor and Thurman J. White (Norman: Univ. Oklahoma Press, 1976); Edward H. Beardsley, *The Rise of the American Chemistry Profession, 1850–1900* (Univ. Florida Monographs, 23) (Gainesville: Univ. Florida Press, 1964); and Arnold Thackray, Jeffrey L. Sturchio, P. Thomas Carroll, and Robert F. Bud, *Chemistry in America, 1876–1976: Historical Indicators* (Dordrecht/Boston: Reidel, 1985).

[3] Most influential in directing attention toward these units of organization have been Charles Rosenberg, "On the Study of American Biology and Medicine: Some Justifications," *Bulletin of the History of Medicine*, 1964, *38*:364–376; Avraham Zloczower, "Career Opportunities and the Growth of Scientific Discoveries in Nineteenth-Century Germany" (M.A. diss., Hebrew Univ., Jerusalem, 1966); and J. B. Morrell, "The Chemist Breeders: The Research Schools of Liebig and Thomas Thomson," *Ambix*, 1972, *19*:1–46.

[4] Robert E. Kohler, *From Medical Chemistry to Biochemistry: The Making of a Biomedical Science* (Cambridge/New York: Cambridge Univ. Press, 1982).

bedded in the medical context. The clinical style of American biochemistry, Kohler suggests, helps explain why American biochemists were generally caught unprepared for the revolutionary advances in molecular biology of the late 1940s and early 1950s. Kohler's work is fabulously rich in information about specific departments of biochemistry; it looks beyond the university and explores, for example, the largely unsuccessful efforts to make the hospital a site of biochemical research; and it avoids the pitfall of attempting to spin the history of biochemistry entirely out of American sources by giving serious attention to German and British research traditions.

A model in many ways, it does fall short in one respect: Kohler denies any significant role to scientific ideas as causal agents in his analysis. He sees a certain minimum level of achievement as necessary if a specialty is to flourish, but refrains from specifying what this minimum might be. Many would-be disciplines attain the level, but comparatively few prosper; hence, Kohler assigns overriding importance to social and institutional context. The rigor with which Kohler pursues this contextual analysis makes his book methodologically intriguing, but sometimes at the expense of clarity in exposition. It causes him some difficulties, for instance, in making clear exactly what scientific style might be; more important, it obscures that sense of intellectual excitement and enthusiasm which plays an important role in attracting talented students to new specialities. We may learn much from Kohler's powerful and challenging analysis, but institutional opportunity and marketing success, important as they often are, do not by themselves explain the genesis and growth of specialties.

The work of Margaret Rossiter on agricultural chemistry and Jean-Claude Guédon on chemical engineering, although differing greatly in style, come closest to combining a sensitivity for conceptual developments and a concern with ambient social conditions. Rossiter's *Emergence of Agricultural Science* deftly outlines the major themes in agricultural research during her period, shows how conditions in New England's farming districts and universities were related to the development of opportunities for agricultural chemists, and follows the careers of three prominent agricultural chemists who sought to create in America institutions for agricultural research. Her work is especially notable for its analysis of the ways in which institutions and ideas may be altered in the process of transmission from one society to another, as happened when American scientists sought to import Justus von Liebig's science, research ideal, and laboratory method of instruction into America. Rossiter's book is not, strictly speaking, a history of agricultural chemistry in America; she ends her account just at the point when the first agricultural research stations were established in the United States. Nevertheless, her book constitutes a solid foundation for further study of this specialty.[5]

Likewise, Guédon has not written a comprehensive survey of the history of chemical engineering in America, but in a series of articles he has, with Gallic precision, illuminated certain crucial aspects of that history. In particular, his

[5] Margaret W. Rossiter, *The Emergence of Agricultural Science: Justus Liebig and the Americans, 1840–1880* (New Haven: Yale Univ. Press, 1975). Other valuable sources on agricultural chemistry are Charles Rosenberg's elegant articles on agricultural research stations, in *No Other Gods: On Science and American Social Thought* (Baltimore: Johns Hopkins Univ. Press, 1976) pp. 133–210; and Oscar E. Anderson, Jr., *The Health of a Nation: Harvey S. Wiley and the Fight for Pure Food* (Chicago: Univ. Chicago Press for the Univ. Cincinnati, 1958).

work has focused on the origins and evolution of the notion of unit operations—the idea that the typical procedures of industrial chemistry can be analyzed in terms of a limited number of fundamental operations such as crystallization, distillation, and filtration. Perhaps most intriguing is his study of the question of why the concept of unit operations has played a central role in the development of chemical engineering in the United States but was adopted only gradually in Europe. His answer is that this notion was an outcome of a local process of tacit negotiations involving the chemical industries, the universities, and professional industrial chemists. Only in America, where universities were still flexible in their organization, where a booming chemical industry was concerned with rationalizing its operations, and where a vigorous community of industrial chemists had taken shape—were the conditions appropriate for the widespread adoption of the concept of unit operations. In England, France, and Germany, local institutional conditions gave rise to other solutions to the problem of integrating science into industrial practice. His analysis illustrates the advantages of treating American and European developments in science comparatively, and, like Rossiter's work, Guédon's displays a discriminating sensitivity to the interaction of conceptual and social factors.[6]

Although the works of Kohler, Rossiter, and Guédon are among the most impressive representatives of this genre, other recent books and articles merit mention. R. G. A. Dolby, Arturo Russo, and John W. Servos have written about a field in which Americans made early and impressive contributions, namely, physical chemistry. Although they emphasize different periods in the history of that specialty, all three authors seek to explain episodes of controversy through reference to the persistence of distinct physical and chemical orientations among physical chemists. Servos discerns a similar tension between physical and chemical traditions in his study of the foundation of the Geophysical Laboratory of the Carnegie Institution of Washington, a laboratory which, despite its name, was a center for the development of the specialty of geochemistry. The contrast between physical and chemical research traditions is no less important to Lawrence Badash's *Radioactivity in America*, which treats the history of a specialty to which both chemists and physicists initially laid claim.[7] Taken together, these contributions suggest that additional research on the history of interdisciplinary specialties is merited. Borderlands between established scientific disciplines

    [6] Jean-Claude Guédon, "Conceptual and Institutional Obstacles to the Emergence of Unit Operations in Europe," in *History of Chemical Engineering*, ed. William F. Furter (Washington, D.C.: American Chemical Society, 1980), pp. 45–75; Guédon, "Chemical Engineering by Design: The Emergence of Unit Operations in the United States" (unpublished MS); and Guédon, "From Unit Operations to Unit Processes: Ambiguities of Success and Failure in Chemical Engineering," in *Chemistry and Modern Society: Historical Essays in Honor of Aaron J. Ihde*, ed. John Parascandola and James C. Whorton (Washington, D.C.: ACS, 1983), pp. 43–60.
    [7] R. G. A. Dolby, "The Transmission of Two New Sciences from Europe to North America in the Late 19th Century," *Annals of Science*, 1977, *34*:287–310; Arturo Russo, "Mulliken e Pauling: Le due vie della chimica-fisica in America," *Testi e Contesti*, 1982, *6*:37–59; John W. Servos, "A Disciplinary Program That Failed: Wilder D. Bancroft and the *Journal of Physical Chemistry*, 1896–1933," *Isis*, 1982, *73*:207–232; Servos, "G. N. Lewis: The Disciplinary Setting," *Journal of Chemical Education*, 1984, *61*:5–10; Servos, "To Explore the Borderland: The Foundation of the Geophysical Laboratory of the Carnegie Institution of Washington," *Historical Studies in the Physical Sciences*, 1984, *14*:147–185; Servos, "The Intellectual Basis of Specialization: Geochemistry in America, 1890–1915," in *Chemistry and Modern Society*, ed. Parascandola and Whorton, pp. 1–19; and Lawrence Badash, *Radioactivity in America: Growth and Decay of a Science* (Baltimore: Johns Hopkins Univ. Press, 1979).

have presented scientists with unusual opportunities for creative achievement. Nevertheless, because they are explored by workers with disparate backgrounds, methods, and motives, they also afford ample occasion for conflict. Practitioners of the specialty may view such conflicts as embarrassments, but historians can profit from the insights they afford into both the conceptual structure and social workings of science.

This brief discussion of the disciplinary genre has not been exhaustive. Several recent publications of the American Chemical Society contain valuable essays on the histories of chemical specialties written by practitioners, and Terry Reynolds's history of the American Institute of Chemical Engineers is a superior example of official institutional history.[8] But rather than treat these and related works in any detail, it might be useful to review several topics that have yet to receive the treatment they deserve. Three stand out, namely, the histories of organic, colloid, and polymer chemistry.

The history of organic chemistry is of special interest because, by comparison with a field like physical chemistry, its development in America appears somewhat retarded, and surely the study of a specialty that was slow to prosper under American conditions will tell us no less about the dynamics of growth than the study of those that expanded rapidly. America produced several organic chemists of note in the late nineteenth and early twentieth centuries, but none of them established a research school comparable to the great schools of Germany. Neither James Mason Crafts nor Arthur Michael had much contact with advanced American students. John Ulric Nef had access to graduate students at the University of Chicago, but was too much the martinet to attract many disciples. Moses Gomberg's laboratory at the University of Michigan produced several fine organic chemists, but his influence as a teacher was not commensurate with his importance as the discoverer of the free triphenylmethyl radical—perhaps because the American academic system was not entirely hospitable to Russian Jews in the early twentieth century. Neither Claude S. Hudson, who was perhaps the finest student of the structure of sugar molecules in the generation after Emil Fischer, nor Walter A. Jacobs, who did much to determine the structure of the biologically active substances in digitalis, had opportunity to build true research schools, since neither held academic posts.

Ira Remsen, the first professor of chemistry at the Johns Hopkins University, would seem to represent an exception to this pattern. He did train many graduate students. Indeed, he sent scores of them into careers at American colleges and universities. But Remsen was not as original a chemist as Michael or Gomberg, and only a handful of his students did significant scientific work. Remsen was an educator who happened to be a chemist, and a chemist who happened to be interested primarily in organic chemistry. His students reflected these characteristics. They were influential in organizing chemical departments and programs and in propagating the laboratory method of instruction, but they typically were not dedicated heart and soul to organic research. They established the foundations for a research community in America but did not challenge

[8] Terry S. Reynolds, *Seventy-five Years of Progress: A History of the American Institute of Chemical Engineers* (New York: AIChE, 1983); Furter, ed., *History of Chemical Engineering* (cit. n. 6); William F. Furter, *A Century of Chemical Engineering* (New York: Plenum Press, 1982); and Herman Skolnik and Kenneth M. Reese, eds., *A Century of Chemistry: The Role of Chemists and the American Chemical Society* (Washington, D.C.: ACS, 1976).

German chemists for leadership in the study of the structure and synthesis of organic substances.

The period between the 1870s and 1920 or so constituted an extended apprenticeship for organic chemistry in the United States; American organic chemists who possessed the skill to establish centers of excellence generally did not have the desire or opportunity to do so. It was not until the mid-1920s that the field began to flourish in this country, a development that was probably related to the expansion of opportunities during the interwar era in the drug, petroleum, and chemical industries. One line of improvement was in classical organic chemistry, and especially the study of the structure and synthesis of natural products. Here the research school of Roger Adams was preeminent during the 1920s and 1930s. Working at the University of Illinois, Adams trained more than two hundred Ph.D. and postdoctoral students, and, together with his colleague Carl S. Marvel, started *Organic Syntheses,* an annual volume of methods for preparing organic compounds. During the postwar era, when a raft of new instruments (nuclear magnetic resonance, mass and infrared spectroscopy) revolutionized the study of organic structure, Robert B. Woodward of Harvard assumed leadership through his synthesis of such substances as quinine, strychnine, cholesterol, and chlorophyll.

A second line of development, commencing in the 1920s, was in the application of the methods of physical chemistry to the study of organic reactions. By contrast with research on the structure of organic molecules, in which Americans were still seeking to equal their European colleagues, the field of physical-organic chemistry was one in which Americans quickly assumed positions of leadership. Influential here were Morris Kharasch at the University of Chicago, James B. Conant and his pupil Paul D. Bartlett at Harvard, and Howard J. Lucas at the California Institute of Technology.

Parts of the history of organic chemistry in America have received attention of late. D. Stanley Tarbell contributed an excellent survey to the ACS Centennial History issue of *Chemical and Engineering News* in 1976; more recently, he and his wife have completed a fine biography of Roger Adams. P. Thomas Carroll has treated the history of the chemistry department at the University of Illinois, which was for many years the center of organic chemistry in America, in his doctoral dissertation. Ira Remsen's career, the subject of a biography by his student, Frederick Getman, has been critically reexamined by Owen Hannaway. Leon Gortler is currently working on the history of physical-organic chemistry in the United States.[9] Nevertheless, there is great need of both a general history of this specialty in America and of particular episodes or events in that history. Moses Gomberg's work on free radicals deserves historical atten-

[9] D. Stanley Tarbell, "The Past 100 Years in Organic Chemistry," *Chemical and Engineering News,* 1976, *54*:110–123; D. Stanley Tarbell and Ann Tracy Tarbell, *Roger Adams: Scientist and Statesman* (Washington, D.C.: ACS, 1981); see also D. Stanley Tarbell, "Organic Chemistry, 1876–1976," in *A Century of Chemistry,* ed. Skolnik and Reese, pp. 339–350; P. Thomas Carroll, "Perspectives on Academic Chemistry in America, 1876–1976: Diversification, Growth, and Change" (Ph.D. diss., Univ. Pennsylvania, 1982); Frederick H. Getman, *The Life of Ira Remsen* (Easton, Pa.: Journal of Chemical Education, 1940); Owen Hannaway, "The German Model of Chemical Education in America: Ira Remsen at Johns Hopkins," *Ambix,* 1976, *23*:145–164; and Leon B. Gortler, "Louis Hammett, Paul Bartlett and the Origins of Physical Organic Chemistry," paper presented at the Symposium on Chemistry in America, 13th Middle Atlantic Regional Meeting, American Chemical Society, March 1979.

tion, as does the research of Arthur Michael on the theory of organic reactions.[10] Although neither chemist founded a flourishing research school, their researches foreshadow in some respects the subsequent work of physical-organic chemists on free radicals and on the thermodynamics of organic reactions. Perhaps the most interesting questions are of a comparative nature. For example, it would be rewarding to compare the prolonged and stuttering beginnings of an organic research tradition in America with the far more rapid expansion of significant research in physical chemistry. Likewise, it might be worthwhile to study the question of why organic chemistry played the role of locomotive in the development of the German chemical community but that of the caboose in America.[11]

If one believes that the history of failures, no less than successes, can illuminate the past, then no topic is as intriguing as the history of colloid chemistry. A number of American scientists were active in the study of colloids during the first decade of the twentieth century, but explosive growth appears to have been ignited by a lecture tour that Wolfgang Ostwald, son of the prominent physical chemist, Wilhelm Ostwald, made of the United States in 1913–1914. His lectures (fifty-six in seventy-four days!) stressed the scientific and technological possibilities opened up by the study of what Ostwald called "the world of neglected dimensions," that is, the study of aggregates of matter intermediate in size between the large masses of the physicist and the atoms and molecules of the classical chemist.[12]

Beginning as a branch of physical chemistry devoted to the study of the properties of finely divided matter, the field expanded until, during the 1920s and early 1930s, efforts were made to establish departments or entire schools of colloid science.[13] Enthusiasts argued that matter in a finely divided state displayed phenomena unknown in the study of either coarse aggregates or molecular systems and suggested that unknown laws governed the behavior of matter in such a state. Since protoplasm is a colloidal suspension of proteins, these scientists argued that their field held the key to understanding the special properties of living matter. Their expectations of practical advances in biology and medicine were similarly extravagant. Disease might be a result of an excessive degree of coagulation or of dispersal of bodily colloids and might be cured by the administration of agents that restored tissues to their natural state. The process of aging itself might be due to the coagulation of cellular colloids and might be reversed just as the coagulation of albumin in a test tube could be reversed through the application of a dispersing agent.

But while colloid chemistry was in its most vigorous stage of expansion, during the late 1920s, a serious blow was struck to it as persuasive evidence began to accumulate indicating that proteins were not agglomerates governed by peculiar laws, as many colloid chemists argued, but were instead molecules that

---

[10] Albert B. Costa has made a start on the latter in his "The Meeting of Thermodynamics and Organic Chemistry," *J. Chem. Educ.*, 1976, 48:243–246.

[11] Martha Moore Trescott's *The Rise of the American Electrochemicals Industry, 1880–1910: Studies in the American Technological Environment* (Westport, Conn.: Greenwood, 1981) contains some relevant *aperçus*.

[12] Wolfgang Ostwald's lectures were later fashioned into a book, *An Introduction to Theoretical and Applied Colloid Chemistry*, trans. Martin H. Fischer (New York: Wiley, 1917).

[13] See, e.g., Julius Stieglitz to Max Mason, 28 Dec. 1925, Box 16, Folder 2, Presidents' Papers, Univ. Chicago.

obeyed the laws of classical chemistry. There ensued a protracted period of decline, and although colloid chemistry still exists as a specialty, its practitioners no longer entertain the grand ambitions of some of its founders.

Research on the history of this specialty would be instructive on several counts. Colloidal models were widely used to explain the mechanism of drug action during the 1920s and even into the 1930s, and without better knowledge of the history of colloid chemistry it is difficult to make sense of an entire generation of research in pharmacology and in cellular metabolism more generally. Colloid chemists were widely employed by industry, and the influence of their theories on research on such products as paint and pigments, photographic emulsions and films, rubber, and cement deserves study. And while we have many examples of studies that treat the dynamics of exponentially growing specialties, we have very few that treat the process of decay.[14]

Closely related to both the history of organic and colloid chemistry is the development of polymer chemistry. The literature on this topic is woefully meager in comparison to the importance of polymers in twentieth-century history. Polymers have altered the look and texture of our material civilization, and the study of the reactions by which polymers are produced was an essential prerequisite to the emergence of molecular biology. Yet the development of polymer science, to which American chemists made signal contributions, has only recently begun to attract historians' interest.[15] The debates over the nature of macromolecules that enlivened the 1920s, the investigations undertaken during the emergency effort to develop synthetic rubber in World War II, and the postwar articulation of a theory capable of relating the structural characteristics and physical properties of polymers—all stand in need of unraveling. Prospects for research on this topic should improve dramatically in the near future, however, since the Center for History of Chemistry is embarking on a major project that will include a program of interviews with chemists and chemical engineers, both in unversities and industry, who were involved in polymer developments from the 1920s to the 1960s. The Center is also seeking to locate, preserve, and make available archival sources relating to the field, taking as one of their models the exemplary work of John T. Edsall and David Bearman on sources in the history of biochemistry and molecular biology.[16]

THE SOCIAL AND ECONOMIC GENRE

Chemistry may be considered as a body of concepts and techniques evolving in response to internal and external demands, but also as a science that alters the society in which it exists. It does so, not directly, but always through inter-

[14] Robert E. Kohler discusses other facets of the history of colloid chemistry in "The History of Biochemistry: A Survey," *Journal of the History of Biology,* 1975, 8:275–318, esp. pp. 290–291.

[15] Robert D. Friedel, *Pioneer Plastic: The Making and Selling of Celluloid* (Madison: Univ. Wisconsin Press, 1983); Yasu Furukawa, "Staudinger, Carothers, and the Emergence of Macromolecular Chemistry" (Ph.D. diss., Univ. Oklahoma, 1983); and John K. Smith and David A. Hounshell, "Carothers and Fundamental Research at Du Pont," paper presented at the annual meeting of the History of Science Society, Norwalk, Conn., October 1983; see also Raymond B. Seymour, ed., *History of Polymer Science and Technology* (New York: Marcel Dekker, 1982).

[16] David Bearman and John T. Edsall, *Archival Sources for the History of Biochemistry and Molecular Biology: A Reference Guide and Report* (Boston: American Academy of Arts and Sciences; Philadelphia: American Philosophical Society, 1980).

mediaries, among them industrial technology, the agencies of government, and medicine.

It is through the medium of industrial technology that chemistry has had its most significant impact on society. Chemists have been involved in the industrial development of the United States since at least 1801, when Robert Hare invented the oxyhydrogen blowtorch, but, aside from the occasional invention of this sort, American chemists generally performed only routine analytical functions for industry until the end of the nineteenth century. They examined raw materials for purity, tested the quality of finished products, and occasionally made improvements in processes of production, but typically neither the chemist nor the businessman believed chemistry capable of transforming industries through the development of new products. This attitude began to change rapidly around the turn of the century, in response both to examples of successful innovation in the dyestuffs industry abroad and to competition within the new photographic and electrical industries at home. Additional stimulus to the development of industrial research resulted from the massive shortages of fine chemicals, nitrates, pharmaceuticals, optical glass, and other products caused by the interruption of trade with Germany during World War I, and during and after the war businesses by the hundreds initiated research and development programs. Since chemical transformations are involved in most industrial processes, chemists were among the major beneficiaries of this movement.

Quite naturally, the dramatic expansion of research and development work has attracted the attention of historians, both historians of science and technology and economic and business historians. The work they have accomplished suggests that the science of chemistry and American industry have had four principal meeting grounds since the end of the nineteenth century: within the corporation itself, both in the plant and in the industrial research laboratory; in the university, through the medium of cooperative projects supported by business; in consulting firms; and in special laboratories doing contract research for individual firms or trade associations.

It is the first point of contact, the in-house laboratory, that is currently receiving the greatest attention from historians. Martha Moore Trescott's work affords us voluminous information about the genesis and development of research in the electrochemicals industry, one of the first sites of in-house corporate research. The gradual institutionalization of research on photographic processes forms a motif in Reese Jenkins's treatment of the American photographic industry. In two superb articles, George Wise has used the early history of the General Electric Research Laboratory to discuss the transfer of both academic chemists and academic chemistry to industry. John L. Enos's *Petroleum Progress and Profits,* a book that does not receive the attention it deserves, relates the history of innovations in processes for cracking petroleum and focuses especially on assessing the financial rewards that accrue to companies pursuing successful research strategies. The history of Du Pont's research and development program has been the subject of several recent papers, and John K. Smith and David A. Hounshell are currently preparing a comprehensive history of that company's research work.[17]

---

[17] Trescott, *Rise of the American Electrochemicals Industry* (cit. n. 11); Reese V. Jenkins, *Images and Enterprise: Technology and the American Photographic Industry, 1839–1925* (Baltimore: Johns

Although dealing with a multiplicity of topics in a variety of styles, several basic questions may be extracted from these works: how and why was research first incorporated into the American corporation, how did businessmen and laboratory directors assign priorities in the laboratory, how did scientists perceive industrial research as a career, and what contribution did research and development work make to industrial profits? In treating these questions it has been quite natural to give attention first to such large and successful corporate laboratories as those of General Electric, Eastman Kodak, and Du Pont and to the careers of such accomplished scientists as Irving Langmuir and William D. Coolidge. There is nothing wrong with this, and indeed, there is need for more work along these lines. It is, for instance, astonishing that we know so little about the transformation of the major American pharmaceutical houses from makers of a handful of standard products to centers for research and development capable, in some cases, of making significant advances in therapeutics. Nevertheless, valuable as such work is, it is also important that we learn more of those industrial laboratories that were not showcases and of those journeymen scientists, many of them chemists, who did not earn Nobel prizes. By 1921 over 500 industrial research laboratories, employing nearly 4,000 chemists, existed in the United States, and these numbers would more than double during the next decade—we know little or nothing of most of these institutions and the chemists who worked in them.[18]

In-house laboratories answered the research needs of most American firms in the twentieth century, but other avenues were explored and continue to be used by some companies, among them, cooperative arrangements with university departments, the use of consultants or consulting firms, and contracts with private research institutes funded by groups of firms or trade associations. John W. Servos has treated the relations between industry and the chemical engineering program at MIT, and P. Thomas Carroll those between industry and the chemistry department at the University of Illinois. Both Jeffrey Sturchio's overview of the historical relations between chemists and American industry and Arnold Thackray's recent synthetic essay on this subject contain useful ideas and information about these alternative mechanisms of corporate research.[19] Nevertheless, none has been adequately studied by historians. For example, we lack histories of the Arthur D. Little consulting firm and the Mellon Institute, nor do

Hopkins Univ. Press, 1975); George Wise, "A New Role for Professional Scientists in Industry: Industrial Research at General Electric, 1900–1916," *Technology and Culture*, 1980, *21*:408–429; Wise, "Ionists in Industry: Physical Chemistry at General Electric, 1900–1915," *Isis*, 1983, *74*:7–21; John L. Enos, *Petroleum Progress and Profits: A History of Process Innovation* (Cambridge, Mass.: MIT Press, 1962); Jeffrey L. Sturchio, "Chemistry and Corporate Strategy at Du Pont," *Research Management*, 1984, *27*:10–18; and Smith and Hounshell, "Carothers and Fundamental Research at Du Pont."

[18] Thackray, Sturchio, Carroll, and Bud, *Chemistry in America* (cit. n. 2), p. 364; Williams Haynes's *American Chemical Industry, A History,* 6 vols., (New York: Van Nostrand, 1945–1954) remains the best source.

[19] John W. Servos, "The Industrial Relations of Science," *Isis*, 1980, *71*:531–549; Carroll, "Perspectives on Academic Chemistry" (cit. n. 9); Jeffrey L. Sturchio, "Chemists and Industry in Modern America: Studies in the Historical Application of Science Indicators" (Ph.D. diss., Univ. Pennsylvania, 1981); and Arnold Thackray, "University-Industry Connections and Chemical Research: An Historical Perspective" in *University-Industry Research Relationships: Selected Studies* (Annual Report of the National Science Board) (Washington, D.C.: National Science Foundation, 1982).

we know very much about the attitudes of academic chemists toward consulting work and contract research. Perhaps most striking is our ignorance of the motives and attitudes of the businessmen who were making choices as to how and where to spend their research funds. A study of the evolution of research policies at the Standard Oil Company of New Jersey would be most revealing in this respect, since that firm, at various times in its history, made use of all of the above-mentioned research strategies.

If industry has been a major consumer of chemists' services during the past century, so too has government. Several aspects of the chemist's role in government have received satisfactory treatment. The history of food and drug regulation, a story in which chemists and chemistry have played leading roles, has attracted the greatest attention. Oscar Anderson's *Health of a Nation* remains the best treatment of the circumstances leading up to passage of the Pure Food Act of 1906. Anderson's work deals intelligently not only with Wiley's flamboyant investigations of food adulteration but also with such topics as his research on behalf of the American sugar industry, the history of the Bureau of Chemistry of the U.S. Department of Agriculture, and the linkage between progress in analytical chemistry and the regulatory process. James Whorton's *Before Silent Spring* is a fine account of the development and use of pesticides in America prior to the introduction of DDT and of efforts to impose controls on their use, a job that was subsequently assumed by the Food and Drug Administration and still later by the Environmental Protection Agency. Complementing Whorton's book is Thomas R. Dunlap's excellent *DDT: Scientists, Citizens, and Public Policy*. These, together with a number of other works treating the role of the chemist in the FDA and its predecessor, the Bureau of Chemistry, make this a comparatively well studied topic.[20]

The activities of chemists in those outgrowths of World War I, the Chemical Warfare Service and the Fixed Nitrogen Research Laboratory, and in the synthetic rubber, synthetic fuels, and Manhattan Projects of World War II have also been discussed in recent books and dissertations. Nevertheless, chemists' roles in such peacetime agencies as the Geological Survey and Forestry Service have been neglected.[21]

---

[20] Anderson, *The Health of a Nation* (cit. n. 5); James Whorton, *Before Silent Spring: Pesticides and Public Health in Pre-DDT America* (Princeton: Princeton Univ. Press, 1974); Thomas R. Dunlap, *DDT: Scientists, Citizens, and Public Policy* (Princeton: Princeton Univ. Press, 1981); see also James Harvey Young, *The Medical Messiahs: A Social History of Health Quackery in Twentieth Century America* (Princeton: Princeton Univ. Press, 1967); Young, *The Toadstool Millionaires: A Social History of Patent Medicines in America Before Federal Regulation* (Princeton: Princeton Univ. Press, 1961); Young, "Sulfanilamide and Diethylene Glycol," in *Chemistry and Modern Society*, ed. Parascandola and Whorton (cit. n. 7), pp. 105–126; Sheldon Hochheiser, "The Establishment of Synthetic Food Color Regulation in the United States, 1906–1912," *ibid.*, pp. 127–146; Paul M. Priebe and George B. Kauffman, "Making Government Policy under Conditions of Scientific Uncertainty: A Century of Controversy about Saccharin in Congress and the Laboratory," *Minerva* 1980, *18*:556–574; and Peter Temin, *Taking Your Medicine: Drug Regulation in the United States* (Cambridge, Mass.: Harvard Univ. Press, 1980).

[21] Daniel P. Jones, "The Role of Chemists in Research on Poison Gases During World War I" (Ph.D. diss., Univ. Wisconsin, 1969); Jones, "American Chemists and the Geneva Protocol," *Isis*, 1980, *71*:426–440; Margaret Jackson Clarke, "The Federal Government and the Fixed Nitrogen Industry, 1915–1926," (Ph.D. diss., Oregon State Univ., 1977); William M. Tuttle, Jr., "The Birth of an Industry: The Synthetic Rubber 'Mess' in World War II," *Technol. Cult.*, 1981, *22*:35–67; Joseph Marchese, "Government and Energy: The Demonstration Program for Synthetic Liquid Fuels, 1944–1955," (Ph.D. diss., Princeton Univ., 1983); and Richard G. Hewlett and Oscar E. Anderson, Jr.,

Chemists in the Drug Division of the Bureau of Chemistry of the U.S. Department of Agriculture devising new methods to analyze drugs for purity (circa 1920). From the National Archives, courtesy of the Center for History of Chemistry, Philadelphia.

Spotty as this literature is, however, it appears rich by comparison with the work that has been done on chemists' roles in state and local government. The rapid growth of American cities and the expansion of urban industry during the late nineteenth century made "civic chemistry" a growth field, and as Jeffrey Sturchio has observed, many of this country's leading chemists were intimately involved in civic and state activities prior to World War I.[22] Ira Remsen advised the city of Baltimore regarding its water supply; Charles F. Chandler of Columbia University served on the Metropolitan Board of Health of New York City; Harvey Wiley was a chemist for the state of Indiana before joining the federal government. It became less common to find chemists of this caliber involved in such work after the war—many procedures had been routinized and taken over by technicians, and other duties were assumed by bacteriologists and engineers. Nevertheless, for a period of some decades during the late nineteenth and early twentieth centuries, chemists had a significant impact on the lives of millions through their service to the state and municipal governments and were responsible for innovations in water supply and purification, smoke abatement, housing codes, and zoning regulations that today are taken for granted. Cursory inspection of a book like Charles Baskerville's *Municipal Chemistry,* published in 1911, suggests a range of such topics awaiting historical study.[23]

The New World, 1939/1946: A History of the United States Atomic Energy Commission (University Park, Pa.: Pennsylvania State Univ. Press, 1962) Vol. I.

[22] Jeffrey L. Sturchio, "Civic Chemistry in Metropolitan New York, 1870–1910: The Chandler Circle," paper presented at the 17th Middle Atlantic Regional Meeting of the American Chemical Society, White Haven, Pennsylvania, 8 Apr. 1983.

[23] Charles Baskerville, ed., *Municipal Chemistry: A Series of Thirty Lectures by Experts on the Applications of Chemistry to the City, Delivered at the City College of New York, 1910* (New York: McGraw-Hill, 1911). On the social and political setting of the work of municipal chemists, see Martin V. Melosi, *Garbage in the Cities: Refuse, Reform, and the Environment, 1880–1980* (College Station, Tex.: Texas A & M Univ. Press, 1981); and Melosi, ed., *Pollution and Reform in American Cities, 1870–1930* (Austin: Univ. Texas Press, 1980).

The relationships between chemistry and medicine and between the chemist and the medical doctor form another area that deserves far more attention than it has so far received. Academic chemistry, both in this country and earlier in Europe, had one of its roots in the medical school; it had another in the apothecary shop. And if chemistry has owed much to the stimulation and opportunities afforded by medicine, so too has medicine great debts to chemistry. At no time and place have those debts been so obvious as in twentieth-century America. Yet the literature treating this subject is dwarfed by that dealing with such topics as the reform of American medical schools.

In a survey of the historical intersections of chemistry and medicine, several topics stand out as inviting areas for research. We know little about the reception accorded Ehrlich's side-chain theory of immune response by American chemists or about that accorded the physico-chemical theory propounded by Svante Arrhenius and Thorvald Madson. Research on antiseptics and on the theory of antiseptic action, stimulated by the needs of army physicians in World War I, deserves study, as does research on anesthetics, also stimulated by the war.[24] American chemists were intimately involved in the isolation of both hormones and vitamins during the early twentieth century, and while some good work has been done on these topics, there is opportunity for more. John Parascandola has discussed Lawrence J. Henderson's research on the mechanisms regulating the acid-base balance in blood, but again there is opportunity for additional work, especially on the development of those diagnostic tests upon which clinical hematology rests.[25]

Perhaps the most intriguing point of contact between chemistry and medicine is chemotherapy. Ehrlich's synthesis of Salvarsan shortly before World War I prompted many chemists and medical scientists, in America as well as abroad, to seek other chemical agents capable of destroying pathogenic bacteria and gave rise to a period of intense speculation regarding the mechanisms of both immune response and drug action.[26] Programs of research were begun in pharmacology departments at American medical schools, in some drug firms, and in the laboratories of academic chemists. But the history of these efforts is largely untold. In particular, we need case studies of specific lines of research on drugs for such diseases as tuberculosis, research on the professional relations between physicians and chemists, and work on the history of drug manufacture and the role of chemists in that industry. There is nothing specifically American about these topics, but the importance of American work on these matters in the twentieth century makes such sources as the John Joseph Abel Papers at the Institute for the History of Medicine in Baltimore or the archives of the Eli Lilly Company strategic sites for research.

---

[24] E.g., the development of mercurochrome and hexyl-resorcinol by chemists at the Johns Hopkins University and of butacaine by Roger Adams and Ernest H. Volwiler at the University of Illinois.

[25] See Jane H. Murnaghan and Paul Talalay, "John Jacob Abel and the Crystallization of Insulin," *Perspectives in Biology and Medicine*, 1967, *10*:334–380; Charles E. Rosenberg, "Science Pure and Science Applied," in *No Other Gods*, pp. 185–195; and John Parascandola, "Lawrence J. Henderson and the Concept of Organized Systems" (Ph.D. diss., Univ. Wisconsin, 1968).

[26] For a starting point to work on reception of European theories, see John Parascandola and Ronald Jasensky, "Origins of the Receptor Theory of Drug Action," *Bull. Hist. Med.*, 1974, *48*:199–220; and Parascandola, "Arthur Cushny, Optical Isomerism, and the Mechanism of Drug Action," *J. Hist. Biol.*, 1975, 8:145–165.

CONCLUSION

The foregoing discussion of research accomplishments and needs is hardly definitive. The beginnings of chemistry in America have been ignored, although there is much to be learned about the development of textbooks and instruments in America and about the motives of those who cultivated the subject. Likewise, little has been said of the period following World War II, when chemistry, a science practiced by small groups using fairly inexpensive apparatus, has faced special funding problems in an era of billion-dollar particle accelerators. Nor has special attention been given the transmission of the laboratory method of instruction to America and its rapid expansion here, a topic on which there is a small literature but opportunity for much more.[27]

Nevertheless, what has already been said is sufficient to demonstrate that the history of chemistry in America is an exceptional vehicle for treating problems of interest both to the historian of science and the American social and economic historian. The growth of chemistry in America was dependent on the transmission of ideas and institutions from one national context to another; thus it raises the general issue of how such transmission is effected. That growth took place through a continuing process of specialization, and thus it involves the general issues of how specialties coalesce, grow, and, in some instances, decay. Comparative study of American and European work in chemistry affords opportunity to treat the problem of how local circumstances mold national styles in science, a topic of more than parochial interest. Finally, the study of the work of individual chemists and their research schools promises to yield further insight into the ways in which conceptual and social circumstances affect the development of scientific knowledge. It is well also to recall that chemistry prospered in America because of its ability to answer the needs of consumers found in other scientific disciplines, industry, government, and medicine. Research into chemistry's relations with these consumers has and should continue to inform us about the relationship between science and technology, the evolution of American medicine and industry, and the peculiar problems inherent in using science to shape public policy. Chemistry may, or may not be, as Bancroft claimed, the central science, but the history of chemistry should surely occupy a central place in the telling of the larger story of science and its uses in modern America.

[27] See esp. Rossiter, *Emergence of Agricultural Science* (cit. n. 5), and Hannaway, "Ira Remsen at Johns Hopkins" (cit. n. 9).

History of Biology

By Jane Maienschein*

WIDESPREAD INTEREST in American biology is a fairly recent development. Historians, using a variety of methods and representing a range of approaches, have begun to study significant aspects of the subject, though we lack as yet a comprehensive history of biology in the United States. History of science meetings offer numerous symposia each year, and articles and books appear at an increasing rate. The conviction that the field is a fertile and important one has stimulated this inquiry, but what constitutes the field and what is the best way to approach it are unresolved questions. This survey suggests that a diversity of approach and method, relying on a variety of resources and examining a wide range of subjects, remains the best way to deal with both the above questions.

GENERAL TRENDS: WHAT IS AMERICAN BIOLOGY?

Conscious exploration of American biology as a separate subject antedates the current explosion in historical examination. With the establishment in the late nineteenth century of professional biological institutions—the Johns Hopkins University, the University of Chicago, and the Marine Biological Laboratory (MBL)—came considerable discussion about what biology in the United States would be like.[1] This very discussion implied that the participants' convictions that national setting might have an effect on scientific ideas and activity, so that there might be a self-consciously American biology. The very earliest historical studies, however, did not explicitly address the question of how the particular setting might have influenced the science.

In fact, one of the earliest subjects of interest, and the only subject that has consistently attracted scholarly attention, is the notorious mid-nineteenth-century race to collect dinosaur remains. Not always professional histories of science, some of these studies nonetheless stirred the imagination and raised serious questions about the scientific developments that had inspired the fossil seekers. In particular, the energetic debates between Othniel Marsh and Edward Cope over interpretations of the evolutionary history of fossil animals provided such a focus. And such later works as Wayne Hanley's *Natural History in America* place the dinosaur races in the context of natural historical work and ideas more generally, revealing the emphasis on those alternative evolutionary theories.[2]

* Department of Philosophy, Arizona State University, Tempe, Arizona 85287.

[1] See, e.g., Charles Otis Whitman, "Specialization and Organization: Companion Principles of All Progress—The Most Important Need of American Biology," *Biological Lectures*, 1891, *1890*:1–26. See generally the Gilman Collection, Johns Hopkins University Manuscripts Collection; Whitman Papers, University of Chicago Archives; and Agassiz Collection, Museum of Comparative Zoology Archives, Harvard University.

[2] Wayne Hanley, *Natural History in America, from Mark Catesby to Rachel Carson* (New York:

The 1940s and 1950s brought an interest in documenting histories of institutions and groups of individuals. Several series, in *Bios* and the *Turtox News,* for example, appeared to detail the emergence of laboratories or biological departments. Such institutional studies typically focused on personalities or major events and did not seriously examine the intellectual or political roots of the institutions. Nevertheless, these chronologies provide useful information about what individuals contributed, and they do underline the particular local features of the field. In addition, by the 1950s a few biologists had begun to examine the contributions and ideas of individual scientists who had remained theretofore little known. Ralph Dexter provided the most notable example of such work with his study of the group of naturalists located around Salem, Massachusetts, including biographical and group sketches of E. S. Morse, Alpheus Hyatt, A. S. Packard, Frederic W. Putnam, and others.[3]

American biology began to be treated as a serious subject within the context of American history in the 1950s and 1960s. The period brought works that examined the lives of major figures in American biology. From that period, A. Hunter Dupree's biography of botanist Asa Gray and Edward Lurie's biography of zoologist Louis Agassiz remain classics. Each paints a clear and striking portrait of the individual and his life in science. These pioneering scholarly works in the emerging field of history of American biology also provide considerable insight into what role the national setting played in shaping the science, illustrating that the subjects' American context did in fact shape the character of the scientific work done. These studies exemplify the move from chronicling to concern with deeper historical questions.[4]

The 1960s saw historians of science, as distinct from American historians, begin to treat biology in the United States. Their work grew out of a burgeoning interest in the philosophy of science and in the rise of modern science generally, and it reflected the professionalization of history of science, which traditionally had emphasized intellectual history and scientific ideas. Because of this emphasis, the studies from this period generally treat the American setting as of secondary or virtually negligible importance. In his widely read *The Structure of Scientific Revolutions,* Thomas Kuhn argued that science entails community, as well as individual practices. But biology itself—as ideas held by a community—rather than its place in the United States occupied center stage for Kuhn and for these historians.[5] Accordingly, they explored scientific ideas, the relative roles of theory and empirical observation, the role of experimentation, and the

Quadrangle, 1977). In addition, Url Lanham, *The Bone Hunters* (New York: Columbia Univ. Press, 1973), presents a tale of the rough-and-ready, distinctly American competition; and Nathan Reingold, ed., *Science in Nineteenth Century America: A Documentary History* (New York: Hill & Wang, 1964), includes letters that illustrate the debates about bones and evolution.

[3] Ralph Dexter, "The Annisquam Seaside Laboratory of Alpheus Hyatt," *Scientific Monthly,* 1952, 74:112–116; Dexter, "Three Young Naturalists Afield—The First Expedition of Hyatt, Shaler, and Verrill," *Sci. Mo.,* 1954, 79:45–51; Dexter, "Excerpts from Alpheus Hyatt's Log of the *Arethusa*," *Essex Institute Historical Collections,* 1954, 90:229–260; and Dexter, "The 'Salem Secession' of Agassiz Zoologists," *Essex Inst. Hist. Coll.,* 1965, 101:27–39.

[4] A. Hunter Dupree, *Asa Gray* (Cambridge, Mass.: Harvard Univ. Press, 1959); Edward Lurie, *Louis Agassiz: A Life in Science* (Chicago: Univ. Chicago Press, 1960); and Lurie, *Nature and the American Mind: Louis Agassiz and the Culture of Science* (New York: Science History, 1974).

[5] Thomas Kuhn, *The Structure of Scientific Revolutions* (Chicago: Univ. Chicago Press, 1962).

advent of modern science, taking for granted the setting in which these developments were unfolding.

There are other works in the intellectual tradition which directly treat biology that took place in the United States, though they do not address whether it is peculiarly American biology. Elof Carlson's *The Gene: A Critical History*, written in the 1960s, examines a special area of biology and emphasizes work done in the United States. More recent efforts in this vein include Donna Haraway's imaginative study of the use of metaphor by three embryologists, one of whom (Ross Granville Harrison) was American. Likewise, Scott Gilbert's study of Thomas Hunt Morgan's developmental biology concentrates on ideas in embryology and genetics that happened to appear in the United States, although he does not identify them as particularly American in any way. Garland Allen's early studies of Thomas Hunt Morgan, Edward Manier's look at Morgan's experimental approach, Alice Levine Baxter's examination of Edmund Beecher Wilson's cytology, and John Farley's several chapters on largely American work in *Gametes and Spores* further exemplify this tradition. Two overviews that also remain firmly in the tradition of intellectual history are Ernst Mayr's opus *The Growth of Biological Thought*, with its instructive historiographic introduction, and Allen's *Life Science in the Twentieth Century*. The latter author reviews modern biology, much of which is American, but neither he nor his critics, who have advanced alternative interpretations of the early twentieth-century morphological tradition and the other scientific ideas and methods he discusses, have addressed the American setting in particular.[6]

Extreme examples of this approach have been labeled "internalist" history of science and have been criticized for ignoring nonintellectual factors which, critics argue, also direct science. While a history of ideas alone obviously cannot uncover all aspects of the history of science, so long as the intellectual historian understands that he or she is in effect looking through a narrowing lens and thus allowing much of the context of science to remain out of focus, then the approach can be acceptable. But a more subtle limitation is that by emphasizing ideas such studies have often tended to address periods of dramatic scientific change and hence, in the history of biology, to concentrate on the post-Darwinian era while neglecting earlier contributions. Charles Rosenberg's overview

---

[6] Elof Alex Carlson, *The Gene: A Critical History* (Philadelphia: Saunders, 1966); Carlson, *Genes, Radiation, and Society: The Life and Work of H. J. Muller* (Ithaca, N.Y.: Cornell Univ. Press, 1981); Donna J. Haraway, *Crystals, Fabrics, and Fields: Metaphors of Organicism in Twentieth-Century Developmental Biology* (New Haven: Yale Univ. Press, 1976); Scott Gilbert, "The Embryological Origins of the Gene Theory," *Journal of the History of Biology*, 1978, *11*: 307–351; Garland Allen, "Thomas Hunt Morgan and the Problem of Sex Determination, 1903–1910," *Proceedings of the American Philosophical Society*, 1966, *110*:48–57; Allen, "T. H. Morgan and the Emergence of a New American Biology," *Quarterly Review of Biology*, 1969, *44*:168–188; Edward Manier, "The Experimental Method in Biology: T. H. Morgan and the Theory of the Gene," *Synthese*, 1969, *20*:185–205; Alice Levine Baxter, "Edmund Beecher Wilson and the Problem of Development: From the Germ Layer Theory to the Chromosome Theory of Inheritance" (Ph.D. diss., Yale Univ., 1974); Baxter, "E. B. Wilson's 'Destruction' of the Germ-Layer Theory," *Isis*, 1977, *68*:363–374; Baxter, "Edmund B. Wilson as Preformationist: Some Reasons for His Acceptance of the Chromosome Theory," *J. Hist. Biol.*, 1976, *9*:29–57; John Farley, *Gametes and Spores: Ideas about Sexual Reproduction, 1750–1914* (Baltimore: Johns Hopkins Univ. Press, 1982); Ernst Mayr, *The Growth of Biological Thought: Diversity, Evolution, and Inheritance* (Cambridge, Mass.: Harvard Univ. Press, 1982); Allen, *Life Science in the Twentieth Century* (New York: Wiley, 1975); and Jane Maienschein, Keith Benson, and Ronald Rainger in "Special Section on American Morphology at the Turn of the Century," *J. Hist. Biol.*, 1981, *14*:83–158.

of internalist and externalist approaches in the history of science, outlining the strengths and weaknesses of each, is a sensible and useful guide to these broad traditions.[7]

The externalist tradition also finds its representatives in the history of biology. That approach emerged explicitly in the late 1960s and early 1970s as part of the more general move within history toward social history. It concentrated primarily on external, or institutional and social, settings. This approach also has its critics, who have argued that much institutional and social history deals only secondarily with the intellectual content, as though neither theories nor empirical studies mattered and science were nothing more than the product of its setting. Fortunately historians of biology have generally avoided such extreme externalism.

Probably the largest area of inquiry that has continued to attract primarily social and political rather than intellectual analysis is eugenics. Since particular eugenics arguments so clearly exhibit national differences and depend on social and political factors, historians have accepted an external approach to the subject as appropriate. Major studies of American eugenics include Mark Haller's *Eugenics* and Kenneth Ludmerer's *Genetics and American Society,* both of which concentrate on the social uses of science. Written from the context of American social history, these books also examine the science of genetics behind social discussions and policy decisions. In a related study, Hamilton Cravens examines changing attitudes to the nature-nurture controversy in the United States. A number of recent works have explored various elements of the eugenics movements as well, most notably Daniel Kevles's highly praised series in the *New Yorker.*[8]

Another recent provocative study of the way in which extraintellectual factors can direct science is Diane Paul's examination of genetics textbooks. She reports that even though the infamous twin studies by psychologist Cyril Burt have been resoundingly discredited (as detailed in Leslie Hearnshaw's biographical assessment of Burt), many textbooks continue to cite them. Furthermore, a majority of textbooks, most of them American, still maintain Burt's conclusion that intelligence is strongly heritable, even relying on his discredited results as evidence. Paul's emphasis on nonintellectual factors such as the politics of textbook writing and the politics of inheritance theories is appropriate in this case, where the extrascientific factors seem clearly to have dictated what is presented in textbooks as the currently best scientific ideas.[9]

In another example, Robert Kohler has been accused by some of adopting an overly externalist view in his history of biochemistry, *From Medical Chemistry to Biochemistry.* He does at times sound as though he were disregarding the ideas of science. Yet a careful reading proves the criticism unfair. Indeed,

---

[7] Charles Rosenberg, "On the Study of American Biology and Medicine: Some Justifications," *Bulletin of the History of Medicine,* 1964, *38*:364–376.

[8] Mark H. Haller, *Eugenics: Hereditarian Attitudes in American Thought* (New Brunswick, N.J.: Rutgers Univ. Press, 1963); Kenneth Ludmerer, *Genetics and American Society: A Historical Appraisal* (Baltimore: Johns Hopkins Univ. Press, 1972); Hamilton Cravens, *Triumph of Evolution: American Scientists and the Heredity-Environment Controversy* (Philadelphia: Univ. Pennsylvania Press, 1978); and Daniel Kevles, "Annals of Eugenics," *New Yorker,* 1984, 8 Oct., pp. 51–115, 15 Oct., pp. 52–125, 22 Oct., pp. 92–151, and 29 Oct., pp. 51–117.

[9] Diane Paul, "Textbook Treatments of the Genetics of Intelligence," *Quart. Rev. Biol.,* 1985, forthcoming; Leslie Hearnshaw, *Cyril Burt, Psychologist* (Ithaca, N.Y.: Cornell Univ. Press, 1979).

Kohler provides an unusual example of what a good and original institutional history can be. His work is complex and occasionally dry with detail, but it nonetheless yields an important perspective on a specialty within American biology and chemistry. Kohler staunchly maintains that scientific disciplines are political institutions that rise and fall as other political institutions do. They do not reflect essential, fixed categories in nature, and thus they may vary greatly from one setting to another. For example, he argues that American biochemistry's close alliance with medicine made theoretical changes in molecular biology more difficult. For Kohler, institutional factors strongly limit and direct science, a view very different from that of the intellectual historian. But he does not claim that ideas are irrelevant.[10]

One unusual study that embraces both internalist and externalist approaches considers British rather than American biology. Gerald Geison's examination of Michael Foster's school of physiology provides a model in weaving together analysis of such factors as the individual scientist, scientific ideas, the roles of other participants working together in the school, the establishment of a discipline that was successful depite its lack of brilliant or useful results, the institutional setting, and the political climate.[11] Historians of science in America are only beginning to embrace similarly sophisticated and productive approaches, providing examples of diverse approaches to the history of American biology.

These examples of internal and external approaches, all representing history of biology that occurred in America, demonstrate a robust diversity. Yet history of American biology can also be cut up in different ways as historians regard one or another interpretation of American biology as more appropriate. For some, American biology is implicitly treated as that set of biological ideas which appeared in the United States; for them intellectual history is the appropriate approach. For others, who regard American biology as that biology practiced by important figures, biographical studies seem the proper point of attack. Still others see leading institutions that direct science as the significant factor; they therefore pursue institutional history at various levels. For yet others, American biology is an activity pursued in characteristically different ways in its American setting and influenced by that setting; for them, social and political history will figure centrally in the story of American biology. Biographies, institutional histories, and histories of disciplines thus emerge as the major representative categories of historical works, with other less traditional works appearing as well. Diversity of approach remains productive and desirable as long as the question about what constitutes American biology remains open to a variety of answers.

SOURCES

Some historians of science have begun to pursue traditional subjects through heretofore underutilized sources of information. For example, William Provine's extensive study of Sewall Wright draws heavily on interviews, notebooks, and reprint collections, seeking a system to grapple with the wealth of materials available for living (and recently deceased) figures in science. As F. L. Holmes

---

[10] Robert E. Kohler, *From Medical Chemistry to Biochemistry: The Making of a Biomedical Discipline* (Cambridge/New York: Cambridge Univ. Press, 1982).

[11] Gerald Geison, *Michael Foster and the Cambridge School of Physiology: The Scientific Enterprise in Late Victorian Society* (Princeton, N.J.: Princeton Univ. Press, 1978).

has pointed out, there is room to explore the "fine structure of scientific creativity."[12] Generally, however, interest in twentieth-century biology has not prompted the use of oral history, on which historians of physics and chemistry have begun to rely, so living biologists remain a major resource as yet largely untapped.

Printed materials are still the central source of historiographic information. Yet reprint collections such as that at the MBL and other collections in libraries are often overlooked. They can, obviously, prove especially valuable when they are annotated by a known hand. In addition, many universities and the American Philosophical Society have amassed valuable archival collections of American biologists. Librarians and administrators are becoming more aware of the worth of their archival resources and the need to make these materials available to scholars, so more collections have been opened. *The Mendel Newsletter,* edited by Frederick Churchill, and David Bearman and John Edsall's *Archival Sources* can help point the way to manuscripts, as can the collections of letters edited by Nathan Reingold and Ida Reingold. For biographical data to guide research, the *Dictionary of Scientific Biography* continues to add supplements, while such works as Joseph Ewan and Nesta Ewan's *Biographical Dictionary of Rocky Mountain Naturalists* provide information on more specialized groups of individuals.[13]

Historians of biology can also profit greatly from the reissue of important texts. As out-of-print books become increasingly expensive and inaccessible and as new graduates move away from major libraries, availability becomes problematic. Thus the reissue by Columbia University Press and Yale University Press of such works as Theodosius Dobzhansky's *Genetics and the Origin of Species,* Mayr's *Systematics and the Origin of Species,* and Richard Goldschmidt's *Material Basis of Evolution* is important. Similarly, the volumes produced by Arno Press, such as the series on the history of paleontology edited by Steven Jay Gould and Thomas Schopf and the volumes in natural history edited by Keir Sterling, represent valuable but underutilized materials.[14]

[12] William Provine, *Sewall Wright: Geneticist and Evolutionist* (Chicago: Univ. Chicago Press, forthcoming); F. L. Holmes, "The Fine Structure of Scientific Creativity," *History of Science,* 1981, *19*:60–70.

[13] Frederick B. Churchill, ed., *The Mendel Newsletter* (Philadelphia: Library of the American Philosophical Society); David Bearman and John T. Edsall, eds., *Archival Sources for the History of Biochemistry and Molecular Biology: A Reference Guide and Report,* 1980 (Boston: American Academy of Arts and Sciences; Philadelphia: American Philosophical Society, 1980); Nathan Reingold and Ida Reingold, eds., *Science in America: A Documentary History, 1900–1939* (Chicago: Univ. Chicago Press, 1981); Nathan Reingold, ed., *Science in Nineteenth Century America* (cit. n. 2); Charles Gillispie, ed., *Dictionary of Scientific Biography* (New York: Scribners, 1970–1980); Supplement II, forthcoming, ed. F. L. Holmes; and Joseph Ewan and Nesta Ewan, *Biographical Dictionary of Rocky Mountain Naturalists: A Guide to the Writings and Collections of Botanists, Zoologists, Geologists, Artists, and Photographers, 1682–1932* (The Hague/Boston: W. luwer, Boston, 1981).

[14] The Arno reprint series includes works by Alpheus Hyatt, Alfred Sherwood Romer and Llewellyn I. Price, George Gaylord Simpson, Henry Fairfield Osborn, Louis Agassiz, Edward Drinker Cope, William King Gregory, and Joseph Leidy; see also Keir Sterling, *Last of the Naturalists: The Career of C. Hart Merriam* (New York: Arno, 1977); Sterling, ed., *Contributions to the History of American Natural History* (New York: Arno, 1974); Theodosius Dobzhansky, *Genetics and the Origin of Species* (New York: Columbia Univ. Press, 1970); Dobzhansky, *Dobzhansky's Genetics of Natural Populations* ed. R C. Lewontin, John A. Moore, William B. Provine, and Bruce Wallace (New York: Columbia Univ. Press, 1981); Ernst Mayr, *Systematics and the Origin of Species* (New York: Columbia Univ. Press, 1982), intro. by Niles Eldredge; and Richard Goldschmidt, *The Material Basis of Evolution* (New Haven: Yale Univ. Press, 1982), intro. by Stephen Jay Gould.

Just as histories of biology tend to group together under the broad categories of biographies, institutional histories, and disciplinary histories, depending on each historian's definition of the problem at hand, so different resources can prove particularly useful. For biographies, therefore, oral histories, notebooks, and reprint collections can provide an extra dimension of information. Institutional histories are often strengthened by archival sources and administrative records, though these often remain difficult to locate. Disciplinary histories can profit from the increased availability of a wide range of primary resources through reprints and manuscripts spanning the disciplinary area.

BIOGRAPHIES

Mary Alice Evans and Howard Ensign Evans's study of Harvard entomologist William Morton Wheeler is an excellent portrait of the man as scientist within a particular social and political setting. It details the development of Wheeler's work on wasps, at the same time discussing the way in which that work emerged against a background of Wheeler's education at Clark University and positions at the University of Chicago, the University of Texas, and the American Museum of Natural History, with visits to the Naples Zoological Station and the MBL. The Evanses combine an understanding of the scientific contributions with details of Wheeler's life and personality.[15]

Another biography, by Garland Allen, has traced chronologically the development of one man's science, Thomas Hunt Morgan's genetics. Allen defends the thesis that biology underwent a revolution from morphological to experimental and that Morgan contributed centrally to effecting that change. Allen's work, while primarily a study of one man's science rather than a personal history or a thorough consideration of setting, details an important story of intellectual development that reflects both personal and institutional elements.[16]

In a different sort of study, Kenneth Manning provides a marvelously vital portrait of a black scientist in a scholarly biography that includes considerable personal and social history. Manning's study of Ernest Everett Just weaves together institutional history (the Rockefeller Foundation, MBL, Howard University), personal details (family life, friendships, love affairs), social history (southern upbringing, race, the Spingarn prize, Howard University, connections with Frank Lillie, Jacques Loeb, and the MBL), national differences in scientific style (American, German, French), and Just's science. By portraying one scientist, Manning believes he will reveal a great deal about the institution of science. Was Just denied opportunity in the United States because he was black? Clearly, yes. How good a scientist was he and how different should his professional position have been? Manning suggests that Just was an excellent scientist, but he does not explicitly argue the point. Instead, he draws on an impressive range of sources to illustrate the pressures, demands, and limitations Just confronted as a black scientist in the United States. Throughout this book, Manning raises questions about how different factors influence scientific ideas and institutions that historians of biology should pursue. He stimulates the reader to

---

[15] Mary Alice Evans and Howard Ensign Evans, *William Morton Wheeler, Biologist* (Cambridge, Mass.: Harvard Univ. Press, 1970).

[16] Garland Allen, *Thomas Hunt Morgan: The Man and His Science* (Princeton, N.J.: Princeton Univ. Press, 1978).

consider related problems as well. In addition, Manning's powerful narrative underlines the value of writing history well.[17]

Evelyn Fox Keller's biography of Barbara McClintock is a fairly straightforward description of McClintock's life and work. Keller does not inquire into scientific change or the individual's place in a complex historical setting. Rather, she wants to convince her readers that McClintock has faced peculiar problems because she is a woman, a thesis that surely has merit although McClintock herself insists that she is different, that she savors the exceptions rather than the rules in nature, and that therefore her own situation cannot be explained as the problem of a class of people—namely women—generally. Keller's treatment of McClintock as an exemplar thus weakens her study. The work also demonstrates the advantages and the problems of writing about a living scientist. In brief, while Keller's biography has reached a wider audience than most histories of biology, the work reveals a lack of objectivity resulting from the author's necessary reliance on interviews and personal acquaintance with her subject.[18]

Autobiographies offer similar problems of objectivity. Nevertheless, such delightful works as G. Evelyn Hutchinson's *Kindly Fruits of the Earth* and such reflections as Mayr's "How I Became a Darwinian," George Gaylord Simpson's *Concession to the Improbable,* and Edwin Colbert's *A Fossil Hunter's Notebook: My Life with Dinosaurs and Other Friends* provide valuable insights into the workings of biology and its setting. They can serve as both primary and secondary sources, for they often provide historical interpretations of the ideas and settings of an earlier time.[19]

Other problems come with studying relatively unknown or undocumented scientists. For example, Marilyn Ogilvie and Clifford Choquette had to probe local records and contact relatives and friends for information on cytologist Nettie Stevens. Though some of Stevens's work on chromosomes has been discussed, for example, in Stephen Brush's article on her work on sex determination, personal details have generally not been available. Ogilvie's recent study of Alice Boring called for similar tactics. The resulting picture of individuals who have contributed important scientific ideas while remaining in relative personal obscurity is invaluable.[20] Many other such life stories remain to be constructed from oral histories and local or family records.

INSTITUTIONS

Biological institutions provide another subject of serious study. Historians have begun to move beyond chronologies toward more sophisticated analyses of the

[17] Kenneth B. Manning, *Black Apollo of Science: The Life of Ernest Everett Just* (New York: Oxford Univ. Press, 1983).

[18] Evelyn Fox Keller, *A Feeling for the Organism: The Life and Work of Barbara McClintock* (San Francisco: Freeman, 1983).

[19] G. Evelyn Hutchinson, *Kindly Fruits of the Earth: The Development of an Embryo Ecologist* (New Haven, Conn.: Yale Univ. Press, 1979); Ernst Mayr, "How I Became a Darwinian," in *The Evolutionary Synthesis: Perspectives on the Unification of Biology,* ed. Mayr and William Provine (Cambridge, Mass.: Harvard Univ. Press, 1980), pp. 413–423; and George Gaylord Simpson, *Concession to the Improbable: An Unconventional Autobiography* (New Haven, Conn.: Yale Univ. Press, 1978).

[20] Marilyn Ogilvie and Clifford Choquette, "Nettie Marie Stevens (1861–1912): Her Life and Contributions to Cytogenetics," *Proc. Amer. Phil. Soc.,* 1981, *125*:292–311; Stephen Brush, "Nettie M. Stevens and the Discovery of Sex Determination by Chromosomes," *Isis,* 1978, *69*:163–172; and

nature and significance of various institutions. Mary P. Winsor's examination of Louis and Alexander Agassiz and the Museum of Comparative Zoology (MCZ) at Harvard University offers one example.[21] The MCZ's role as the primary academic center for natural history and hence biological research in the mid-nineteenth century, as well as the powerful and long-term influence of the Agassiz father and son, indicate the ability of individuals to shape the scientific work within an organization under tight administrative control. The rich materials in the well-organized MCZ archives (some on microfilm) make that institution a promising source of materials and subject for further study.

The Johns Hopkins University biology program has also received attention, largely because of its critical importance in attracting a group of American investigators to solid morphological work, as well as in setting new directions in biological research in the 1880s and 1890s. Philip Pauly has looked at the relations of psychology to biology in Hopkins's early years under G. Stanley Hall, for example, and has discovered a lack of communication between practitioners of the two areas of study. Also looking at the Johns Hopkins, Dennis McCullough and Keith Benson have examined the impact of William Keith Brooks on morphological work there in the 1890s.[22] Unfortunately, the roots of physiological investigation in the biology program at Hopkins have remained largely unexamined, as has the period after the "golden age" of the nineteenth century.

The University of Chicago has also received some attention but warrants much more. Lincoln Blake, for example, has provided an unsophisticated introduction to science at Chicago, with a bare sketch of the chronology of what happened in the early years.[23] Despite the lack of a large organized archival collection on the early years, the University of Chicago and the Lillie and Whitman Collections at the MBL do contain a number of documents that should inform a much richer study of Chicago and the principals involved than has yet appeared. In particular Charles Otis Whitman, first head of biology at Chicago and first director of the MBL, had strong ideas about what biology meant: morphology and physiology, zoology and botany, descriptive and experimental work. In organizing both institutions, he stubbornly held to his convictions about biology but stepped aside from administrative control in the early 1900s. Whitman's student Frank Lillie took over both the MBL and the Chicago department, in turn imposing his own strong ideas about biology. Both men were so closely tied with these institutions that biography and institutional history need to be written together. In addition, each man in his own way stressed the importance of the whole organism, examination of which promises significant

---

Ogilvie, "Alice M. Boring: An American Scientist in China," paper presented to the History of Science Society, Norwalk, Conn., Oct. 1983.

[21] Mary Pickard Winsor, "Louis Agassiz and the Species Question," *Stud. Hist. Biol.*, 1979, *3*:89–117.

[22] Philip Pauly, "G. Stanley Hall and His Successors: A History of the First Half-Century of Psychology at Johns Hopkins University," paper prepared for the G. Stanley Hall Centennial, Johns Hopkins University, 1983; Dennis McCullough, "W. K. Brooks's Role in the History of American Biology," *J. Hist. Biol.*, 1969, *2*:411–438; Keith Benson, "William Keith Brooks (1848–1908): A Case Study in Morphology and the Development of American Biology" (Ph.D. diss., Oregon State Univ., 1979); and Benson, "American Morphology in the Late Nineteenth Century: The Case of the Biology Department of Johns Hopkins University," *J. Hist. Biol.*, forthcoming.

[23] Lincoln C. Blake, "The Concept and Development of Science at the University of Chicago, 1890–1905" (Ph.D. diss., Univ. of Chicago, 1966), esp. Ch. 4 on Charles Otis Whitman and biology.

*The Marine Biological
Laboratory (MBL) at
Woods Hole, Massa-
chusetts. Courtesy
of the MBL.*

results since Chicago turned out a number of major scientists (including Just)
concerned with the organism as a whole.

A thorough study of Henry Fairfield Osborn's role in establishing Columbia
University's biology program, which became the home for Morgan's genetics
and Wilson's cytology, might likewise illuminate the interaction between institutions and the practice of biology with particular styles. So would a closer look
at other university departments and research laboratories, revealing interactions
of individuals, problems, methods, and settings.

Marine laboratories—the Marine Biological Laboratory, Friday Harbor, Cold
Spring Harbor, the Bermuda Experimental Station, and Scripps—have also
played an important part in American biology. Attention has been directed to
the MBL, both in Dexter's study of Alpheus Hyatt's and Louis Agassiz's an-
tecedent laboratories and in a conference dedicated to the history of the MBL
and the Naples Zoological Station. And Paul Galtsoff and Dean Allard have
examined the United States Fish Commission, which was closely tied to the
MBL in the early years. Aside from Lillie's limited documentary history of 1944,
however, there is no systematic study of later MBL history and the laboratory's
very considerable impact on American science. The approaching centennial (in
1988) has already begun to stimulate interest, the MBL has a growing archives
to provide material, and administrators and librarians there are welcoming historical interest with unusual enthusiasm and support. Historians of biology
should avail themselves of this opportunity. As for other marine laboratories,

Keith Benson has traced the early history of the University of Washington's laboratory at Friday Harbor; an early article by Ernst Dornfeld establishes the few available facts about the short-lived but important Allis Lake Laboratory; and materials from the Carnegie Institution's Cold Spring Harbor Laboratory have begun to receive some consideration.[24] A fuller picture of the motivations behind, and works carried on at, these marine laboratories, their antecedents, and other similar research facilities including summer camps will add to our understanding of the setting and constraints on American biological research.

Other institutions, including public and private museums and societies have likewise shaped the science of biology in the United States. Ronald Rainger has begun an examination of the American Museum of Natural History and of Henry Fairfield Osborn's role there. He demonstrates that scientific convictions influenced the public presentation of specimens and that, in turn, the need for public presentation at times influenced the scientific ideas. Charles Coleman Sellars's account *Mr. Peale's Museum* illustrates the earlier desire to collect natural curiosities and to make sense of them. Philip Kopper's beautifully illustrated book on the holdings, though not really the history, of the National Museum of Natural History exemplifies the interest in natural history in our "nation's attic." Jeffrey Stott's study of American zoological parks underlines the American fascination with public lands and the way in which protection supports science. Wardell Pomeroy's study of the Kinsey Institute for Sex Research points to a different sort of natural history work, with its collection of largely American sex histories, of interest to historians of biology. Sally Kohlstedt's examination of the Boston Society of Natural History illustrates the depth of local concern with nature study.[25] Further examples of museums, other public enterprises concerned with natural history or biology, and biological societies abound, providing opportunities for further study and for comparative assessments.

---

[24] Ralph Dexter, "From Penikese to the Marine Biological Laboratory at Woods Hole—The Role of Agassiz's Students," *Essex Inst. Hist. Coll.*, 1974, *110*:151–161; Dexter, "The Annisquam Seaside Laboratory of Alpheus Hyatt, Predecessor of the Marine Biological Laboratory at Woods Hole, 1880–1886," in *Oceanography: The Past*, ed. Mary Sears and Daniel Merriam (New York: Springer, 1980), pp. 94–100; "The MBL and the Stazione Zoologica: A Century of History," conference at Ischia, Italy, Oct. 1984, with proceedings to appear in *Biological Bulletin*, June 1985; Detlev W. Bronk, "Marine Biological Laboratory: Origins and Patrons," *Science*, 1975, *189*:613–617; Paul Galtsoff, *The Story of the Bureau of Commercial Fisheries Biological Laboratory, Woods Hole, Mass.* (Washington, D.C.: U.S. Dept. of the Interior, 1962); Dean Conrad Allard, Jr., *Spencer Fullerton Baird and the U.S. Fish Commission* (Ph.D. diss, George Washington Univ., 1967; New York: Arno, 1978); Frank Rattray Lillie, *The Woods Hole Marine Biological Laboratory* (Chicago: Univ. Chicago Press, 1944); Keith Benson, "A History of the Laboratory at Friday Harbor" (forthcoming); Ernst J. Dornfeld, "The Allis Lake Laboratory," *Marquette Medical Review*, 1956, *21*:115–144; and James Ebert, article on Cold Spring Harbor and the Carnegie Foundation, *Biological Bulletin*, 1985, forthcoming.

[25] Ronald Rainger, "Fossils for Knowledge and Enlightenment: Vertebrate Paleontology at the American Museum, 1890–1910," paper presented to the History of Science Society, Chicago, Dec. 1984; Charles Coleman Sellars, *Mr. Peale's Museum: Charles Willson Peale and the First Popular Museum of Natural Science and Art* (New York: Norton, 1980); Philip Kopper, *The National Museum of Natural History* (New York: Abrams/Smithsonian, 1982); Jeffrey R. Stott, "The Historical Origins of the Zoological Park in American Thought," *Environmental Review*, 1981, *5*:52–65; Sally Gregory Kohlstedt, "From Learned Society to Public Museum: The Boston Society of Natural History," in *The Organization of Knowledge in Modern America, 1860–1920*, ed. Alexandra Oleson and John Voss (Baltimore: Johns Hopkins Univ. Press, 1979), pp. 386–406; Kohlstedt, "The Nineteenth Century Amateur Tradition: The Case of the Boston Society of Natural History," in *Science and Its Public: The Changing Image*, ed. Gerald Holton and William A. Blanpied (Dordrecht/London: Reidel, 1976), pp. 173–190.

One such comparative assessment comes in Pauly's examination of the relations between medicine and biology around 1900. Those universities with strong medical programs generally failed to establish effective and progressive biology programs in part, Pauly argues, because the required service to medicine drew resources away from active efforts to define and enliven the biological research. This thesis needs further articulation and development but suggests similar comparative questions. The relations between development of biology programs and the availability of strong or weak museums, herbaria, or journals might illuminate how nonintellectual factors have influenced the development of biology. Such factors might also prove to have influenced the ideas of science, as Rainger has shown they did at the American Museum.[26]

In addition, William Haas has raised questions about the differences between American and Chinese attitudes toward botany and taxonomy, indicating that botanists have exchanged materials and visits but generally not approaches in the twentieth century. His study highlights the issue of how biology in the United States is in at least some ways a distinctly American enterprise. It may be that particular techniques or laboratory styles or individual schools have affected the national character of American biology in unique and significant ways. It may also be that there are broader underlying ideological assumptions that influence the national character of science, as Haas suggests.[27]

Other sorts of institutions such as university museums, herbaria, journals, and funding agencies have received little attention. Both Pnina Abir-Am and Robert Kohler have studied the Rockefeller Foundation's approach to molecular biology and thus provide examples of how one foundation influenced biology.[28] Manning has examined funding by the Rockefeller Foundation of Just's biology. Many other questions remain to be explored, and many of these institutions have archival collections and resources of their own awaiting historical study.

DISCIPLINES

Disciplinary studies are enjoying considerable popularity. Rainger's and Stephen Jay Gould's work on the history of paleontology and taxonomy and Toby Appel's study of Jeffries Wyman and evolution return to subjects neglected in the recent rush to understand modern experimental biology. Charlotte Porter's work serves as a reminder of the vast untouched materials in the history of natural history. John C. Greene's admirable survey of American science in the age of Jefferson offers rich chapters on zoology, botany, and related themes. Greene demonstrates that this early and neglected period of American biology deserves attention even though the various American activities in science did not always fall together into a neatly connected story. More specialized studies include Pa-

[26] Philip Pauly, "The Appearance of Academic Biology in Late Nineteenth Century America," *J. Hist. Biol.*, 1984, *17*:369–397; Rainger, "Fossils for Knowledge and Enlightenment" (cit. n. 25).

[27] William Haas, "Botany in Republican China: The Leading Role of Taxonomy," paper presented at the Rockefeller Institution.

[28] Pnina Abir-Am, "The Discourse of Physical Power and Biological Knowledge in the 1930s: A Reappraisal of the Rockefeller Foundation's 'Policy' in Molecular Biology," *Social Studies of Science*, 1982, *12*:341–382; and Robert E. Kohler, "Warren Weaver and the Rockefeller Foundation Program in Molecular Biology: A Case Study in the Management of Science," in *The Sciences in the American Context: New Perspectives*, ed. Nathan Reingold (Washington, D.C.: Smithsonian Institution Press, 1979), pp. 249–293.

tricia Gossel's examination of the origins of American bacteriology as a scientific discipline; Gossel surprisingly reveals that American bacteriology initially found its place as a part of botany. And Michael Sokal has shown that psychology deserves attention as a field closely related to biology in the early years of this century.[29] Oceanography, agriculture, ecology, embryology, and evolutionary biology also have begun to receive more careful attention.

Susan Schlee's studies of oceanography are a case in point. Though she does not define oceanography or worry explicitly about what is included and what is not, she presents a useful descriptive history of studies, including American studies, of the sea. Government patronage played a critical role in setting up oceanographic ventures, so Schlee describes details of funding and political hopes, along with the realities of various expeditions and laboratories. Schlee's accounts are both well written and full of suggestions for deeper and more analytical studies. Her history of the *Atlantis* exemplifies one creative direction for further work, for it provides essentially a biography of one important American research vessel. A collection of historical papers originally presented at the Woods Hole Oceanographic Institution suggests other approaches to the history of oceanography as different authors tackle a variety of problems.[30]

Agriculture maintains close relations with biology in some respects, but few historians of science have examined these connections, with the notable exceptions of Margaret Rossiter, in several different contexts, and Charles Rosenberg, in articles. Rossiter's book-length study of scientific training, agricultural laboratories, and educational institutions from 1840 to 1880 documents the way in which Americans embraced the important science of agricultural chemistry. Her more recent essay serves as a tantalizing introduction to the decades 1860–1920, but no one has seriously studied later periods. As Rossiter points out in a separate and useful bibliography, histories of agriculture have tended to stick with the ABCs (Johnny Appleseed, Luther Burbank, and George Washington Carver), even though many biological contributions began or continued in agricultural settings—for example, studies of sex production and sexual reproduction.[31]

---

[29] Ronald Rainger, "Paleontology and Philosophy: A Critique," *J. Hist. Biol.*, forthcoming; Rainger, "The Understanding of the Fossil Past: Paleontology and Evolution Theory, 1850–1910" (Ph.D. diss, Indiana Univ., 1982); Stephen Jay Gould, *Ontogeny and Phylogeny* (Cambridge, Mass.: Harvard Univ. Press, 1977); Toby Appel, "A Little Too Modest: Jeffries Wyman, Philosophical Anatomy and Evolution," paper presented to the History of Science Society, Norwalk, Conn., Oct. 1983; Charlotte Porter, "The Concussion of Revolution: Publications and Reform at the Early Academy of Natural Sciences, Philadelphia, 1812–1842," *J. Hist. Biol.*, 1979, *12*:273–292; Porter, "The Excursive Naturalists or the Development of American Taxonomy at the Philadelphia Academy of Natural Sciences, 1812–1842" (Ph.D. diss., Harvard Univ., 1976); John C. Greene, *American Science in the Age of Jefferson* (Ames: Univ. Iowa Press, 1984); Patricia Gossel, "The Species Problem in Bacteriology: A Technical and Diagnostic Dilemma," paper presented to the Joint Atlantic Seminar for the History of Biology, Washington, D.C., Apr. 1984; Michael Sokal, ed., *An Education in Psychology: James McKeen Cattell's Journal and Letters from Germany and England, 1880–1888* (Cambridge, Mass.: MIT Press, 1981); and Sokal and Patrice Rafail, comps., *A Guide to Manuscript Collections in the History of Psychology and Related Areas* (Millwood, N.Y.: Kraus International, 1982).

[30] Susan Schlee, *The Edge of an Unfamiliar World: A History of Oceanography* (New York: Dutton, 1973); Schlee, *On Almost Any Wind: The Saga of the Oceanographical Vessel* Atlantis (Ithaca, N.Y.: Cornell Univ. Press, 1978); and Sears and Merriam, eds. *Oceanography: The Past.*

[31] Margaret Rossiter, *The Emergence of Agricultural Science: Justus Liebig and the Americans, 1840–1880* (New Haven, Conn.: Yale Univ. Press, 1975); Rossiter, "The Organization of the Agricultural Sciences," in *Organization of Knowledge*, ed. Oleson and Voss (cit. n. 25), pp. 211–248;

Trained as a sociologist, Rachel Volberg has examined networks of communication that surrounded the development of a professional discipline of botany in the United States from 1880 to 1920. She argues that as botany became professionalized, researchers eliminated problems and disciplines within botany not amenable to experimentation. Volberg stresses the nonintellectual and nontheoretical convictions that have dictated what was considered "acceptable" science. Although Volberg emphasizes details about the way science works that may seem odd or extraneous to the more theoretically inclined, this work is an important examination of botany's relations to agriculture, funding agencies, institutions, and other disciplines. The historian of science can gain valuable insight about the role of communication networks and shared assumptions from this and other sociological studies.[32]

Ecology has begun to receive a great deal of attention recently, as demonstrated by the range of studies Frank Egerton discusses in a two-part article. The first reviews the field to date and is not strictly limited to ecology in the United States. The second part surveys the history of applied ecology in North America. Researchers are focusing on such questions as what ecology is and how it emerged and has developed in relation to other biological disciplines. Donald Worster's more popular treatment, Ronald Tobey's look at grasslands ecology, Sharon Kingsland's study of laboratory and experimental work, Eugene Cittadino's discussion of professionalization in botany in the United States, James Collins's work on evolutionary ecology, William Coleman's work on ecology prior to the evolutionary synthesis, and Joel Hagen's examination of plant ecology and taxonomy, among others, serve to illustrate this healthy and productive diversity of questions and approaches.[33]

Embryology and genetics have also attracted increasing interest. For example, Jeffrey Werdinger's doctoral dissertation studying embryology at the MBL demonstrates the importance of the community there in facilitating the exchange of ideas. At a time when experimental embryology was changing rapidly, this exchange proved central to the emergence of the field and its separation from the

Rossiter, comp., "A List of References for the History of Agricultural Science in America" (University of California at Davis Agricultural History Center, 1980); and Charles Rosenberg, "Rationalization and Reality in the Shaping of American Agricultural Research, 1875–1914," *Soc. Stud. Sci.*, 1977, 7:401–422; see also Rosenberg, *No Other Gods: On Science and American Social Thought* (Baltimore: Johns Hopkins Univ. Press, 1976), Chs. 8–12.

[32] Rachel Volberg, "Constraints and Commitments in the Development of American Biology, 1880–1920" (Ph.D. diss., Univ. California at San Francisco, 1983).

[33] Frank N. Egerton, "The History of Ecology: Part I, Achievements and Opportunities," *J. Hist. Biol.*, 1983, *16*:259–310; "Part II, Applied Ecology in North America," *ibid.*, forthcoming; Ronald C. Tobey, *Saving the Prairies: The Life Cycle of the Founding School of American Plant Ecology, 1895–1955* (Berkeley/Los Angeles: Univ. California Press, 1981); Sharon Kingsland, "The Refractory Model: The Logistic Curve and the History of Population Ecology," *Quart. Rev. Biol.*, 1982, 57:29–52; Kingsland, "Modelling Nature: Theoretical and Experimental Approaches to Population Ecology, 1920–1950" (Ph.D. diss., Univ. Toronto, 1981); Eugene Cittadino, "Ecology and the Professionalization of Botany in America, 1890–1905," *Stud. Hist. Biol.*, 1980, *4*:171–198; James Collins, "Evolutionary Ecology and the Changing Role of Natural Selection in Ecological Theory," paper presented to the conference on history, philosophy, and biology, Denison University, July 1983; William Coleman, "Evolution into Ecology," paper presented at the conference "Reflections on Ecology and Evolutionary Biology," Arizona State University, March 1985; Joel Hagen, "Experimental Taxonomy, 1930–1950: The Impact of Cytology, Ecology, and Genetics on Ideas of Biological Classification" (Ph.D. diss., Oregon State Univ., 1982); and Hagen, "Experimentalists and Naturalists in Twentieth-Century Botany: Experimental Taxonomy, 1920–1950," *J. Hist. Biol.*, 1984, *17*:249–270.

study of heredity, to the extent that such a separation really occurred. Scott Gilbert's analysis of Morgan's embryological work also examines the emergence of genetics and embryology as distinct disciplines, as does a recent paper by Garland Allen. Jan Sapp is exploring lines of biological research in the United States and France that continued to emphasize development, with his study of work on cytoplasmic inheritance. And Jane Maienschein's survey of studies of sex determination points to an area lying between embryology and genetics, an area that has begun to attract increasing historical attention from biologists as well. Numerous studies of the rise of the field of genetics have appeared or are in preparation, including Horace Judson's masterfully written *Eighth Day of Creation* and Kevles's survey of genetics in the United States and Britain.[34]

Studies of evolutionary biology include several examinations of the relationship between biology and religion, with attention to issues of creationism. Others focusing more directly on biological work, particularly in the nineteenth century, include Cynthia Russett's study of the reception of Darwinism in the United States and essays by Michele Aldrich and Edward Pfeifer in Thomas Glick's *Comparative Reception of Darwinism*. Peter Bowler has extended the discussion of evolutionary ideas after Darwin. Bowler's *Eclipse of Darwinism*, for example, classifies such Americans as Louis Agassiz, Edward Drinker Cope, Henry Fairfield Osborn, and William Berryman Scott as anti-Darwinians and discusses the various versions of anti-Darwinism. Ernst Mayr and William Provine have extended the discussion to more recent evolutionary biologists with the *Evolutionary Synthesis*.[35]

Among the subjects closely related to history of biology but not traditionally considered part of biology is the environment, the subject of several studies. Stephen Pyne's *Fire in America* provides an excellent and imaginative definition of a problem in this area, examining approaches to forest and fire management, bureaucratic control, and factors very closely related to studies of botany and agriculture. Roderick Nash, Thomas Dunlap, and Michael Cohen have also examined issues of conservation and environment in ways that illuminate American concerns with organic nature.[36]

---

[34] Jeffrey Werdinger, "Embryology at Woods Hole: The Emergence of a New American Biology" (Ph.D. diss., Indiana Univ., 1980); Gilbert, "Embryological Origins of Genetics" (cit. n. 6); Garland Allen, "T. H. Morgan and the Split Between Embryology and Genetics, 1910–1926," paper presented to the British Society for Developmental Biology, Nottingham, 1983; Jan Sapp, "Cytoplasmic Inheritance and the Struggle for Authority in the Field of Heredity, 1891–1981" (Ph.D. diss., Univ. Montreal, 1984); Sapp, "The Field of Heredity and the Struggle for Authority: Some New Perspectives on the Rise of Genetics," *J. Hist. Biol.*, 1983, *16*:311–342; Jane Maienschein, "What Determines Sex? A Study of Converging Approaches, 1880–1916," *Isis*, 1984, *76*:457–480; and Horace Judson, *The Eighth Day of Creation: Makers of the Revolution in Biology* (New York: Simon & Schuster, 1979); Daniel J. Kevles, "Genetics in the United States and Great Britain, 1890–1930: A Review with Speculations," in *Biology, Medicine, and Society, 1840–1940* (Cambridge: Cambridge Univ. Press, 1981), pp. 193–215.

[35] On creationism see, e.g., Dorothy Nelkin, *The Creation Controversy: Science or Scripture in the Schools* (New York: Norton, 1982); see also Cynthia Russett, *Darwin in America: The Intellectual Response, 1865–1912* (San Francisco: Freeman, 1976); Michele Aldrich, "United States: Bibliographic Essay," in *The Comparative Reception of Darwinism*, ed. Thomas F. Glick (Austin: Univ. Texas Press, 1974), pp. 207–226; Edward Pfeifer, "The United States," *ibid.*, pp. 168–206; Pfeifer, "The Genesis of American Neo-Lamarckism," *Isis*, 1965, *56*:156–167; Peter Bowler, *The Eclipse of Darwinism: Anti-Darwinian Evolution Theories in the Decade around 1900* (Baltimore: Johns Hopkins Univ. Press, 1983); and Richard Burkhardt, "Lamarckism in Britain and the United States," in *Evolutionary Synthesis*, ed. Mayr and Provine (cit. n. 18).

[36] Stephen Pyne, *Fire in America: A Cultural History of Wildland and Rural Fire* (Princeton, N.J.:

Despite the growing number of studies on subareas of biology and despite the expansion into related concerns, relatively few scholars have moved to later or to more technical disciplines, such as biochemistry or molecular biology. Few historians of science have pursued philosophically informed questions about how science has developed in the twentieth century, questions about the nature of biology or the ways in which biology has changed, for example. Further, only a handful have examined the way in which specialties arose because of, or with the support of, institutional factors. The diversity of studies to date makes the history of biology in America a healthy and robust field, but a further abundance of problems, questions, and approaches obviously remains to be explored as well.

CONCLUSION

On the whole, the historical study of biology in the United States is a robust, exciting field. While historians concerned with recent developments in particular will need to update their technical knowledge, many other historians would profit from a more sophisticated understanding of biology and what it is. Moreover, some will find it useful to explore philosophical and theoretical issues, while others should acquire more sophisticated tools of social analysis in order to assess the impact of factors external to scientific ideas. Others will begin to explore the way in which or extent to which biology in America has also been a distinctly American biology. They will begin to answer those questions about how to define the field of history of American biology and how best to approach it. As historians of biology investigate new issues, they will begin to divide up their problems in new ways, yielding novel perspectives. The continued willingness to accept diverse approaches should allow the history of biology in America to remain a particularly lively, productive, and inviting field of study.

Princeton Univ. Press, 1982); Roderick Nash, *The American Environment: Readings in the History of Conservation* (Reading, Mass.: Addison-Wesley, 1968); Nash, *Wilderness and the American Mind* (3rd. ed.; New Haven, Conn.: Yale Univ. Press, 1982); Thomas Dunlap, *DDT: Scientists, Citizens, and Public Policy* (Princeton, N.J.: Princeton Univ. Press, 1981); and Michael P. Cohen, *The Pathless Way: John Muir and American Wilderness* (Madison: Univ. Wisconsin Press, 1984).

History of Physics

By Albert E. Moyer*

HENRY DAVID THOREAU expressed it simply: "The life which men praise and regard as successful is but one kind. Why should we exaggerate any one kind at the expense of the others?"[1] In recent years, we—those writing on the history of American physics—have exaggerated the life of the twentieth-century physicist-politician. This is the man (few have been women) whose success in physics has been linked to American political and military history, especially the two world wars. We have heaped attention on—at times, even praised—not only the personal events and circumstances of his life but also his particular research program and home institution. His research program has typically involved some branch of atomic, nuclear, or high-energy physics; his home institution has usually been a prominent university affiliated with a big-science laboratory funded by a major philanthropic or federal agency. Although the study of the physicist-politician is legitimate and perhaps even primary in any full analysis of American physics, we have, as I intend to show, exaggerated this one kind of life at the expense of the others.

A convenient date for beginning a survey of recent writings on American physics is 1976, about the time Daniel Kevles was completing *The Physicists*. This is a convenient date not only because of the unrivaled classic stature of *The Physicists* but also because of the summational nature of this well researched book; Kevles drew on all the principal prior studies.[2] A first glance at the many monographs, editions, and articles published since 1976—excluding for the moment Kevles's own book, which appeared in 1978—indicates that they cover myriad topics from diverse perspectives. On closer examination, nevertheless, a pattern emerges. Most authors share either an explicit or an implicit concern with the issue of the United States' rise to world leadership in physics. Authors attempt to describe or explain the dramatic transition during roughly the first half of the twentieth century from international scientific subservience to dominance. What gives meaning and even urgency to this line of historical inquiry is the striking success of physicists in creating atomic weapons during World War II. In fact, for many writers the experience of physicists at Los Alamos and other wartime laboratories has become a touchstone for evaluating research topics. That is, a topic's relevance to the wartime experience has become a popular criterion for certifying its historical legitimacy.

The preference for twentieth-century topics of this type contrasts sharply with

* Department of History, Virginia Polytechnic Institute and State University, Blacksburg, Virginia 24061.
[1] Henry David Thoreau, *Walden, or Life in the Woods* (1854; rpt. New York: Harper & Row, 1965), p. 15.
[2] Daniel J. Kevles, *The Physicists: The History of a Scientific Community in Modern America* (New York: Knopf, 1978), pp. 435–464.

the relative lack of interest in topics dealing with the entire period prior to about 1900. Admittedly, there have been new works pertaining to, for example, J. Willard Gibbs, Edwin Hall, Joseph Henry, Albert Michelson, Henry Rowland, and John Trowbridge. However, except for Nathan Reingold and his colleagues' ongoing editorial project involving the Henry Papers—five volumes of documents extending into Henry's Princeton years have been completed to date— these studies simply lack the bulk and coherence that characterize the body of writings covering the later period, especially World War II.[3]

There is, of course, an obvious sense in which the experience of physicists in World War II has dominated recent work on the history of American physics. The past few years have seen a surge of inquiries into the individuals, institutions, and ideas involved *directly* in the Manhattan Project. Recall, for example, the rush of research regarding J. Robert Oppenheimer and Los Alamos. This direct interest extends, moreover, beyond the war years themselves to include prewar roots and postwar ramifications. Historians like to trace, by examining organizations such as the Lawrence Berkeley Laboratory, ways in which the Manhattan Project was not only the culmination of prewar institutional styles and research trends but also a source of postwar patterns.

But there is a more subtle and fundamental manner in which the physicists' wartime experience has dominated recent historical work. By focusing on the Manhattan Project as a critical historical episode, scholars have tended to favor a select subset of the diverse questions traditionally asked about American physics. In other words, interest in the atomic weapons project has helped to fix attention on a particular cluster of basic questions; the answers to these questions have helped, in turn, to generate a distinctive view of the history of American physics.

Two questions stand out. First, how did the American physics community reach maturity in the decades immediately prior to the Manhattan Project? Here the implicit assumption is that the physics community, in order to have succeeded so spectacularly in the weapons program, must have come of age sometime during the 1920s or 1930s; a corollary is that up through World War I the community was in its adolescence, although maturing rapidly in its institutional and research capabilities. Second, what effect did the Manhattan Project have upon the evolving character of physics as pursued in a democracy? The implicit assumption here is that there is something distinctive about how American physicists, both individually and collectively, responded to the heightened politicization of physics that occurred through the advent of atomic bombs. This second question, incidentally, reflects a broader concern among historians of American

[3] Martin J. Klein, "The Scientific Style of Josiah Willard Gibbs," in *Springs of Scientific Creativity: Essays on Founders of Modern Science*, ed. R. Avis, H. T. David, and R. H. Stuewer (Minneapolis: Univ. Minnesota Press, 1983), pp. 142–162; Katherine R. Sopka, "The Discovery of the Hall Effect: Edwin Hall's Hitherto Unpublished Account," in *The Hall Effect and Its Applications*, ed. C. L. Chien and C. R. Westgate (New York: Plenum, 1980), pp. 523–545; Nathan Reingold, ed., *The Papers of Joseph Henry*, 5 vols. to date (Washington, D.C.: Smithsonian Institution Press, 1972–present); Loyd S. Swenson, Jr., "The Michelson-Morley-Miller Experiments and the Einsteinian Synthesis," *Astronomische Nachrichten*, 1982, *303*:39–45; A. D. Moore, "Henry A. Rowland," *Scientific American*, 1982, *246*(2):150–161; Gerald Holton, "How the Jefferson Physical Laboratory Came to Be," *Physics Today*, 1984, *37*(12):32–37. On the importance of studying American science prior to about 1900, see Nathan Reingold, "Reflections on 200 Years of Science in the United States," in *The Sciences in the American Context: New Perspectives*, ed. Nathan Reingold (Washington, D.C.: Smithsonian Institution Press, 1979), pp. 9–20.

physics with the overall nature of physics in a democracy; to a large degree, this concern underpins scholars' general preoccupation with physicist-politicians as viewed against the full backdrop of twentieth-century political and military history.

The more general preoccupation is illustrated well by the growing body of scholarship on Robert Millikan—a key twentieth-century physicist-politician outside the immediate sway of World War II. Historians of physics have begun a dialogue about the proper interpretation of Millikan's career as an educator, researcher, fundraiser, and (including during World War I) policymaker. For example, Robert Kargon has labeled Millikan "a conservative in the midst of a revolutionary world," an interpretation that has been questioned by other Millikan scholars. Still other historians have been engaging in a more technical dialogue about Millikan's research orientation.[4] To be sure, studies such as these have sharpened our vision of physicist-politicians outside the circumscribed world of the Manhattan Project. Nevertheless, the emphasis since 1976 has remained on those physicist-politicians with ties to World War II.

This characterization of recent writings arises through a review of literature that excludes Kevles's *Physicists*. Nevertheless, the characterization also fits that book, which features the "leaders of the Los Alamos generation" while surveying a host of nineteenth- and twentieth-century scientists. In reviews, Lewis Pyenson and Kargon described the book as primarily a story of "the ruling elite of American physics" and of "external politics." David Hollinger has similarly pointed to Kevles's reliance on "a succession of physicist-politicians" presented within the context of "the succession of wars and Presidential administrations that are the staple of most courses in 20th-century American history." These observations led the three reviewers to complain that the book is not, as its subtitle states, "the history of a scientific community in modern America," but rather the history of one special segment of that community. "Attention is lavished on twentieth-century figures who lightened philanthropic and ultimately public coffers during campaigns for high-energy machines," Pyenson has written, "but scant remarks are directed to less glamorous fields of physics and their rank-and-file exponents."[5]

The similarity between Kevles's book and other current writings raises the possibility that the book has influenced the direction of recent research. Most likely, however, *The Physicists* and other writings are all part of an ongoing

[4] Robert H. Kargon, *The Rise of Robert Millikan: Portrait of a Life in American Science* (Ithaca, N.Y.: Cornell Univ. Press, 1982), p. 74; John L. Michel, "Millikan: A Conservative in Revolutionary Times?" review of Kargon, *The Rise of Robert Millikan, Physics Today*, 1982, 35(10):79–80; Robert W. Seidel, "Millikan and His Era," review of Kargon, *The Rise of Robert Millikan, Science*, 1982, 216:851–852; Nathan Reingold, review of Kargon, *The Rise of Robert Millikan, Isis*, 1983, 74:143–144; and Daniel J. Kevles, "Robert A. Millikan," *Scientific American*, Jan. 1979, 240(1):142–151. See also Gerald Holton, "Subelectrons, Presuppositions, and the Millikan-Ehrenhaft Dispute," *Historical Studies in the Physical Sciences*, 1978, 9:161–224; Alan D. Franklin, "Millikan's Published and Unpublished Data on Oil Drops," *ibid.*, 1981, 11:185–201; Harvey Fletcher, "My Work with Millikan on the Oil-Drop Experiment," *Phys. Today*, 1982, 35(6):43–47; and Peter Galison, "The Discovery of the Muon and the Failed Revolution against Quantum Electrodynamics," *Centaurus*, 1983, 26:262–316, esp. pp. 305–308.

[5] Kevles, *The Physicists*, pp. ix, 334, 367, 394, 423–424. Lewis Pyenson, "Physics in America from the Top Down," review of Kevles, *The Physicists, Phys. Today*, 1978, 31(3):63–64; Robert Kargon, "Physics in the United States," review of Kevles, *The Physicists, Science*, 1978, 199:524–525; and David A. Hollinger, "The Powerful Interplay of Physics and Society," review of Kevles, *The Physicists, Chronicle of Higher Education*, 20 Mar. 1978, pp. 27–28.

historiographic trend—one that Charles Rosenberg characterizes as "an ever-increasing interest in the twentieth century and especially in the role of big science."[6] The existence of this trend helps to explain why, for example, few authors bother to cite Kevles's broader themes in justifying their own corresponding analyses. When they do turn to Kevles (a surprisingly infrequent occurrence), it is usually only for facts and details to support their own seemingly self-generated analyses.

In identifying the particular historiographic viewpoint of Kevles and other writers of late, I am not denigrating their work. On the contrary, as I remarked earlier, a concern with physicist-policymakers and all that their careers entail is not only legitimate but also perhaps primary in any comprehensive account of American physics; indeed, during recent decades, physicists in essentially every branch of the profession have had some stake in politics or weaponry. Even the few reviewers who strongly disapproved of aspects of *The Physicists* acknowledged that the book is a major contribution to historical scholarship. What I am doing is calling attention to this prevailing historiographic viewpoint so that persons either working or reading in the field can become more aware of the presuppositions and limitations of the research. In addition, future scholars might be able to complement this hitherto dominant line of research with alternative studies, aiming for a wider understanding of American physics.

LEGACY OF THE MANHATTAN PROJECT

The alliance between the scientific and the military communities in the United States has strengthened considerably in the past half century. As Alex Roland documents in his essay in this volume, historians are beginning to give the relationship its deserved attention. Among historians concentrating on American physics, an outright boom has occurred in research regarding World War II, especially the Manhattan Project. Receiving the fullest attention has been J. Robert Oppenheimer, the physicist-politican par excellence. Research on Oppenheimer, in fact, has achieved the status of a scholarly industry serving a wide market. There have been biographies by James Kunetka and Peter Goodchild, a volume of letters and recollections edited by Alice Kimball Smith and Charles Weiner, and a new collection of Oppenheimer's essays titled *Uncommon Sense*. Complementing these large-scale projects has been a steady outpouring of articles, ranging from Gerald Holton's sketch of Oppenheimer's youth to Barton Bernstein's exceptionally balanced account of the 1954 security hearing. Balance, in fact, is what most of the new works have brought to the Oppenheimer story. Lately, even Oppenheimer's old colleagues have been easing Oppenheimer out of the mythic realm. Herbert York, for example, in tracing the development of hydrogen bombs by the Americans and the Soviets, offers a first-hand description of Oppenheimer's role in the genesis of these weapons; Hans Bethe and Edward Teller have augmented this description with their own, often contradictory revelations.[7]

---

[6] Charles Rosenberg, "Science in American Society: A Generation of Historical Debate," *Isis*, 1983, 74:356–367, on p. 359; see also pp. 363–364.

[7] James W. Kunetka, *Oppenheimer: The Years of Risk* (Englewood Cliffs, N.J.: Prentice-Hall, 1982); Peter Goodchild, *J. Robert Oppenheimer: Shatterer of Worlds* (Boston: Houghton Mifflin, 1981); J. Robert Oppenheimer, *Robert Oppenheimer: Letters and Recollections*, ed. Alice Kimball

The fascination with the wartime development of bombs has extended to Oppenheimer's Los Alamos lieutenants and other soldiers in the Manhattan campaign—not only publicly conspicuous physicists such as Bethe and Teller but also less prominent men such as Philip Morrison, Leo Szilard, and Robert Wilson. A separate essay would be required merely to summarize the range of scholarly analyses and anecdotal reminiscences written by or about this group of men. In his recent "Nuclear Age Bibliography," Karl Hufbauer needed twenty-two pages simply to list the historical studies concerning nuclear weapons published since 1975. A satirist writing in *Physics Today* concluded that the phrase "nuclear proliferation" really refers to "the seemingly endless memoirs, textbooks, polemics and articles" churned out by veterans of the Manhattan Project, so labeled "because so many literary agents and publishers were located in that part of New York."[8]

Lawrence Badash and two colleagues have brought a degree of order to this varied historical enterprise by helping to assemble a series of lectures by physicists such as Norris Bradbury, Edwin McMillan, and John Manley into a unified volume, *Reminiscences of Los Alamos, 1943–1945*. Jane Wilson has compiled a similar volume based on a series of personal accounts initially published in the *Bulletin of the Atomic Scientists*.[9] The most ambitious project, however, is currently under way at Los Alamos. Lillian Hoddeson is heading a small team investigating the technical development of the first fission bombs. Although squarely within the historiographic tradition that takes as its reference point the Manhattan Project, Hoddeson's investigation promises to provide a corrective to studies that overemphasize the larger policy decisions and personal intrigues of Oppenheimer and other Los Alamos leaders. This comprehensive project also promises to provide a corrective to those personal reminiscences that are of limited or narrow historical scope.

In addition to being studied for their own sake, the physicists' wartime experiences also serve historians as filters through which prewar elements pass and then emerge altered in the postwar years. This orientation leads to the historical genre that I alluded to earlier: scholars operate within the boundaries of general American history to trace the development of particular institutions and to describe their associated physicist-politicians and research programs. The genre is characterized by continuity and compatibility: continuity with wartime events and compatibility with American military and political history. Given the

Smith and Charles Weiner (Cambridge, Mass.: Harvard Univ. Press, 1980); J. Robert Oppenheimer, *Uncommon Sense*, ed. N. Metropolis *et al.* (Cambridge, Mass.: Birkhäuser, 1984); Gerald Holton, "Young Man Oppenheimer," *Partisan Review*, 1981, 48:380–388; Barton J. Bernstein, "In the Matter of J. Robert Oppenheimer," *Hist. Stud. Phys. Sci.*, 1982, 12:195–252; Hans A. Bethe, "Comments on the History of the H-Bomb," *Los Alamos Science*, Fall 1982, pp. 43–53; Edward Teller and Bethe, "Letters: Hydrogen Bomb History," *Science*, 1982, 218:1270; and Herbert York, *The Advisors: Oppenheimer, Teller, and the Superbomb* (San Francisco: Freeman, 1976).

[8] Karl Hufbauer, "Nuclear Age Bibliography: Historical Studies Published since 1/1/1975" (MS, June 1984, available from the Institute on Global Conflict and Cooperation, Central Office, Q-060, University of California–San Diego, La Jolla, CA 92093); see also Stephen G. Brush and Lanfranco Belloni, comps., *The History of Modern Physics: An International Bibliography* (New York: Garland, 1983); and Henry Petroski, "A Short History of Nuclear Editing," *Phys. Today*, 1983, 36(11):9, 110.

[9] Lawrence Badash, Joseph O. Hirschfelder, and Herbert P. Broida, eds., *Reminiscences of Los Alamos, 1943–1945* (Boston: Reidel, 1980); and Jane Wilson, ed., *All in Our Time: The Reminiscences of Twelve Nuclear Pioneers* (Chicago: Bulletin of the Atomic Scientists, 1975).

particular texture of the Manhattan Project—the primary historical filter—it follows that the institutions are usually big-science prototypes or projections, the physicist-politicians are usually university based, and the research programs characteristically involve nuclear or high-energy physics. For convenience, I call this genre "Manhattan-style history."

Of late, the primary vehicle for Manhattan-style history has been the institutional study. Varying degrees of attention have been directed toward all the major nuclear reactor and particle accelerator facilities: Argonne National Laboratory, Brookhaven National Laboratory, Fermi National Accelerator Laboratory, W. K. Kellogg Radiation Laboratory, Lawrence Berkeley Laboratory, Oak Ridge National Laboratory, and the Stanford Linear Accelerator Center. Current issues of the two history of physics newsletters, published by the American Institute of Physics Center for History of Physics and the American Physical Society Division of History of Physics, are replete with reports of historical conferences or archival projects involving the laboratories.

Two of the most noteworthy conferences to shed light on these laboratories, particularly through the recollections of founders from the Manhattan generation of physicists, were actually organized around specific research fields. Nuclear physics was the topic of Roger Stuewer's 1977 symposium at the University of Minnesota; contributors included Hans Bethe, Otto Frisch, Edwin McMillan, Rudolf Peierls, Emilio Segrè, Robert Serber, and Robert Wilson. Three years later, particle physics was the topic of Laurie Brown and Lillian Hoddeson's symposium at Fermilab; among the participants were Herbert Anderson, Robert Marshak, Bruno Rossi, Robert Serber, and Victor Weisskopf. These conferences and the published proceedings are noteworthy not only because of their systematic coverage of subjects but also because of their meshing of the insights of physicists and professional historians. The historians have supplied the interpretive framework often lacking in the physicists' otherwise revealing recollections.[10]

Although the symposia on nuclear and particle physics frequently emphasized American contributions, they were international in scope. This serves to remind us, as Charles Weiner has stressed in an essay on the early history of nuclear physics, that the national scientific community is often merely reflecting the more dominant international community. On the other hand, as Hoddeson has established in a comparison of Fermilab with a Japanese high-energy physics laboratory, national differences do account for the distinctive icing on pieces of the international cake. Hoddeson attributes the eventual divergence in the development of Fermilab and its Japanese counterpart to "different cultural norms (for example, American individualism vs. Japanese conformity), different economic realities, and different social circumstances."[11] Clearly, it is important

[10] Roger H. Stuewer, ed., *Nuclear Physics in Retrospect: Proceedings of a Symposium on the 1930s* (Minneapolis: Univ. Minnesota Press, 1979); Laurie M. Brown and Lillian Hoddeson, eds., *The Birth of Particle Physics* (Cambridge: Cambridge Univ. Press, 1983); and Brown and Hoddeson, "The Birth of Elementary-Particle Physics," *Phys. Today,* 1982, *35*(4):36–43. For more on conferences of this type, see, e.g., "Conferences Highlight New Interest in National Labs History," *Center for History of Physics Newsletter,* 1979, *11*(2):1–2.

[11] Charles Weiner, "Institutional Settings for Scientific Change: Episodes from the History of Nuclear Physics," in *Science and Values: Patterns of Tradition and Change,* ed. Arnold Thackray and Everett Mendelsohn (New York: Humanities, 1974), pp. 187–212, on pp. 187–188, 206–207; and Lillian Hoddeson, "Establishing KEK in Japan and Fermilab in the U.S.: Internationalism, Nationalism, and High Energy Accelerators," *Social Studies of Science,* 1983, *13*:1–48, on p. 34.

One element of the physics community: Nobel laureates attending a physics conference in 1952 at the University of Rochester. Left to right: *Hideki Yukawa (visiting professor, Columbia), Edwin McMillan (Berkeley), Carl Anderson (Caltech), and Enrico Fermi (Chicago). Courtesy of Robert Marshak.*

that we remain cognizant of the delicate distinction between physics in the United States, which merely happens to be done by Americans, and American physics, which is peculiarly American.[12]

Historians of science have been exceptionally attentive to both sorts of physics at the universities and associated big-science laboratories in California. For example, John Heilbron, Robert Seidel, and Bruce Wheaton have been making increasingly discriminating studies of the accelerator laboratory at the Berkeley campus of the University of California. These studies document the centrality in the laboratory's development of yet another quintessential physicist-politician, Ernest Lawrence; they also reveal the centrality in the laboratory's development of interdisciplinary research executed within a flexible organizational structure. Heilbron, Seidel, and Wheaton distinguish themselves by blending internal and external history, integrating technical advances with biographical and institutional elements. This blend has produced Manhattan-style history at its best.[13]

Whereas histories bearing the Manhattan pedigree usually follow the fortunes

---

[12] David A. Hollinger, "American Intellectual History: Issues for the 1980s," *Reviews in American History,* 1982, *10*:306–317, on pp. 311–314.

[13] John L. Heilbron, Robert W. Seidel, and Bruce R. Wheaton, *Lawrence and His Laboratory: Nuclear Science at Berkeley* (Berkeley: Lawrence Berkeley Laboratory and Office for History of Science and Technology, Univ. California, 1981); Heilbron, Seidel, and Wheaton, "A Strong Interaction between Science and Society, 1931–1981," *CERN Courier,* 1981, *21*:335–344; see also Seidel, "The Origins of Academic Physics Research in California," *Journal of College Science Teaching,* 1976, *6*:10–23; and Seidel, "Accelerating Science: The Postwar Transformation of the Lawrence Radiation Laboratory," *Hist. Stud. Phys. Sci.,* 1983, *13*:375–400.

of physicist-politicians affiliated with universities, they do not always have this academic orientation. One of the most discerning studies concerning big-science institutions involved in a branch of nuclear physics and linked to postwar American history is Joan Bromberg's analysis of government scientists working on magnetic fusion reactors. The line of demarcation between academic and governmental scientists has become blurred with the burgeoning of federal funding since World War II; nevertheless, in contrast to high-energy physics, with its academic alliance, "magnetic fusion research bears above all the imprint of the government laboratories." In tracing the increasing centralization of decision making in Washington from the 1950s through the 1970s, Bromberg identifies three periods of fusion research and relates them to specific periods in American political and military history. The result is another example of Manhattan-style history at its best.[14]

COMING OF AGE IN A DEMOCRACY

Historians' fixation on the Manhattan Project has also had a more subtle impact on the direction of historical research. This fixation, as I explained earlier, has helped focus attention on two basic questions that have in turn helped color the general perception of the history of American physics. The questions concern the physics community's attainment of maturity prior to the Manhattan Project and the project's effect upon the character of physics in a democracy. Both questions, when expressed most generally, hark back to the nineteenth century, if not earlier. The issue of maturity—or lack thereof—dates to at least the era beginning with Joseph Henry's eye-opening European tour in 1837 and extends through Henry Rowland's pessimistic "Plea for Pure Science" in 1883. As Reingold, Kevles, and others have documented, physical scientists of this era perceived themselves to be lagging behind their European colleagues in professional standing and standards; consequently, they campaigned to upgrade American educational programs, professional societies, research journals, and other essential institutional components. The related issue of science in a democracy is similarly well-established, dating back to at least the time of Alexis de Tocqueville. Reingold has explained, for example, how Victorian commentators associated American democratic values with a purported propensity for applied rather than basic research.[15]

Interest in the Manhattan Project has served mainly to direct increased attention to the questions of maturity and democracy. The maturity question gains importance in the historiography of American physics through the desire to account for the stunning success of the atomic weapons project. The democracy

[14] Joan L. Bromberg, *Fusion: Science, Politics, and the Invention of a New Energy Source* (Cambridge, Mass.: MIT Press, 1982), quoting p. 253; Bromberg, "TFTR: The Anatomy of a Programme Decision," *Soc. Stud. Sci.*, 1982, 12:559–583.

[15] Reingold, ed., *Papers of Joseph Henry* (cit. n. 3), Vol. III: *January 1836–December 1837: The Princeton Years*, pp. xviii–xix; *ibid.*, Vol. IV: *January 1838–December 1840: The Princeton Years*, pp. xvi–xvii; Kevles, *The Physicists* pp. 43–44; Kevles, "On the Flaws of American Physics: A Social and Institutional Analysis," in *Nineteenth-Century American Science: A Reappraisal*, ed. George H. Daniels (Evanston, Ill.: Northwestern Univ. Press, 1972), pp. 133–151, on pp. 133, 151; Reingold, "Reflections on 200 Years" (cit. n. 3), pp. 10–11; and Reingold, "Definitions and Speculations: The Professionalization of Science in America in the Nineteenth Century," in *The Pursuit of Knowledge in the Early American Republic*, ed. Alexandra Oleson and Sanborn C. Brown (Baltimore: Johns Hopkins Univ. Press, 1976), pp. 33–69, on pp. 33–34.

question becomes more critical as a result of the heightened politicization of American physics that accompanied the weapons program. In his classic *Science in the Federal Government,* A. Hunter Dupree wrote: "By the time the bombs fell on Hiroshima and Nagasaki, the entire country was aware that science was a political, economic, and social force of the first magnitude."[16] Dupree could have explicitly included historians among those with this intensified awareness.

Recently, the first question has spawned a multifaceted body of literature on the so-called maturation, flowering, or coming of age of American physics—on the shift of the international center of gravity to the United States. This literature explores primarily two issues: first, the relative importance of native-born versus European refugee physicists in advancing the American community's institutional and research capabilities during the 1920s and 1930s and, second, the transmission of quantum mechanics to the United States during these same decades and its role in adding theoretical physicists to a community previously dominated by experimentalists. Both issues are germane to the coming-of-age question as asked from the perspective of the Manhattan Project. The refugee physicists, after all, were integral to the breakthroughs at Los Alamos and Chicago. But could they have performed as effectively without an established, vigorous community in which to immerse themselves? In addition, the leaders of the Manhattan Project tended to be the same venturous physicists who had embraced quantum mechanics a decade or so earlier. Could the research behind the bombs have progressed so swiftly without a community already bound together by, and well versed in, the theoretical as well as the experimental aspects of quantum mechanics?

The refugee and quantum issues derive, of course, from more comprehensive topics that are usually studied independently of the Manhattan Project. That is, general historians have been curious for a number of decades about the migration of European intellectuals to the United States; similarly, historians of science have been inquisitive about the rise of quantum theory for its own sake. The recent focus on the coming-of-age question within the context of the Manhattan Project has simply made elements of two already interesting topics even more interesting.

In *The Physicists,* Kevles deals with both the issue of refugee versus native physicists and the question of the transmission of quantum mechanics. As with other recent writers, his interpretation is partially informed by the views of Charles Weiner and Stanley Coben. In 1969 Weiner argued, contrary to popular opinion, which attributed the flowering of American physics to the European refugees, that "the rapid growth of physics in America in the 1930's was under way before the refugees arrived." Coben, following Weiner's lead, argued in 1971 that one of the main preconditions for the "surprising" appearance of theoretical physics among Americans during the late 1920s was the reception of quantum mechanics. Since the publication of Weiner's and Coben's essays, a number of scholars have extended or adjusted the original arguments. Regarding the reception of quantum mechanics, the most exhaustive analysis is Katherine Sopka's *Quantum Physics in America, 1920–1935.* After meticulously examining the institutional and intellectual fiber of the American physics community,

---

[16] A. Hunter Dupree, *Science in the Federal Government: A History of Policies and Activities to 1940* (Cambridge, Mass.: Harvard Univ. Press, 1957), p. 369.

she draws conclusions basically in line with those of Weiner and Coben. The development of quantum physics added "a strong theoretical component to the already well-established tradition of excellence in the experimental domain," thus providing "the crucial element in the timing of the 'coming of age' of physics in America." Consequently, "the European physicists who came here could function as effectively as they did because physics in America had already attained maturity."[17]

Paul Hoch suggests a more central role for the European physicists in his essay contrasting the obstacles that confronted the refugees in the Soviet Union, Britain, and the United States. He claims that the theoretical bent of the refugees complemented the American experimental bent, thus providing "the crucial missing element necessary for rapid progress," as exemplified in the "successful atom bomb project." Gerald Holton tempers this claim in a reconstruction of the refugee story traced through the particular experiences of Albert Einstein. When describing the "remarkable symbiosis" between the Americans and the Europeans, Holton is careful to acknowledge "the excellence of the early theoretical contributions of American physicists." Various scholars besides Holton, responding to the 1979 centennial of Einstein's birth, have weighed the symbolic and substantive impact on American physics of Einstein's immigration. In a centennial essay particularly mindful of Einstein's ties to the Institute for Advanced Study, Harry Woolf agrees with Paul Langevin's contemporary appraisal of Einstein's relocation in 1933: "It's as important an event as would be the transfer of the Vatican from Rome to the New World. The Pope of physics has moved and the United States will now become the center of the Natural Sciences."[18]

In 1979, an international group of scholars met in Italy for a workshop on the growth of quantum mechanics in Germany and the United States. Papers included Robert Seidel's account of the transmission of quantum mechanics as facilitated by funds from the Rockefeller Foundation and similar philanthropic agencies—"the institutionalized largess of American capitalism." The most imaginative paper was John Heilbron's revisionist approach to the issue of quantum transmission. In contrast with Coben, Sopka, and other current historians, Heilbron denies that the transmission altered American physics by adding a previously missing theoretical component: "In 1920 the United States did not lack men competent in theoretical physics; rather, it lacked men able to

[17] Kevles, *The Physicists*, pp. 169, 221, 282–283; Charles Weiner, "A New Site for the Seminar: The Refugees and American Physics in the Thirties," in *The Intellectual Migration*, ed. Donald Fleming and Bernard Bailyn (Cambridge, Mass.: Harvard Univ. Press., 1969), p. 191; Stanley Coben, "The Scientific Establishment and the Transmission of Quantum Mechanics to the United States, 1919–32," *American Historical Review*, 1971, 76:442–460, on pp. 442, 450–452; Katherine R. Sopka, *Quantum Physics in America, 1920–1935* (New York: Arno, 1980), pp. xiii–xiv, 4.1, 5.3; see also Gerald Holton's preface, *ibid.*, pp. i–ix.

[18] Paul K. Hoch, "The Reception of Central European Refugee Physicists of the 1930s: U.S.S.R., U.K., U.S.A.," *Annals of Science*, 1983, 40:217–246, on pp. 218–219, 236; see also "Discussion," in *Nuclear Physics in Retrospect*, ed. Stuewer, pp. 306–322, on 312–320; Gerald Holton, "The Formation of the American Physics Community in the 1920s and the Coming of Albert Einstein," *Minerva*, 1981, 19:569–581, on pp. 570–572; a variant of this article appeared as "The Migration of Physicists to the United States," in *The Muses Flee Hitler*, ed. Jarrell C. Jackman and Carla M. Borden (Washington, D.C.: Smithsonian Institution Press, 1983), pp.169–182; and Harry Woolf, "Albert Einstein: Encounter with America," in *Some Strangeness in the Proportion: A Centennial Symposium to Celebrate the Achievements of Albert Einstein*, ed. Woolf (Reading, Mass.: Addison-Wesley, 1980), pp. 32–33 (quoting Langevin); see also Abraham Pais, *"Subtle Is the Lord . . .": The Science and the Life of Albert Einstein* (Oxford: Clarendon, 1982), pp. 449–454, 473–478.

compete with those few Europeans who were advancing the quantum theory of the atom." He adds, "We must not allow the great subsequent success of quantum mechanics or the grand claims of its expositors . . . to mislead us into judging its transmission to the United States as much more than the addition of a new speciality—modern German atomic theory—to an already imposing American competence in physics." For evidence of early competence in important areas of theoretical physics, Heilbron points to the fields of X-ray spectroscopy and modern acoustics.[19]

Did Americans display competence in theoretical physics prior to the appearance of quantum mechanics? Four recent works help to answer this question. A negative reply comes from Stanley Goldberg's study of the American response to, and assimilation of, Einstein's special theory of relativity. With their "emphasis on the experimental and practical," American physicists created a "hostile environment" for relativity theory. Goldberg goes so far as to elevate this emphasis on the practical into an American national style—a concept, by the way, that Reingold called into question over a decade ago in his sweeping reappraisal of the thesis of "American indifference to basic research." Another negative answer emerges from Lawrence Badash's book on the study of radioactivity in the United States during the opening two decades of the century. In chronicling the rapid "growth and decay" of this science, he remarks that its practitioners "came more in the long-honored tradition of experimentally-oriented data gatherers." Unlike Goldberg, however, Badash puts little stock in this "national aversion to theory," concluding that in the sparsely populated field of radioactivity "personal happenstance was more a determinant." My own research leads to a positive answer. Examining the conceptual orientations of American physicists immediately preceding the advent of the relativity and quantum theories, I have found that Americans were at least well informed about the latest electromagnetic, thermodynamic, and statistical-mechanical theories taking shape in the international community. Finally, John Servos reorients the entire question by examining it from the perspective of the history of mathematics. He maintains that advanced training in applied mathematics is a prerequisite for engaging in advanced physical theory and that such training became available in the United States only after World War I.[20] This, of course, was about the time Americans began to display a mastery of theoretical quantum mechanics.

Interest in the Manhattan Project has also induced historians to pay increased attention to the fundamental question of the evolving character of physics as

---

[19] The papers from the workshop are published in *Fisica e società negli anni '20* (Milan: Clup-Clued, 1980); see Robert Seidel, "Aspetti istituzionali della transmissione della meccanica quantistica agli Stati Uniti," *ibid.*, pp. 189–214; John L. Heilbron, "La fisica negli Stati Uniti subito prima della meccanica quantistica," *ibid.*, pp. 135–158 (I thank the authors for English translations).

[20] Stanley Goldberg, *Understanding Relativity: Origin and Impact of a Scientific Revolution* (Cambridge, Mass.: Birkhäuser, 1984), pp. 263, 265–266; Lawrence Badash, *Radioactivity in America: Growth and Decay of a Science* (Baltimore: Johns Hopkins Univ. Press, 1979), pp. 4, 260, 273; Albert E. Moyer, *American Physics in Transition: A History of Conceptual Change in the Late Nineteenth Century* (Los Angeles: Tomash, 1983), pp. 153, 157, 172. John W. Servos, "Mathematics and the American Scientist, 1880–1930" (rev. version of paper presented at the Annual Meeting of the History of Science Society, Norwalk, Conn., Oct. 1983); see also Nathan Reingold, "Refugee Mathematicians in the United States of America, 1933–1941: Reception and Reaction," *Ann. Sci.*, 1981, *38*:313–338. For background, see Reingold, "American Indifference to Basic Research: A Reappraisal," in *Nineteenth-Century American Science*, ed. Daniels, pp. 38–62.

pursued in a democracy. Unquestionably, Kevles's *The Physicists* is the proper point of embarkation for any tour of recent works on this question. Although the book is foremost a narrative history—a chronological, nonjudgmental description of specific institutions, individuals, and ideas—it does advance a thesis of broad import. Kevles holds that from the late nineteenth century through the mid-1970s the physics community was animated by a tension between traditional elitist leanings and American democratic ideals. Specifically, this recurrent tension involved two types of elitism. "Best-science elitism" is the community's tendency not only to follow loyally but also to give fullest rewards and opportunities to an aristocracy of leading physicists in the few top universities and other organizations at the peak of the American "institutional pyramid." Whereas best-science elitism concerns the internal character of the profession, "political elitism" concerns the community's attitude toward outside, federal patronage of physics. Political elitism is the belief that the primary power to fashion and to control tax-funded research should lie with scientists themselves, not with elected officials or other politically accountable agents. During the 1950s, Kevles concludes, the leaders of the Los Alamos generation firmly established both best-science and political elitism. With the scientific criticisms and reevaluations of the Vietnam era, however, pluralist and democratic viewpoints reappeared.[21]

A few reviewers have been quick to challenge the basic assumptions of Kevles's thesis, particularly his understanding of American democracy. The critic with the most political savvy is David Joravsky. Zeroing in on Kevles's contrast between democracy and elitist science, Joravsky suggests that Kevles translate these "ideological terms into plain talk: 'democracy' means top bosses, 'elitist science' means their technical servants." More generally, Joravsky judges Kevles to be "the court historian of an American establishment" who "obscures the moral responsibility and political impotence of the physicists." In like manner, Robert March detects in Kevles a "faith in the uniquely creative tension between elitist science and the democratic society it serves"—a faith indicative of Kevles's sympathetic and optimistic view of the physicists' contribution to society. This leads March to a strong indictment of the book: "Had this been the history of American accountants, or lawyers, or even the medical profession, it would have been derided as a shameless apologia." In a milder vein, David Hollinger laments that the themes of best-science and political elitism have not been "developed more systematically" and that further "analytic constructs" have not been offered. For example, Kevles could have used his narrative to test the claim that "the growth of professional physics has been a powerful agent in the consolidation of American capitalism." In fairness to Kevles, however, we should not fault him simply for parting company with recent revisionist historians—historians such as David Noble, the bugbear of American scientific technology and corporate capitalism. Kevles's contribution is an expansive initial vision of the history of American physics, not a revision of the more subtle political questions.[22]

[21] The preceding summary is from Albert E. Moyer, review of Kevles, *The Physicists, Isis*, 1978, 69:634.

[22] David Joravsky, "Scientists as Servants," review of five books, including Kevles, *The Physicists, New York Review of Books*, 28 June 1979, pp. 34–39, on pp. 34–35; Robert March, review of Kevles, *The Physicists, Bulletin of the Atomic Scientists*, March 1979, 35:61–64, on pp. 61, 64;

In the past few years, an increasing number of scholars have been addressing the more subtle questions. Among these scholars are not only historians of science but also American political historians and a variety of social and political scientists. In this volume, Sally Kohlstedt, Margaret Rossiter, and Sharon Gibbs Thibodeau highlight the new studies that deal directly with the topics of institutions, public policy, and government. The new studies range from those that integrate particular political events with specific technical developments to those that isolate policy issues from the workaday research context. Many of the works reviewed in this essay offer the former type of integration. An example is Joan Bromberg's history of magnetic fusion research. Thaddeus Trenn's long-winded but earnest survey of the science advisory network since World War II illustrates the latter, more generic study.

Bromberg, as mentioned earlier, identifies three basic periods of fusion research and relates them to corresponding periods in American political and military history. Her detailed investigation implicitly demonstrates the difficulty in generalizing about democratic values and physics. For example, at the same time that the nation's fusion research program was becoming more responsive to the public and seemingly more democratic, it was becoming more controlled by centralized managers in Washington—by an elite group of what Joravsky would call "top bosses." For the most part, Bromberg simply avoids characterizing these types of national trends with problematic labels such as "democratization."[23]

Trenn, on the other hand, freely employs such labels. Like Kevles, he judges that during the late 1960s through the 1970s there occurred a "democratization of science advice" in the United States. But is Trenn any more able than Kevles to define "democratization" in a way that can rationalize the continued existence of an exclusive circle of "top bosses"? "Democracy is not public participation," he explains at one point; "rather it is a system where leaders are chosen. . . . democratization does not necessarily require public participation on a one-to-one basis. Democratic participation could very well be delegating knowledgeable individuals to make decisions on behalf of the larger society." The problem, of course, with definitions of this sort is that they are so general that they could serve equally well to define a range of governmental systems instituted at different periods within the United States and, indeed, around the world.[24]

Clearly, historians face a dilemma in attempting to characterize those aspects of the physics profession that have been conditioned by the American political context. If they choose to write explicitly about democratic values and the like, they face the danger of lapsing into ideological rhetoric lacking in semantic precision. But if they avoid such language and limit themselves merely to describing the interactions of particular interest groups, then they face the danger of

---

Hollinger, "Powerful Interplay" (cit. n. 5), p. 28; see also Samuel H. Day, Jr., "Gods of Big Science," review of Kevles, *The Physicists, The Progressive,* May 1978, *42:*42–43; David F. Noble, *America by Design: Science, Technology, and the Rise of Corporate Capitalism* (Oxford: Oxford Univ. Press, 1977), pp. 4, 147; see also Nathan Reingold, "Clio as Physicist and Machinist," *Rev. Amer. Hist.,* 1982, *10:*264–280, on pp. 271–272.

[23] Bromberg, *Fusion,* pp. 248–251; Bromberg, "TFTR," pp. 577–578.

[24] Thaddeus J. Trenn, *America's Golden Bough: The Science Advisory Intertwist* (Boston: Oelgeschlager, Gunn & Hain, 1983), pp. 6, 174; see also *ibid.,* "Appendix A," pp. 233–234, for the etymology of the term "scientist-politician."

creating a narrative so narrow as to be devoid of analytic constructs and general conclusions. Perhaps comparative studies of the postwar physics communities in the United States and the Soviet Union would help to clarify those traits of American physics that can be attributed to American political values. Clarification might also come from examining the disjunction between physicists' perceptions and the actual circumstances of their political environments.

SOLID ALTERNATIVES

The coming-of-age and the physics-in-a-democracy questions lose their immediacy when we shift our focus away from university-based physicist-politicians and the context of American political and military history. Investigations of the Manhattan Project and Manhattan-style history more generally, with its emphasis on nuclear or high-energy research in big-science facilities, also become less central. But how do we actually achieve this reorientation? To what alternative topics do we shift our focus? The most promising subject at the moment is the American community of solid-state physicists.

Even if investigations of the solid-state community were not potentially revealing of new patterns in the history of American physics, there would still be compelling reasons for undertaking the studies. In recent years, solid-state physics—or, as it is coming to be known, condensed-matter physics—has grown into the most heavily populated field in American physics. In addition, a good case can be made that through its associated technologies such as the transistor, it is having a larger immediate impact on society than even nuclear physics. Be that as it may, historians of science have paid scant attention to this field in comparison with nuclear or even high-energy physics. This began to change around 1980, when Hoddeson joined with physicists and historians such as Spencer Weart in forming the "American team" of the International Project in the History of Solid State Physics. Hoddeson had done some earlier work in this line of historical research. Concentrating on Bell Laboratories and the discovery of the transistor, she had initially provided a narrative account of the basic conceptual and institutional features of perhaps the main branch of solid-state history—semiconductors. Later, under the auspices of the project, she tackled the more elusive history of superconductivity. At the same time, Weart began to analyze the senses in which solid-state physicists constituted a distinct community, bound together intellectually and socially and sharing viewpoints. These and similar components of research by other scholars are beginning to coalesce, providing an altered image of the history of American physics.[25]

One inescapable feature of the new image is the firmness of the bond between

[25] Lillian H. Hoddeson, "The Roots of Solid-State Research at Bell Labs," *Phys. Today*, 1977, *30*(3):23–30, on p. 23; "Introducing the Project," *Newsletter: International Project in the History of Solid State Physics*, May 1982, No. 1, pp. 1–3; and Theodore H. Geballe, "This Golden Age of Solid-State Physics," *Phys. Today*, 1981, *34*(11):132–143, on p. 132. See also Hoddeson, "The Entry of the Quantum Theory of Solids into the Bell Telephone Laboratories, 1925–40: A Case-Study of the Industrial Application of Fundamental Science," *Minerva*, 1980, *18*:422–447; Hoddeson, "The Discovery of the Point-Contact Transistor," *Hist. Stud. Phys. Sci.*, 1981, *12*:41–76; Hoddeson, "The Flowering of Solid State Physics in the Aftermath of World War II," paper presented to the American Physical Society Symposium of the Division of History of Physics, Baltimore, Apr. 1983; and Spencer R. Weart, "Solid State Physics as a Community," a 102-page typescript of Ch. 10 of a jointly authored book on the history of solid-state physics (forthcoming), pp. 68–69 (I thank the author for permission to cite this draft).

physics and industry. Historians of American science and technology frequently mention the institutional and intellectual aspects of this bond, but seldom investigate them in any detail. There are a few exceptions, including George Wise in his study of physicists at General Electric and Arturo Russo and Richard Gehrenbeck in their research into Clinton Davisson and Lester Germer's famous discovery of electron diffraction at Bell Laboratories. Weart speculates that this relative neglect reflects the fact that the majority of historical writing is done by academics who are most comfortable holding forth on their own colleagues—university-based physicists who are well insulated from the practical and financial contingencies of industrial research.[26] What the solid-state initiative does is place the interplay between physics and industry on center stage in the historiography of American physics. Sensitivity to the interplay sharpens our awareness of the great extent to which industry has provided employment for professional physicists in the United States and, more generally, of the great extent to which the evolution of American physics has been joined to American economic history. It also alerts us to individual physicists who have played weighty roles in the history of American physics but who have been upstaged by their Los Alamos peers; few historians of science have even heard of such Bell Lab luminaries from the mid-1940s as John Bardeen, Conyers Herring, and their director, Mervin Kelly.[27]

Another feature of the new image of American physics is the combined modesty and autonomy of individual research programs. As late as the 1960s solid-state physics was the domain of small groups of researchers working on independent projects in unpretentious facilities with limited budgets. This is in sharp contrast, Weart explains, "with high-energy physics, notorious for its 'big-science' of monster accelerators and papers with dozens of co-authors." Weart quotes Bardeen as saying that solid-state physics "is 'little physics,' requiring only modest outlays for equipment." Weart also quotes Frederick Seitz, author of the definitive solid-state textbook, as recalling that the field was one in which "a person could still be himself and not a slave to a machine."[28]

Whereas traditional historiographic issues lose their immediacy in light of the new research on the solid-state community, they do not become irrelevant. Hoddeson finds the rise of quantum mechanics in the United States to be an essential opening chapter in the story of the rise of solid-state physics. Additionally, she finds that wartime experiences, especially at the MIT Radiation Laboratory, accelerated postwar research on superconductivity. Krzysztof Szymborski comes to a similar conclusion regarding the postwar discovery of the transitor. In recent years, moreover, solid-state physicists—like many other physicists—have become ensconced in the military establishment. Nevertheless,

---

[26] George Wise, "Science at General Electric," *Phys. Today*, 1984, *37*(12):52–61; Arturo Russo, "Fundamental Research at Bell Laboratories: The Discovery of Electron Diffraction," *Hist. Stud. Phys. Sci.*, 1981, *12*:117–160; Richard K. Gehrenbeck, "Davisson and Germer," in *Fifty Years of Electron Diffraction*, ed. Peter Goodman (Dordrecht: Reidel, 1981), pp. 12–27; and Spencer R. Weart, "Obstacles to a History of Industrial Science" (MS, Jan. 1984), pp. 2, 5–6, 7–12 (I thank the author for permission to cite this draft).

[27] Weart, "Obstacles," pp. 3, 9; Hoddeson, "Entry of the Quantum Theory," pp. 429, 441–442; see also Hoddeson, "The Emergence of Basic Research in the Bell Telephone System, 1875–1915," *Technology and Culture*, 1981, *22*:512–544; and Hoddeson, "Roots of Solid-State Research," p. 27.

[28] Weart, "Solid State Physics," p. 78; see also Frank Herman, "Elephants and Mahouts—Early Days in Semiconductor Physics," *Phys. Today*, 1984, *37*(6):56–63, on p. 62.

when contrasted with nuclear and high-energy physics, solid-state physics passed through its early decades relatively independent of American political and military history. Hoddeson has established, for instance, that the development of the transistor was well under way prior to World War II.[29]

The solid-state community is not the only alternative worthy of study. The group of American physicists who helped develop lasers constitutes another promising subject. Joan Bromberg, with the assistance of Robert Seidel, is currently laying the groundwork for understanding this group through the Laser History Project. Bromberg and Seidel are interviewing a number of laser pioneers, industrial as well as military contributors—weaponry again appearing as a primary preoccupation of postwar physicists. Historians might also find promise in more comprehensive subjects that embrace the individual subdisciplines of physics. The evolution of physics education, for example, is a topic that scholars have only begun to investigate systematically. Melba Phillips has as yet little company in her surveys of laboratory teaching and the American Association of Physics Teachers. And Stanley Goldberg remains somewhat isolated in his attempt to read the history of American physics through the textbooks of successive generations of physics students.[30]

Although these various studies provide alternatives to the more familiar studies of physicist-politicians, nearly all of the works mentioned to this point are alike in one sense: they share a common methodological heritage. Specifically, they reflect the traditional methodologies of institutional, intellectual, and biographical history fostered by the major graduate programs and history of science organizations across the country. In other words, the mental tools and techniques used to probe the solid-state community are usually the same ones employed to probe the Los Alamos community. This need not be the case. In fact, researchers have lately brought other methodological perspectives to bear on the history of American physics. This reorientation mimics in many ways the recent movement of general American historians away from traditional political history toward social history.[31] The new perspectives, as with the solid-state inquiries, enable investigators to avoid the presuppositions usually associated with studies of physicist-politicians. A list of alternative methodological perspectives includes the sociological (e.g., Andrew Pickering's findings on the hunt for the quark at, among other places, Stanford University), anthropological (e.g., Sharon Traweek's expeditions to the tribal villages of high-energy physicists), and philosophical (e.g., Maila Walter's effort—and the endeavors of participants in a 1982 centenary meeting—to clarify the meaning of Percy Bridgman's mus-

[29] Hoddeson, "Entry of the Quantum Theory," pp. 445–447; Hoddeson, "Flowering of Solid State Physics," pp. 19–21; Hoddeson, "Roots of Solid-State Research," p. 32; and Krzysztof Szymborski, "World War II and American Solid State Physics," paper presented at the Annual Meeting of the History of Science Society, Chicago, Dec. 1984.

[30] See, e.g., Melba Phillips, "Early History of Physics Laboratories for Students at the College Level," *American Journal of Physics*, 1981, *49*:522–527; Phillips, ed., *AAPT Pathways: Proceedings of the Fiftieth Anniversary Symposium of the AAPT* (Stony Brook, N.Y.: American Association of Physics Teachers, 1981); Goldberg, *Understanding Relativity*, pp. 276–293; see also Albert E. Moyer, "Physics Teaching and the Learning Theories of G. Stanley Hall and Edward L. Thorndike," *Physics Teacher*, 1981, *19*:221–228.

[31] For methodological trends among general historians, see Felix Gilbert and Stephen R. Graubard, eds., *Historical Studies Today* (New York: Norton, 1972); see also Reingold, "Clio as Physicist and Machinist," pp. 266–272.

ings).[32] Three other perspectives requiring additional comment also come to mind.

Statistical Surveys. Quantitative methods are providing increasingly concrete information about the physics community. The statistical survey made by Paul Forman, Heilbron, and Weart suggests that, by about 1900, academic physicists in the United States were on a par with those in France, Germany, and England in terms of expenditures, volume of publications, and number of practitioners. Kevles, in another scan of the decades bracketing 1900, has detailed the demographic contrasts between the community of American physicists and their colleagues in mathematics and chemistry. Most recently, Weart has carried out a "statistical reconnaissance" of the American community in the period between the two world wars, including the Depression; the resulting data support the contention of strong ties between industrial and academic physics.[33]

Citation Analyses. Henry Small's publication of the *Physics Citation Index, 1920–1929* has opened the door to convenient analyses of the citation patterns displayed in the leading international journals of the 1920s. For example, Katherine Sopka has teased from these data an inventory of the most cited theoretical papers written by Americans. Similar uses were reported at a 1982 workshop on the historical applications of the index. Tentative plans now exist to extend the index to the period 1930–1954.[34]

Social-Psychological Investigations. Even a topic as overworked as the inception of the Los Alamos weapons program may profitably be viewed from a perspective that departs radically from American political and military history and the fortunes of physicist-policymakers. Weart has illustrated this in his social-psychological account of the "images in the public mind" that helped propel the physicists and the nation to a truly prodigious weapons project. The "road to Los Alamos," according to Weart, was constructed on a bedrock of "images— ideas of ancient origin, charged with all the power of myth."[35] It is interesting to speculate on the degree to which these same archetypal images, if they exist, have also drawn historians to the nuclear motif.

[32] Andrew Pickering, "The Hunting of the Quark," *Isis*, 1981, 72:216–236; Sharon Traweek, "Laboratory Practice in High Energy Physics," paper presented at the Annual Meeting of the Society for Social Studies of Science, Blacksburg, Va., Nov. 1983; Maila Walter, "P. W. Bridgman and the Privacy of Scientific Knowledge," paper presented at the Annual Meeting of the History of Science Society, Norwalk, Conn., Oct. 1983; and P. W. Bridgman Centenary Symposium, Cambridge, Mass., Apr. 1982 (the papers and a tape recording of the symposium are on file at the Harvard University Archives).

[33] Paul Forman, John L. Heilbron, and Spencer Weart, "Physics *circa* 1900: Personnel, Funding, and Productivity of the Academic Establishments," *Hist. Stud. Phys. Sci.*, 1975, 5:1–185; on p. 5; Daniel J. Kevles, "The Physics, Mathematics, and Chemistry Communities: A Comparative Analysis," in *The Organization of Knowledge in Modern America, 1860–1920*, ed. Alexandra Oleson and John Voss (Baltimore: Johns Hopkins Univ. Press, 1979), pp. 139–172; and Weart, "The Physics Business in America, 1919–1940: A Statistical Reconnaissance," in Reingold, *Sciences in the American Context* (cit. n. 3), pp. 295–358, on pp. 305, 327.

[34] Henry Small, *Physics Citation Index, 1920–1929*, 2 vols. (Philadelphia: Institute for Scientific Information, 1981); and "Reports: 1920–29 Citation Index," *History of Physics Newsletter*, 1983, 1:44; see also the frequent physics articles in *Scientometrics*; regarding future plans, see "Executive Committee Meeting," *Hist. Phys. Newsletter*, 1984, 2:2.

[35] Spencer R. Weart, "The Road to Los Alamos," in *Colloque international sur l'histoire de la physique des particules* (*Journal de physique*, 1982, 43, colloque C8, suppl. au no. 12) (Les Ulis Cedex: Editions de Physique, 1982), pp. 301–313, on pp. 302, 308; see also Weart, *Nuclear Fear* (New York: Dial, forthcoming).

New topics and new methodologies allow us to avoid many of the presuppositions of prior writing. Another possible way to avoid past constraints is simply to modify the manner in which we present our subject—in particular, to adopt a mode of presentation that emphasizes the visual as well as the verbal. Films, videotapes, and museum exhibits all have the potential for contributing to a fuller view of the history of American physics. Admittedly, visual media are often less able than books and articles to capture the niceties of a historical episode. Nevertheless, visual presentations can offer fresh insights. We see this in the recent films and videotapes on Oppenheimer and other members of the Los Alamos generation. Most notable is the Oppenheimer documentary by Jon Else; nominated for an Academy Award, the film captures nuances of Oppenheimer's career and character by weaving together vintage footage and recent interviews with old colleagues. Public television has also been presenting discerning retrospective profiles of other Los Alamos veterans; Richard Feynman, Victor Weisskopf, and, most recently, I. I. Rabi have joined Bethe and Teller in their own video features. Although historians have served mainly as consultants in these national productions, they have at times been more directly involved. Roger Stuewer produced a videotape of excerpts from the 1977 Minnesota conference on nuclear physics, while Arthur Miller appeared in the *Nova* segment on Einstein to explain key concepts. Of course, not all the productions are equally successful in presenting history. For example, the characters were overdrawn in *Oppenheimer,* the seven-part television dramatization by Peter Goodchild and Peter Prince.[36]

Museum exhibits also stress the visual. Paul Forman has successfully used this approach at the Smithsonian Institution with exhibits on modern physics. His main displays, all germane to physics in the United States, have included "Atom Smashers" (1977–present), "Einstein" (1979–1980), "The Fall of Parity" (1982), and "Atomic Clocks" (1983–present). Forman grants that museum exhibits are weaker in certain senses than media like books and films; he suggests, nevertheless, that exhibits offer a richness absent elsewhere. The actual artifacts, whether they be cyclotrons or certificates of naturalization, "are inexhaustible wells of significances; every distinguishable feature means not *some* thing but *many* things." It is the role of the historian-curator to "point out to the visitor some well-chosen few of those infinite significances."[37]

TOWARD A DYNAMICS OF AMERICAN PHYSICS

The future is challenging but bright for historians studying American physics. We have at our disposal superb supportive facilities and scholarly resources. While international in their scope, these resources and facilities possess strong American components and, while dependent on a variety of patrons, they owe much to the enthusiasm of present-day American physicists. Leading the list is

---

[36] Bruce S. Eastwood, *Directory of Audio-Visual Sources: History of Science, Medicine, and Technology* (New York: Science History, 1979); John Dowling, "Nuclear War and Disarmament: Selected Film/Video List," *Sightlines,* 1982, *15*(3):19–21; and Dowling, coordinator, *A Cinescope of Physics: An AAPT Film Resource Book* (Stony Brook, N.Y.: American Association of Physics Teachers, 1978); unfortunately, all of these guides are slipping out of date.

[37] Paul Forman, "Exhibiting Modern Physics: Harangue Delivered in the National Museum of American History on Tuesday, January 25, 1983" (MS), pp. 2, 5.

the Center for History of Physics of the American Institute of Physics (AIP), where Spencer Weart, Joan Warnow, and John Aubry are collecting, cataloguing, and interpreting a comprehensive range of source materials. Complementing the center is the Office for History of Science and Technology at the University of California, Berkeley, where John Heilbron and Bruce Wheaton have inventoried a mass of documents relating to twentieth-century physics. We are additionally favored by the newsletters issued by the Center for History of Physics and the new Division of History of Physics of the American Physical Society—a significant institutional asset in itself. Also available are study guides and materials to aid teachers in developing courses on American physics. Finally, there is Kevles's *Physicists*, the single most important work in giving visibility to this field of study.[38]

In recent years the wealth of facilities and resources has helped focus a disproportionate amount of historical attention on American physics. An unfortunate result is the tendency to identify all of American science with American physics, elevating physics to the status of the paradigmatic national science. And because studies of physicist-politicians have dominated recent research, there has been the further tendency to identify American physics with nuclear and high-energy physics. Whereas the surge of scholarship regarding other fields of American science has by now eliminated the mistaken notion of physics as the paragon, there persists the notion that American physics is in essence nuclear and high-energy physics—that it is primarily the realm of physicist-policymakers based in universities affiliated with big-science installations. This brings me back to a key issue raised by critics of Kevles's book: what exactly is the community of American physicists?

The issue can be cast in either a positive or a negative form. Consider first the negative form. Perhaps Kevles's reviewers overlook a more basic point when they complain that he has described only a segment of the American community. Perhaps, in reality, there is no single community. Rather than one monolithic group of practitioners through the century, there have been numerous, shifting, independent groups working in a plethora of technical fields and having a diversity of professional affiliations. The fields have ranged from optics and acoustics to nuclear and solid-state physics; the affiliations have been with universities, private foundations, the government, the military, and industry. Many of us possibly have been misled by the existence of an umbrella organization such as the AIP into assuming that there is a cohesive community with shared characteristics and interests. What we forget is that the founders of the AIP in 1931 had few pretensions about the unity of the five professional societies that came together under its auspices. The AIP was foremost a bureaucratic clearinghouse and economical publishing house for the American Physical Society (APS), the Optical Society of America, the Acoustical Society of America, the Society of Rheology, and the American Association of Physics

---

[38] J. L. Heilbron and Bruce R. Wheaton, *Literature on the History of Physics in the 20th Century* (Berkeley: Office for History of Science and Technology, Univ. California, 1981); Heilbron and Wheaton, *An Inventory of Published Letters to and from Physicists, 1900–1950* (Berkeley: Office for History of Science and Technology, Univ. California, 1982); Roger Stuewer, "Physics and Society in Twentieth-Century America: Study Guide," Dept. of Independent Study, Continuing Education and Extension, Univ. Minnesota, 1981; Stephen G. Brush, "Teacher's Guide: History of Modern Science," History Dept., Univ. Maryland, 1984.

Teachers; it also served to discourage industrial physicists from abandoning the academically biased APS in order to organize a more responsive Society of Applied Physics.[39] Thus historians might find it more appropriate to take as their unit of analysis the community of, for example, solid-state industrial physicists rather than the aggregation of American physicists.

The issue can be inverted and expressed positively. The very diversity in field and affiliation may be the distinguishing feature of the American community of physicists. Many historians in recent decades have suggested that the essence of American science, including physics, is actually contained in the symbiosis of universities, the government, the military, private foundations, and industry. According to this view, the American scientific community is pluralistic, not monistic.

If we believe that there is an all-inclusive community that constitutes the essence of twentieth-century American physics, then we are still left with the task of providing a comprehensive characterization of that community. To date, historians have produced numerous descriptions of various segments of the community over particular spans of time. As we have seen, they are now expanding this research into new and revealing areas, often using novel methodologies and modes of presentation. What remain in short supply, however, are synthetic overviews that critically analyze the dynamics of the relationship between, especially, the academic, governmental, military, and industrial realms of American physics. Other than Kevles's book, the only current attempt at an overview is a retrospective essay by Weart that he prepared for the fiftieth anniversary of the AIP.[40] We need further interpretive analyses—fresh reflections on the fundamental dynamics of the broader community.

In the meantime, it will be enough to ask that we become more fully conscious of the presuppositions behind our research. We must realize that even in this relatively young area of scholarship there exist influential but often tacit historiographic traditions of which we sometimes are unknowing members. Only with this self-knowledge will we avoid exaggerating the life of any one kind of American physicist at the expense of the others.

[39] H. William Koch, "AIP Today—Tomorrow," *Phys. Today*, 1981, *34*(11):235–241, on pp. 235–236; Weart, "Physics Business," p. 321.

[40] Spencer R. Weart, "The Last Fifty Years—A Revolution?" *Phys. Today*, 1981, *34*(11):37–49; this issue is devoted to the AIP's 50-year history.

History of the Social Sciences

By Hamilton Cravens*

A FORCEFUL ANSWER to the question why anyone should be interested in the history of the social sciences in America was provided three decades ago. In his classic *The Organization Man* (1956), William H. Whyte, Jr., instructed an entire generation in the realities of organizational life in America. Whyte insisted that the organization employed the ideas and techniques of the social sciences to resolve tensions between the group and the individual. With equal measures of paternalism and insidiousness, the organization established its hegemony in society and culture. The organization and its social ethic were everywhere: in grade schools and colleges, neighborhoods and suburbs, houses of worship and places of recreation, factories and offices, mass culture and popular government, even in the private worlds of family relations and personal friendships. Whyte argued that the organizational ideal demanded the individual's conformity with and acquiescence to the larger group or collective. The organizational ethos branded individualism as subversive. The organization throttled individuality. The individual had no meaningful existence outside the organization or the group. Thus Whyte attributed much of the responsibility for this crisis of American culture and society, not simply to the organization, but to its appropriation of social science thinking for its own purposes.[1]

Whatever the merits of the particulars of Whyte's indictment, his general thesis provides a useful and exciting point of departure for historical inquiry. The social sciences have played an enormous role in science, society, and culture for much of American history. The social sciences have been concerned with people, not things. They have been concerned with us, our society, economy, polity, and culture. Ultimately they have been invented and formulated as responses to the most profound questions of the relations among human beings, above all of the implications of group identity for individuals in the national population. This does not mean that the social sciences have not been sciences, disciplines, and professions. It does, however, signify that much more has been involved. As useful as a traditional disciplinary focus is, scholars should be prepared to use broader perspectives as well.

In the last quarter century the history of the social sciences has slowly become a recognized field for historians of science and American historians. Within the field there are several journals, a newsletter, and a scholarly organization. Increasingly contributions to the history of various disciplines have appeared with the full apparatus of professional historical scholarship. Today it is not uncommon for historians to write about eugenics, sociobiology, mental testing, and the so-called helping and manipulative professions, for example, topics that bid

* Program in History of Technology and Science, Department of History, Iowa State University, Ames, Iowa 50011.
[1] William H. Whyte, Jr., *The Organization Man* (New York: Simon & Schuster, 1956).

fair to transcend disciplinary lines of the past. Historians of the social sciences have also won considerable recognition from such groups as the Organization of American Historians, which has conferred prizes to several for their books. The field has even attracted attention beyond academe, as, for example, when last fall *The New Yorker* serialized Daniel J. Kevles's new book on eugenics.[2]

Scholars interested in phenomena related to the social sciences have helped legitimate and broaden the field. The distinguished pioneer of American intellectual history, Merle Curti, has recently published a major overview of American ideas of human nature. Such prominent historians of medicine and public health as John C. Burnham and Charles E. Rosenberg have worked on problems involving the social sciences, medicine, and public health. Much of the recent literature on the social sciences has been addressed to the difficult, not to say slippery, problem of professionalism. Historians Roy Lubove and Samuel P. Hays have provided thoughtful discussions of professionalism as involving much more than the establishment of disciplines and professions or the deliberations of a community of expert inquirers. Stimulated in different ways by Bernard Bailyn and Lawrence A. Cremin, some historians of education are asking hard questions about the roles of the social sciences in a new social and cultural history of education very different from the old "foundations of education" tradition. In modern times much history of the social sciences has been intertwined with that of higher education. The pioneering work of Merle Curti and Vernon Carstensen on the University of Wisconsin has been followed up by Bruce Kuklick and others for the liberal arts and professional school universities. When scholars turn to the land-grant and technical institutions and build on Earle D. Ross's classic, this will further enhance the field.[3]

[2] The Organization of American Historians bestowed the Frederick Jackson Turner Prize on Edward A. Purcell, Jr., *The Crisis of Democratic Theory: Scientific Naturalism and the Problem of Value* (Lexington: Univ. Kentucky Press, 1973), and Mary O. Furner, *Advocacy and Objectivity: A Crisis in the Professionalization of American Social Science 1865–1905* (Lexington: Univ. Kentucky Press, 1975); and the Merle Curti Prize on Rosalind Rosenberg, *Beyond Separate Spheres: Intellectual Roots of Modern Feminism* (New Haven: Yale Univ. Press, 1982). The *Journal of the History of the Behavioral Sciences* began publishing in 1965, the *Journal of the History of Sociology* and *Knowledge and Society: Studies in the Sociology of Culture Past and Present*, in 1978; the *History of Anthropology Newsletter* in 1971. Cheiron, the International Society of the History of Behavioral and Social Sciences, has held annual meetings since 1969; it has a multidisciplinary and international membership of 360, about one fourth professional historians. See also Daniel J. Kevles, *In the Name of Eugenics: Genetics and the Uses of Human Heredity* (New York: Knopf, 1985), serialized as "Annals of Eugenics" in *The New Yorker*, 1984, 60(34):51–115; (35):52–125; (36):91–151; (37):51–117.

[3] Merle Curti, *Human Nature in American Thought: A History* (Madison: Univ. Wisconsin Press, 1980). On medicine, public health, and social science, see esp. John C. Burnham, "Psychiatry, Psychology and the Progressive Movement," *American Quarterly*, 1960, 12:457–465; Burnham, *Psychoanalysis and American Medicine, 1894–1918: Medicine, Science, and Culture* (Psychological Issues, Vol. 5, No. 4, monograph 20) (New York: International Universities Press, 1967); and Charles E. Rosenberg, *The Trial of the Assassin Guiteau: Law and Psychiatry in the Gilded Age* (Chicago: Univ. Chicago Press, 1968); see also Barbara G. Rosenkrantz, *Public Health and the State: Changing Views in Massachusetts, 1842–1936* (Cambridge, Mass.: Harvard Univ. Press, 1972). On professionalism, see Roy Lubove, *The Professional Altruist: The Emergence of Social Work as a Career 1880–1930* (Cambridge, Mass.: Harvard Univ. Press, 1965); Samuel P. Hays, *Conservation and the Gospel of Efficiency: The Progressive Conservation Movement, 1890–1920* (Cambridge, Mass.: Harvard Univ. Press, 1959). On American public education, see Bernard Bailyn, *Education in the Forming of American Society* (Chapel Hill: Univ. North Carolina Press, 1960); and Lawrence A. Cremin, *The Transformation of the School: Progressivism in American Education, 1876–1957* (New York: Knopf, 1961). On higher education see Merle Curti and Vernon Carstensen, *The University of Wisconsin: A History, 1848–1925*, 2 vols. (Madison: Univ. Wisconsin Press, 1949); Laurence A.

The field presents challenging possibilities for American historians and historians of science. American historians attracted to the "organizational synthesis," for example, can find rich fare indeed in the history of the social sciences. Those interested in modern social thought and the development of public and social policy will find ample new opportunities. Nor does this exhaust the list of potential topics for American historians. The social sciences are no less fascinating for historians of science. The social sciences have interacted with the traditional physical and social sciences, as, for instance, in the complex relations between biological and social science since Darwin. For social and cultural historians of science and technology, not to mention those interested in science, technology, and social studies, the social sciences offer much. In certain respects the social sciences present us with a different range of conceptual issues than other sciences. Arguably the term *social science* is a misnomer. The social sciences could be rechristened the *social technologies*. Most have had a heavily applied orientation. For the most part they have been invented and developed in response to social and public policy concerns.[4]

As a field the history of the social sciences appears undeveloped. There is no paucity of books. Social scientists have been writing the history of their disciplines since the 1920s. Much of this practitioner genre is useful: it can provide information and perspectives difficult to find elsewhere. Furthermore, much recent practitioner literature easily meets the professional historian's standards of research and interpretation. But the available information is skewed in certain ways. The overwhelming majority of contributions have been written from within the framework of traditional disciplinary history. Works on anthropology, psychology, and sociology so dominate the field that it is difficult to find work on such staple social sciences as economics, geography, and political science, not to mention such lesser-known applied social sciences as home economics, urban planning, industrial engineering, agricultural economics, child development, industrial engineering, rural sociology, public administration, or agricultural engineering. Undoubtedly a broad definition of what disciplines have been among the social sciences would be useful. Even more helpful would be many more works than currently exist that transcend disciplinary boundaries and focus on larger problems in science, society, and culture. In a word, the field has not gelled intellectually. There is no framework, descriptive or interpretive. It is no

Veysey, *The Emergence of the American University* (Chicago: Univ. Chicago Press, 1965); Bruce Kuklick, *The Rise of American Philosophy: Cambridge, Massachusetts, 1860–1930* (New Haven: Yale Univ. Press, 1977); Winton U. Solberg, *The University of Illinois, 1867–1894: An Intellectual and Cultural History* (Urbana: Univ. Illinois Press, 1968); Earle D. Ross, *A History of the Iowa State College of Agriculture and Mechanic Arts* (Ames: Iowa State College Press, 1942). On science and technology in the land-grant college setting, see Alan I Marcus and Erik Lokensgard, "Greater Than the Sum of Its Parts: Chemical Engineering, Agricultural Wastes, and the Transformation of Iowa State College, 1920–1940," *Annals of Iowa,* forthcoming.

[4] On the "organizational synthesis" see Samuel P. Hays, *The Response to Industrialism 1885–1914* (Chicago: Univ. Chicago Press, 1957); Robert H. Wiebe, *The Search for Order, 1877–1920* (New York: Hill & Wang, 1967); and Louis P. Galambos, "The Emerging Organizational Synthesis in Modern American History," *Business History Review,* 1970, 44:279–290. On discussions of the interaction between the social sciences and social thought, see Stow Persons, *American Minds: A History of Ideas,* rev. ed. (1958; Huntington, N.Y.: Krieger, 1975); on the biological and social sciences, see, e.g., Cynthia Eagle Russett, *The Concept of Equilibrium in American Social Thought* (New Haven: Yale Univ. Press, 1966); and Hamilton Cravens, *The Triumph of Evolution: American Scientists and the Heredity-Environment Controversy, 1900–1941* (Philadelphia: Univ. Pennsylvania Press, 1978).

simple task to supply authoritative answers to many routine questions of fact. The construction of broader generalizations is accordingly more difficult.

A critical overview of the field can best be realized if it is understood that the issues of disciplinary history, important as they are, are secondary to larger areas of inquiry, such as the invention and use of knowledge in society and culture and the constituencies and merchandisers of such knowledge. Apparently the social sciences have been important because they have been concerned with the most profound questions of social taxonomy and public policy. As such general conceptions have changed over time, so have the social sciences and the history of the various social science "disciplines."

"DEMOCRATIC" SOCIAL SCIENCE

The social sciences first took shape in America in the late 1830s and early 1840s. They emerged as a new and distinctive way of perceiving and defining the nature and arrangement of groups in the social order, and of proscribing the implications of group identity for the individual. As American society and culture underwent a "democratic" transformation for white, middle-class, Protestant males of British ancestry, questions of what democracy meant and how far it extended came to the fore. This transformation of social thought meant recognition of the group in the American social order; prior social thought had focused on the individual as a member of civilization who was free to pursue commerce and enterprise.[5] Many scholars have noted this dramatic shift in American social attitudes, whether indicated in ideas or in actions, from the "individualism" and "egalitarianism" of the Revolutionary era to the hard-bitten group consciousness of the mid-nineteenth century. As many scholars have argued, the mid-century group consciousness assumed that differences in color of skin, religious faith, and gender—differences often signified by physical or biological characteristics and manifested in different behavioral patterns—had enormous consequences for the social order and, therefore, for public and social policy. Abolitionism offers a good example of this change in attitude. Until the 1830s, those abolitionists in the organized movement believed in colonization. They considered all blacks, free or slave, as Africans and, therefore, incapable of ever fitting into the civilized society that was America. On the other hand, the new abolitionists of the 1830s, whether we refer to the Garrisonians or their competitors in the movement, defined slaves as members of a group, blacks, that constituted a natural part of the American social order and whose enslavement was both a moral blight upon the republic's fair reputation and an affront to Christianity. The colonizationists' conception of blacks, free or slave, as alien Africans soon dissipated. By the early 1840s slaves were widely perceived as a group in society that constituted a social problem, just as free blacks in the North found that they had become redefined by the white majority as a troublesome group—with the enactment of many so-called Jim Crow laws in the 1830s and later.[6]

[5] A perceptive account of this transformation of social thought is Alan I Marcus, "In Sickness and In Health: The Marriage of the Municipal Corporation to the Public Interest and the Problem of Public Health, 1820–1870: The Case of Cincinnati," (Ph.D. diss., Univ. Cincinnati, 1979).

[6] See, e.g., Phillip J. Staudenhaus, *The African Colonization Movement, 1815–1861* (New York: Columbia Univ. Press, 1961); Louis Filler, *The Crusade Against Slavery, 1830–1860* (New York: Harper & Row, 1970); and Leon W. Litwack, *North of Slavery: The Negro in the Free States,*

According to the new social thought, group identity had certain implications. Individuals could rise to the full potential of the group to which they belonged. Whether they could transcend their group identity was problematic. Perhaps white male immigrants or non-Protestants could. Certainly, however, woman operated in a different sphere from man. An Uncle Tom might be a Christ-like figure, but he was nevertheless black. As Charles Rosenberg and George W. Stocking, Jr., have noted, the new group consciousness of this period was optimistic about the possibilities of individual improvement, but to a point. If the individual exerted himself or herself so much that the physical constitution of organs were altered, there was a corresponding change in behavior reflecting mental and moral causation. Presumably such changes would be beneficial for the individual, although there was advice aplenty about the permanent consequences of "drink storms" and the "solitary vice." Many Americans arrived at conclusions about individual development and self-improvement that closely resembled certain aspects of ideas popularized by Lamarck.[7] The medium of morality and culture was, then, biological structure, including the brain as an organ. In a self-proclaimed fluid, democratic, and free society, nineteenth-century Americans, as Neil Harris has reminded us from a somewhat different perspective, were fascinated by the facts, routine and bizarre, of biological structure, and their importance for moral behavior.[8]

By any reasonable yardstick, phrenology was a mid-century social science. In one incarnation or another, phrenology lasted in Europe for the better part of a century. It had both great vogue and fascinating affiliations. Phrenologists helped disseminate the idea that the brain was the organ of the mind, one of the nineteenth century's great discoveries and one that has become a part of modern natural and social science.[9] Phrenology came into its own in America only after the late 1830s, but by mid century it was a roaring business, at least for the Fowler brothers, who did so much to merchandise it. As John D. Davies argued thirty years ago, the doctrines of Franz Gall and G. Spurzheim underwent considerable transformation in America. If phrenology did not have the widespread impact that Davies and a more recent historian, Madeline B. Stern, have implied, nevertheless Americans considered it a "social science" devoted to social and individual amelioration. The Fowler brothers and other phrenologists insisted that environmentally induced adaptations in the brain would change

1789–1860 (Chicago: Univ. Chicago Press, 1961). On parallel shifts in majority attitudes towards Indians, immigrants, and women, see Robert F. Berkhofer, *The White Man's Indian: Images of the American Indian From Columbus to the Present* (New York: Knopf, 1978); Ray Allen Billington, *The Protestant Crusade 1800–1860: A Study of the Origins of American Nativism* (New York: Macmillan, 1938); and Carl N. Degler, *At Odds: Women and the Family in America from the Revolution to the Present* (New York: Oxford Univ. Press, 1980).

[7] Charles E. Rosenberg, "The Bitter Fruit: Heredity, Disease, and Social Thought," *Perspectives in American History,* 1974, 8:189–235, rpt. in Charles E. Rosenberg, *No Other Gods: On Science and American Social Thought* (Baltimore: Johns Hopkins Univ. Press, 1976), pp. 25–53; George W. Stocking, Jr., "Lamarckianism in American Social Science: 1890–1915," *Journal of the History of Ideas,* 1962, 23:239–256, rpt. with some changes in Stocking, *Race, Culture and Evolution: Essays in the History of Anthropology* (New York: Free Press, 1968), pp. 234–269.

[8] Neil Harris, *Humbug: The Art of P. T. Barnum* (Boston: Little, Brown, 1973). See also Carroll Smith-Rosenberg and Charles E. Rosenberg, "The Female Animal: Medical and Biological Views of Woman and Her Role in Nineteenth-Century America," *Journal of American History,* 1973, 60:332–356.

[9] See Edwin G. Boring, *A History of Experimental Psychology,* 2nd ed. (New York: Appleton-Century-Crofts, 1950), pp. 50–60.

behavior—Lamarckism, as it were, for the bumps on the head. Phrenology was thus both a method of social analysis based on the anatomy of the brain—one postulating, increasingly, different types of brains for the different groups in society—and a way of reforming the circumstances of the human race.[10]

Phrenology belongs to a group of mid-century "sciences" and "crusades." Some have been studied, like diet reform, others are largely forgotten or ignored by historians, including mesmerism, animal magnetism, and hydropathy (nor can it be said, for that matter, that we have enough studies of phrenology or even of dietary reform). But just because some phenomena in a past age seem, from a present perspective, not to have led to developments in our own time is no reason to slight them. Many of these movements attracted many followers who thought their doctrines eminently "scientific." Here seems an excellent opportunity for examining the problems of "professional" and "popular" science in new ways.[11]

Another mid-century social science was scientific polygenism, a body of doctrines that shared many assumptions with phrenology. Both stressed the importance of biological factors, or the organic and physical basis of mind and culture. Both arose in the later 1830s as fully articulated movements, as responses to the new perception that society was constituted of biologically defined groups. Both assumed Lamarckian formulae. Thanks to William Stanton's work, the existence of American polygenism is widely recognized in the literature. Known then on both sides of the Atlantic as the "American School," American polygenists were led by Samuel G. Morton, a Philadelphia physician and anatomist, whose *Crania Americana* (1839) launched the school. Their fundamental argument was that the various races of man had been created separately. The evidence they used focused on measurements of skulls and, hence, of brains, reminiscent certainly of phrenology. Stanton has insisted that the American school's members, save for the notoriously proslavery Southern doctor Josiah C. Nott, took the positions they did for "scientific" rather than "social" reasons. Yet that is a more difficult distinction to make than might first appear, especially for so complex and emotive an issue as race. There is considerable evidence that "social" beliefs in racial inferiority and "scientific" beliefs in polygenism were more commonly correlated. Louis Agassiz, who joined the group, was repelled by blacks, for example. If Stanton can argue that polygenism was unattractive to proslavery Southerners because it contradicted the Book of Genesis, that contradiction did not prevent Nott and others from peddling polygenism in the South as an intellectual prop for slavery.[12]

[10] See John D. Davies, *Phrenology: Fad and Science: A Nineteenth-Century American Crusade* (1955,; 2nd ed., Hamden, Conn.: Archon, 1971); and Madeline B. Stern, *Heads and Headlines: The Phrenological Fowlers* (Norman: Univ. Oklahoma Press, 1971).

[11] Ronald L. Numbers, *Prophetess of Health: A Study of Ellen G. White* (New York: Harper & Row, 1976); Stephen Nissenbaum, *Sex, Diet, and Debility in Jacksonian America* (Westport, Conn.: Greenwood, 1980); and James Whorton, *Crusaders for Fitness: A History of American Health Reformers, 1830–1920* (Princeton, N.J.: Princeton Univ. Press, 1982). Alice Felt Tyler, *Freedom's Ferment: Phases of American Social History From the Colonial Period to the Outbreak of the Civil War* (New York: Harper & Row, 1944; Torchbook ed., 1962), mentions some of these movements in passing.

[12] William Stanton, *The Leopard's Spots: Scientific Attitudes Towards Race in America 1815–1859* (Chicago: Univ. Chicago Press, 1960); see also Frank Spencer, "Samuel G. Morton's Doctoral Thesis on Bodily Pain: The Probable Source of Morton's Polygenism," *Transactions and Studies of the College of Physicians and Surgeons of Philadelphia*, 1983, Series 5, 5(4):321–338; and, on Agassiz, Stephen Jay Gould, *The Mismeasure of Man* (New York: Norton, 1981), pp. 42–50.

Moreover, modern scholarship has shown that racism was hardly unique to the South, as Stanton seemed to assume. The vast majority of American whites then and later firmly believed in the hierarchy of races, whether they invoked the authority of science or not. Scientific polygenism as support for the pro-slavery argument was in that sense superfluous and presumably used only by those who found the idioms of science congenial. Works by two distinguished historians have shown that a large body of popular racial doctrines, some inspired by contemporary science, some not, emerged after the mid 1830s. In *The Black Image in the White Mind,* George M. Fredrickson has cogently shown that a large number of popularizers besides Nott widely disseminated the American school's arguments on race. He has also demonstrated assumptions in polygenism that paralleled ideas of black inferiority in popular literature and in such political doctrines as Free Soil. More recently, Reginald Horsman has studied the role of popular and scientific racism in the development of American expansionism from the Revolution to the Civil War. Based on meticulous research and precise analysis, his discussions of the American school, of "Anglo-Saxonism," of racial attitudes as disseminated from scientists to the public, and of the mechanisms of popularization are particularly illuminating. Fredrickson and Horsman provide satisfying examples of how the history of science can be integrated with that of the larger culture and society. Fredrickson's work undermines the notion that the "science" of polygenism was somehow isolated from society and culture. Both authors have also shown how polygenism was related to other popular movements. Horsman directly links polygenists and phrenologists. If most polygenists were not avowed phrenologists, many of them, including Morton, were interested in phrenology as a science. And many phrenologists, including the Fowler brothers, believed in polygenism. Polygenist doctrines may have been more widely accepted, especially among mid-century doctors and anatomists, than has been thought. Morton's work may simply have been the most noted of a larger genre of thought that was generated by medical "scientists," especially if they were not "regular" doctors.[13]

Yet another "discipline" that requires more investigation is anthropology, if anthropology indeed was distinct from polygenism. For this period, a disciplinary orientation to historical conceptualization may obscure more than it illuminates. Thus it is unclear whether we can speak of a discipline of anthropology in the mid-nineteenth century. Individuals did address problems that might be defined as anthropological, but this often meant they worked alone, despite the widespread currency of such dogmas as polygenism.[14] Consider the

---

[13] George M. Fredrickson, *The Black Image in the White Mind: The Debate on Afro-American Character and Destiny, 1817–1914* (New York: Harper & Row, 1971); and Reginald Horsman, *Race and Manifest Destiny: The Origins of American Racial Anglo-Saxonism* (Cambridge, Mass.: Harvard Univ. Press, 1981). Of many fine books on white attitudes towards blacks, see, e.g., Winthrop D. Jordan, *White Over Black: American Attitudes Toward the Negro, 1550–1812* (Chapel Hill: Univ. North Carolina Press, 1968); David Brion Davis, *The Problem of Slavery in the Age of Revolution, 1770–1823* (Ithaca: Cornell Univ. Press, 1975); August Meier and Elliott Rudwick, *From Plantation to Ghetto* (New York: Hill & Wang, 1966; rev. ed., 1976); John Hope Franklin, *From Slavery to Freedom* (1947; 5th ed., New York: Knopf, 1980); and Thomas F. Gossett, *Race: The History of an Idea in America* (Dallas: Southern Methodist Univ. Press, 1963).

[14] Thus John C. Greene, in *American Science in the Age of Jefferson* (Ames: Iowa State Univ. Press, 1984), pp. 320–408, has shown there were individuals working on problems that we might retrospectively label physical anthropology, archaeology, and comparative linguistics well before the 1820s. What is unclear is whether this led to the crystallization of a community of investigators who invented research traditions.

example of Lewis Henry Morgan. We know little of Morgan, save that his work was apparently ignored in his own day but has been influential in modern times. George W. Stocking, Jr., insists that Morgan's schemes of mental evolution had a profound impact upon John Wesley Powell and his colleagues at the Bureau of American Ethnology. Two distinguished anthropologists have seen Morgan's contributions to the science of anthropology differently. The British social anthropologist Meyer Fortes has argued that Morgan had a protean, if delayed, influence on British social anthropology. Morgan's greatest scientific discovery, Fortes insisted, was that the custom of designating relatives has scientific significance. When W. H. Rivers "discovered" Morgan's work around 1900, this enabled him and his colleagues to develop British structural or synchronic anthropology, which was quite different from the cultural or diachronic anthropology that Franz Boas established in early twentieth-century America. But Marvin Harris has claimed that Morgan's importance was in underlining the techno-economic environment as the proper subject matter for the science of anthropology, a position that Harris himself has elaborated.[15] What was the importance of individuals such as Morgan, or Henry Schoolcraft, for that matter, and other "anthropologists" in nineteenth-century America? Did they form or create a science? Obviously more work can be done in this area.

A final example of a mid-century social science was the so-called moral treatment used on the insane. Moral treatment had certain affinities with phrenology and polygenism. Advocates of moral treatment assumed that mental disease resulted directly from physical causes, from the disease and distortion of organs, especially the brain. Ultimately the larger causes were mental and moral, within the individual. Champions of moral treatment believed, as a later generation did not, that bad habits or diseased structures could be modified and even cured if the individual exerted himself or herself sufficiently. A systematic program could be effective therapy. Moral treatment thus had important intellectual affinities with Lamarckian ideas.

In contrast to phrenology and polygenism, the history of the mental hospital and of moral treatment has become a topic of considerable controversy in recent years. Unfortunately, more heat than light has been generated. The central issue is whether those in charge of the asylum had good intentions or not. Perhaps the most able interpretation by a modern mental health professional is that of Albert Deutsch, who chronicled the history of American mental hospitals as the victory of enlightened values over the callous, unscientific opinions of the past. Deutsch viewed the mid-nineteenth century asylum as a vast improvement over the past, yet characterized that institution and the therapy of moral treatment as too flawed to survive in the increasingly scientific world that came into existence in the later nineteenth and early twentieth centuries. In *Concepts of Insanity in the United States, 1789–1865* (1964), the historian Norman Dain portrayed the development of what he dubbed psychiatric thought in America. In his account, a new conception of mental illness arose in the mid 1830s. The

---

[15] The standard study is Carl Resek, *Lewis Henry Morgan: American Scholar* (Chicago: Univ. Chicago Press, 1960); see also Stocking, *Race, Culture, and Evolution* (cit. n. 7), pp. 116–117; Meyer Fortes, *Kinship and the Social Order: The Legacy of Lewis Henry Morgan* (Chicago: Aldine, 1969); Marvin Harris, *The Rise of Anthropological Theory: A History of Theories of Culture* (New York: Thomas Y. Crowell Company, 1969), esp. pp. 180–188; and Harris, *Cultural Materialism: The Struggle for a Science of Culture* (New York: Random House, 1979).

mentally ill were no longer individuals; they belonged to a group—the insane. The therapy of moral treatment included somatic analysis, on the assumption, not far removed from the doctrines of phrenology or polygenism, that "insanity was a physical disorder of the brain that manifested itself in psychological symptoms." Furthermore, moral therapy was designed explicitly to help the patient overcome his or her disease by well-defined and regimented physical activity that would in time lead to release (if not cure). Dain's account was in many respects eminently useful, well researched, judicious in its interpretations, and certainly open in pointing to the failures and lapses of the mental hospitals, not simply in the occasionally brutal treatment of any inmates, but also in the systematic discrimination against the poor, nonwhites, and immigrants. And Dain recognized that the origins of moral treatment lay in the larger culture as well as in "science."[16]

More recently social critics have attacked the nineteenth-century asylum, standing Deutsch's enlightened-practitioner interpretation on its head. They see the asylum as an agent of oppression and social control of the deviant, invented by the capitalistic class in its drive for hegemony and dominance. From this point of view, moral therapy receives short shrift as a mere rationalization or sham. It is easy to oversimplify the critics' positions. Genuine qualitative differences exist among them in the sophistication and tightness of their arguments. They are more easily classified by their abhorrence of the asylum than by anything else. Science, medicine, practitioners, helping institutions, and the like are viewed as the problem, not the solution. More crudely, the system is bad; ineluctable social forces produce bad people in positions of authority. Much of the criticism of the nineteenth-century asylum stems from presentist social policy commitments and values.[17]

In the 1970s, historians David J. Rothman and Gerald N. Grob debated this issue from within the framework of the new social history. Rothman saw the asylum as a harsh, repressive instrument of social control invented by those in power, who were frightened by the rapid social changes of Jacksonian society, particularly immigration, urbanization, and industrialization. Grob agreed that the asylum had to cope with those same social problems, which, in the end, distorted and engulfed the asylum. In some sense both Rothman and Grob defined history as a morality play. Rothman's practitioners wore the blackest of hats; Grob's *dramatis personae* had, if not white hats, then certainly ones no darker than medium gray. In most respects Grob had the better of the argument. Rothman interpreted the therapy of moral treatment as harsh, rigid discipline. The fact that in the state asylums this therapy eventually deteriorated into unconscionable and tragic treatment for many inmates, especially the less fortunate

---

[16] Albert Deutsch, *The Mentally Ill in America* (New York: Columbia Univ. Press, 1937; 2nd ed., 1949); and Norman Dain, *Concepts of Insanity in the United States, 1789–1865* (New Brunswick, N.J.: Rutgers Univ. Press, 1964), quoting p. 84. See also Dain, *Disordered Minds: The First Century of Eastern State Hospital in Williamsburg, Virginia, 1766–1866* (Williamsburg, Va.: Colonial Williamsburg Foundation, distributed by Univ. Virginia Press, 1971), which substantiates the general point of the shift from the individual to the group in social thought and action and the particular notions of that world view in terms of the implications of group identity for the individual.

[17] Nancy Tomes, *A Generous Confidence: Thomas Story Kirkbride and the Art of Asylum Keeping, 1840–1883* (New York: Cambridge Univ. Press, 1984), pp. 1–18, ably summarizes the controversy and provides citations to other discussions; see also Lawrence J. Friedman, "The Demise of the Asylum," *Reviews in American History*, 1984, 12:241–247.

and minorities, does not support Rothman's Manichean derivation of the motives of individuals. Grob's research has been more intensive, precise, and wide-
ranging and his analyses more judicious. Moreover, he has not argued with the
past, and he has portrayed individuals as honorably motivated. His criticisms
of the social control argument have been cogent. Yet it may be wondered how
important individual motives are in a historical interpretation that assumes ineluctable social forces.[18]

In *A Generous Confidence: Thomas Story Kirkbride and the Art of Asylum-
Keeping, 1840–1883*, Nancy Tomes has viewed the past on its own terms and
made a superb contribution to our understanding of the private asylum, its relations with its clients and patients, the implementation of moral treatment, the
structural problems and dilemmas of the asylum as institution, treatment of inmates, the profession of practitioners, and the circumstances that led to the deterioration of the asylum. Tomes's richly researched, perceptively argued, and
judiciously constructed interpretation is far too complex to summarize here. For
present purposes, she has discussed and analyzed the therapy, the asylum's reg-
imen, the relations between administrator, doctor, patient, and patient's family,
and related matters in an eminently comprehensible manner. She has moved
beyond the ideological horizons of the controversy and has made the past come
alive as the past. Unlike some social historians active today, she is sensitive to
the nuances of ideas.[19]

Between the 1830s and the 1870s a new style of social thought took shape in
American culture and society. It emphasized two levels of description and analysis that were related to the group and the individual. One stressed the taxo-
nomic hierarchy of superior and inferior groups in the national population. The
other underlined the implications of group identity for the individual. To what
extent Lamarckian formulae were widespread is difficult to say. But evidently
a common way of looking at issues of group and individual worth appeared in
several different "social sciences." Clearly those social sciences had useful applications and therapies, and important implications for public and social policy,
insofar as many Americans of that era were concerned.

---

[18] David J. Rothman, *The Discovery of the Asylum* (Boston: Little, Brown, 1971); Gerald N.
Grob, *The State and the Mentally Ill: A History of the Worcester State Hospital in Massachusetts,
1830–1920* (Chapel Hill: Univ. North Carolina Press, 1966); Grob, *Mental Institutions in America:
Social Policy to 1875* (New York: Free Press, 1973); and Grob, *Edward Jarvis and the Medical World
of Nineteenth-Century America* (Knoxville: Univ. Tennessee Press, 1978). Also of interest is Grob's
devastating response to Rothman, "Rediscovering Asylums: The Unhistorical History of the Mental
Hospital," *Hastings Center Report*, 1977, 7(4):38–41. A critical review of Rothman is Jacques Quen,
The Journal of Psychiatry and Law, 1974, 2(1):105–122. A more concrete and subtle version of the
social-control thesis is Richard Fox, *So Far Disordered in Mind: Insanity in California, 1870–1930*
(Berkeley/Los Angeles: Univ. California Press, 1978); Fox has studied the process of commitment,
using statistical analysis of court records, and concluded that commitment occurred to rid society
of persons of deviant ideas and behavior.
[19] Tomes, *A Generous Confidence*, passim. Tomes's real contribution has been to find a satisfy-
ingly interdisciplinary way of writing social history of medicine and science. On mental asylums in
Britain, see Rosen, *Madness in Society: Chapters in the Historical Sociology of Mental Illness* (New
York: Harper & Row, 1968), pp. 247–330; William Parry-Jones, *The Trade in Lunacy: A Study of
Private Madhouses in England in the Eighteenth and Nineteenth Centuries* (London/Boston: Rout-
ledge & Kegan Paul, 1972); and Andrew Scull, *Museums of Madness: The Social Organization of
Insanity in Nineteenth-Century England* (New York: St. Martin's Press, 1979). See also Michel Fou-
cault, *Madness and Civilization: A History of Insanity in the Age of Reason*, trans. Richard Howard
(New York: Pantheon Books, 1965).

THE PROFESSIONAL IDEA

Most scholars would agree that the social sciences underwent a major transformation in the five decades following Appomattox. They became national professions. In this period all manner of national professional, trade, advocacy, and social policy institutions were founded or recreated. Growing numbers of Americans believed that local institutions and perspectives could no longer solve their problems, and that national agendas, analyses, and remedies were necessary. In the nineteenth century's closing decades, America suddenly seemed no longer local.

These changes affected the social sciences no less than other aspects of American civilization. The social sciences developed the full apparatus of the national professional subculture, not always smoothly to be sure. Usually it was the rapidly changing system of higher education that gave the new professionals their base of operations. Other institutional loci of the social scientists included governmental agencies, business enterprise, charitable organizations, and the new foundations. Professional and learned societies, heretofore national in name at most, now began to function as national organizations. The new professionals founded or reestablished journals, newsletters, and programs of selection and training to serve these larger national goals.[20]

The professionals developed much more than a code of manners or implied programs of social mobility, as Burton Bledstein seems to imply. They sought to achieve power and status in a national arena by claiming license to or, if possible, a monopoly over a defined body of knowledge.[21] First they created their profession, its institutions and intellectual horizons. Then they invented specialized disciplines of knowledge. From the large store of old facts in Europe and America they plucked those most congenial and comprehensible, as if to reaffirm the importance of tradition. They also invented or discovered new facts, to emphasize that knowledge was progressive. Implied in the new notion of a specialized discipline was the assumption that it was independent of and no less scientific than any others.

The social science professionals developed specific positions on the taxonomy of society, and of groups and individuals that seemed rigorous and scientific,

---

[20] On this shift see, for the social science professions, Lubove, *The Professional Altruist,* and Hays, *Conservation and the Gospel of Efficiency* (both cit. n. 3); for the general trend, see Wiebe, *The Search for Order* (cit. n. 4); and John A. Garraty, *The New Commonwealth, 1877–1890* (New York: Harper & Row, 1968). See also Hamilton Cravens, "American Science Comes of Age: An Institutional Perspective, 1850–1930," *American Studies,* 1976, *17*:49–70; Alan I Marcus, "Professional Revolution and Reform in the Progressive Era: Cincinnati Physicians and the City Elections of 1897 and 1900," *Journal of Urban History,* 1979, *5*:183–207; Marcus, "Disease Prevention in America: From a Local to a National Outlook, 1880–1910," *Bulletin of the History of Medicine,* 1979, *53*:184–203; and Martin Lazerson, *Origins of the Urban School: Public Education in Massachusetts, 1870–1915* (Cambridge: Harvard Univ. Press, 1971).

[21] See Burton J. Bledstein, *The Culture of Professionalism: The Middle Class and the Development of Higher Education in America* (New York: Norton, 1976); for the professions as a fitting marketplace model, see Magali Sarfatti Larson, *The Rise of Professionalism: A Sociological Analysis* (Berkeley: Univ. California Press, 1977), a work superior to many other sociological treatments, even that of Joseph Ben-David, *The Scientist's Role in Society* (Englewood Cliffs, N.J.: Prentice-Hall, 1971). On the place of research, see Robert E. Kohler, *From Medical Chemistry to Biochemistry: The Making of a Biological Discipline* (New York: Cambridge Univ. Press, 1982); and Ronald C. Tobey, *Saving the Prairies: The Life Cycle of the Grasslands School of Ecology, 1895–1955* (Berkeley/Los Angeles: Univ. California Press, 1981).

but which in fact borrowed much from public and social policy discussion in society and culture. Questions arose of the segregation of "superior" from "inferior" groups in the national population, as with the backlash against the "new" immigrants, the reassertion of the separate spheres of male and female activity, and with the drawing of the color line. The increasing emphasis on race, nationality, religion, and gender meant an intensification of the notion that groups differed in moral and social value. There was a new emphasis, not simply on biology as a key to culture, but upon pessimistic and deterministic formulations of that bromide. Several historians have argued that in this era doctrines of biological determinism became widely accepted and expressed in social thought and action.[22] The hierarchy of superior and inferior groups in the national population took on determinist implications for immigrants as well as nonwhites. Group identity foretold an individual's fate, and individual self-improvement now appeared virtually impossible. Models of biological and social inheritance invoked Lamarckian idioms, depicting the individual as a member of a group or type with predetermined and shared group characteristics. What seemed somewhat fluid in an earlier age seemed so no longer.

For some historians of the social sciences the most important shift in this period was that from social science to social science*s*. Thomas L. Haskell has argued that the gradual disintegration of the American Social Science Association between 1865 and 1909 represented a change in the locus of social authority from the older genteel elites, who believed in a unified science of society, to the new academic professionals, chiefly academic historians, economists, and sociologists, who insisted that their disciplines had proved the general premise that society is interdependent. In an interdependent society, Haskell argues, only the professional social scientists possessed sufficient expertise to interpret and explain to laity how and why the various parts of society were interdependent and constituted a whole. While Haskell discusses the ASSA's breakup, Mary Furner takes up its consequences. She discussed the development of social-science professionalism largely by examining the tensions generated among the new academics (chiefly economists) over "advocacy" and "objectivity" as they appropriated social science from those who had dominated the ASSA. By tracing a series of academic freedom cases, Furner showed how a conservative reaction against "liberal" economists imposed professional discipline upon the new social science professoriate. Ideological extremes were abandoned. Centrist politics and "objective" social science became the new professional credo.[23]

Haskell and Furner have provided valuable accounts of the shift from the social science movement to the new academic disciplines and professions. More work could be done. Haskell's discovery of the pervasiveness of the notion of the interdependence of the elements of society is helpful. The idea of interdependence, however, was not a generalized abstraction that never changed. Conceptions of the interdependence of the elements, or groups, of society, of the

[22] Persons, *American Minds* (cit. n. 4), pp. 237–365; Stocking, *Race, Culture, and Evolution* (cit. n. 7), pp. 42–68, 110–132, 234–269; and Rosenberg, *No Other Gods: On Science and American Social Thought* (Baltimore: Johns Hopkins Univ. Press, 1976), pp. 25–53.

[23] Thomas L. Haskell, *The Emergence of Professional Social Science: The American Social Science Association and the Nineteenth-Century Crisis of Authority* (Urbana: Univ. Illinois Press, 1977); Furner, *Advocacy and Objectivity* (cit. n. 2). On the gentry, see, e.g., Stow Persons, *The Decline of American Gentility* (New York: Columbia Univ. Press, 1973).

relationships between the parts and the whole, have permeated the social sciences since their inception. Furner's portrait of social scientists donning the cloak of professionalism to practice in society's marketplace as best they could seems more satisfying than Haskell's notion of a community of inquirers. And, according to Furner, some rambunctious or woolly-headed economists apparently required external threats and internal sanctions to be professionally disciplined. Yet instances of dissent, let alone rebellion, were relatively rare in comparison with the great number who accepted and acquiesced. For most professionals, no threats of dismissal or isolation were necessary; working within the system was precisely what the new professionalism was about. A comparison of the fate of the ASSA with the American Association for the Advancement of Science and the engineering societies, moreover, suggests a general trend towards a conception of specialization in one discipline with the professionals in each redefined as members of a national profession.

As Dorothy Ross has suggested, the new professionals were not obscurantists peddling arcane knowledge. They appealed to those popular bromides, scientism and progress, to legitimate their status in society from the relatively safe havens of academe. With the possible exception of the economists, the social scientists had relatively little subject matter of their own to offer. Here we need to know far more about why research became so important to the new professionals, and what that research meant to those who ran colleges and universities, and to its other consumers.[24]

Of all the new or reinvented social sciences of this period, American anthropology had the smallest potential for merchandizing its wares to external constituencies.[25] The issues anthropologists dealt with, such as race, biological and social heredity, and anthropometry, were important to many Americans. And their views on these matters were firmly established, as Franz Boas was to discover repeatedly in his career. In the late nineteenth century, anthropology's institutional base was in museums and in the federal government, including the Smithsonian Institution and the Bureau of American Ethnology, under the leadership of John Wesley Powell and W. J. McGee. In his superlative portrait of anthropology at the Smithsonian, Curtis M. Hinsley, Jr., has shown that its practitioners found anthropology far closer in meaning and purpose to natural history and the wondrous regularities of nature than to a social science. Nor was the government anthropologists' notion of "professionalism" the same as that of the academic social scientists. As Hinsley and George W. Stocking, Jr., have ably shown, it was a commitment to the study of nature and man's place in it, more akin to the localized and general scientific professionalism of an earlier age than to the cosmopolitan or national programs of the new academic specialists.[26]

[24] See Dorothy Ross, "The Development of the Social Sciences," *The Organization of Knowledge in Modern America, 1860–1920*, ed. Alexandra Oleson and John Voss (Baltimore: Johns Hopkins Univ. Press, 1979), pp. 107–138; Ross, "American Social Science and the Idea of Progress," in *The Authority of Experts*, ed. Thomas L. Haskell (Bloomington: Indiana Univ. Press, 1984); and Ross, "Historical Consciousness in Nineteenth-Century America," *American Historical Review*, 1984, 89:909–928; see also Cravens, *The Triumph of Evolution* (cit. n. 4), Chs. 1 and 2.

[25] Anthropology had more potential in other cultures; see, e.g., Henrika Kuklick, "The Sins of the Fathers: British Anthropology and African Colonial Administration," *Research in Sociology of Knowledge, Sciences, and Art*, 1978, 1:93–119.

[26] Curtis M. Hinsley, Jr., *Savages and Scientists: The Smithsonian Institution and the Develop-

Yet late nineteenth-century anthropology had a social message, one drawn from contemporary Lamarckian formulations of evolutionism, as Hinsley, Stocking, and John S. Haller, Jr., have argued. Indeed, as previously noted, the Lamarckian principles of biology and social heredity had hardened considerably since mid century. The evolutionary anthropologists invented schemes of evolutionary racial "progress" in which some groups—most, in fact—could never reach the pinnacle of civilization of the white race. Through his investigation, which included the views of doctors and anatomists as well as anthropologists, Haller especially has expanded our notions of the history of racial anthropology.[27] Curiously for a set of doctrines so in tune with popular attitudes on racial superiority, evolutionary anthropology had few institutional centers and fewer constituencies and patrons.

Franz Boas and his allies led the movement to reorient American anthropology towards the graduate university and cultural anthropology. In time they reinvented anthropology's institutions as part of a national academic profession, but not without bitter controversy. Recently Joan Mark has criticized Boas. Insisting that there was a science and profession of American anthropology long before Boas executed his campaign to change power relations in discipline and profession, Mark considers Boas a self-regarding, ungracious spoiler contemptuous of his American colleagues. Boas was prickly and difficult. As Columbia University president Nicholas Murray Butler ruefully discovered, Boas could be downright uppity. Yet the structuring of a professional revolution is an impersonal process that does not depend on a single individual. Anthropology did not become fully professionalized until the later 1920s, and Mark is on firm ground when she insists that nineteenth-century anthropology needs more examination on its own terms. Much work is needed on its intellectual history. We need not embrace Marvin Harris's positivistic interpretations of Boasian anthropology, for example, to agree that much has not been explored.[28] Why, for example, did Boas and his followers ignore British social anthropology and physical anthropology? What contributions did they make to the science of anthropology? What role did they play in scientific discussions of race? What did culture and cultural relativism mean at various times? Why did recognition and legitimacy only come to the Boasians after the mid 1920s?[29]

If American anthropology had relatively few external constituencies, this cannot be said of psychology. As several scholars have shown, the American proto-psychologists rapidly cast off their heritage of philosophy and German psychology and invented a new mental science appropriate for those questions of group and individual worth that mattered to American culture and not incidentally to American constituencies. Unlike anthropology's social message, psy-

ment of American Anthropology, 1846–1910 (Washington, D.C.: Smithsonian Institution Press, 1981); Stocking, Race, Culture, and Evolution (cit. n. 7); Stocking, ed., The Shaping of American Anthropology, 1883–1911: A Franz Boas Reader (New York: Basic, 1974).

[27] John S. Haller, Jr., Outcasts From Evolution: Scientific Attitudes of Racial Inferiority, 1859–1900 (Urbana: Univ. Illinois Press, 1971).

[28] Joan Mark, Four Anthropologists: An American Science in its Early Years (New York: Science History, 1980); and Marvin Harris, The Rise of Anthropological Theory: A History of Theories of Culture (New York: Thomas Crowell, 1969), Chs. 9–13.

[29] On these issues, esp. anthropology's late professionalization, see, e.g., Stocking, Race, Culture and Evolution, pp. 270–305; and Cravens, The Triumph of Evolution (cit. n. 4), pp. 89–120, 180–190.

chology's was powerful in society and culture. Within the academy psychologists benefited from mushrooming enrollments in the 1890s and after, attributable to some extent to the expansion of programs for public school teachers. But the real point, as John C. Burnham and others have insisted, was that psychologists' evolving concern to develop a psychology of capacity and a psychology of conduct had enormous appeal.[30] Moreover, the new professionals declared, they could accomplish one of professionalism's main purposes—standardization— by creating standards and yardsticks of mind and emotion. This was indeed a powerful message. As Dorothy Ross points out in her fine analysis of one of psychology's founders, G. Stanley Hall, if Hall was a man with serious problems as a scientist, professional, publicist, and administrator, he nevertheless spawned an entire movement of professional and popular child study— almost in spite of himself. The man mattered less than the message or even the medium in which it was cast. And, as Michael M. Sokal has demonstrated in his studies of another of psychology's founders, James McKeen Cattell, the message was the thing. Cattell was a difficult character, a curmudgeon early in life. He had a genius for creating enemies and for isolating himself from mainstream opinion in the scientific community; even his famous research program on individual mental differences, launched in the 1890s, soon fell apart. Yet as Hall stood for the notions that mind has evolved and the child is the father to the man, Cattell asserted that the scientific measurement of individual and group mental differences was just around the corner.[31]

Soon psychologists turned to mental measurement in droves. Some, such as Edward L. Thorndike, did so from within the academy and influenced several generations of psychologists and educators, not to mention those thousands who took the examinations of intellect he and others devised. The real breakthrough in mental testing, however, came when Henry H. Goddard and Lewis M. Terman published their "Americanized" versions of Binet's mental scales, which they redefined as tests of innate intelligence, which was not at all what Binet had assumed or intended.[32] Mental testing was a social technology if there ever

[30] Burnham, "Psychology, Psychiatry, and the Progressive Movement" (cit. n. 5); and Cravens, The Triumph of Evolution, pp. 56–86.

[31] Dorothy Ross, G. Stanley Hall: The Psychologist as Prophet (Chicago: Univ. Chicago Press, 1972); Michael M. Sokal, ed., An Education in Psychology: James McKeen Cattell's Journal and Letters from Germany and England, 1880–1888 (Cambridge, Mass.: MIT Press, 1981); Sokal, "Graduate Study with Wundt: Two Eyewitness Accounts," in Wundt Studies: A Centennial Collection, ed. Wolfgang G. Bringmann and Ryan D. Tweney (Toronto: Hogrefe, 1980), pp. 210–255; Sokal, "James McKeen Cattell and the Failure of Anthropometric Testing, 1890–1901," in The Problematic Science: Psychology in Nineteenth-Century Thought, ed. William R. Woodward and Mitchell G. Ash (New York: Praeger, 1982), pp. 322–345; Sokal, "The Origins of the Psychological Corporation," J. Hist. Behav. Sci., 1981, 17:54–67; Sokal, "James McKeen Cattell and American Psychology in the 1920s," in Explorations in the History of Psychology, ed. Josef Brozek (Lewisburg, Pa.: Bucknell Univ. Press, 1984), pp. 273–323; and Sokal, "Science and James McKeen Cattell, 1894 to 1945," Science, 1980, 209:43–52.

[32] Geraldine Joncich, The Sane Positivist: A Biography of Edward Lee Thorndike (Middletown, Conn.: Wesleyan Univ. Press, 1968); Theta H. Wolfe, Alfred Binet (Chicago: Univ. Chicago Press, 1973); and see Gould, The Mismeasure of Man (cit. n. 11), pp. 146–233, for a good discussion of how scientific, in retrospect, the early American uses of Binet were (this is a polemical rather than a historical discussion). See also, e.g., L. S. Hearnshaw, Cyril Burt, Psychologist (Ithaca: Cornell Univ. Press, 1979), a judicious critique of mental testing, and Leon J. Kamin, The Science and Politics of the I.Q. (Potomac, Md.: Lawrence Erlbaum Associates, 1974), which led to the exposure of Burt's serious methodological problems. The emergence of the "new" hereditarian mental testing came with Arthur R. Jensen, "How Much Can We Boost I.Q. and Scholastic Achievement?" Harvard Educational Review, 1969, 33:1–123. See also Cravens, The Triumph of Evolution, pp. 78–86.

was one. The question of whether mental testing was a tool of the corporate state, as some have charged, may not have been satisfactorily resolved, but clearly testing became a method by which psychologists identified and served client populations and patrons alike.[33]

Nor were the most prestigious institutions immune from the seductive appeals of applied psychology. As Matthew Hale, Jr., notes, the German-born and trained psychologist Hugo Münsterberg spent much of his time at Harvard advocating and encouraging mental measurement and all manner of other psychotechnics to his students and, indeed, to anyone who would listen. These applications included the law, industry, and the clinic. Münsterberg believed that social function was directly related to biological and mental structure. Psychologists even participated in the widespread public discussions of the meaning of work in the industrial age, of whether the routine of factory and office had made the work ethic passé.[34] Psychotechnics, of course, received much support from the behaviorist movement of the 1910s. In a very real sense, behaviorism provided a scientific rationale and legitimacy for applied psychology, for it stressed the measurement, prediction, and control of human behavior. As John M. O'Donnell has argued, perhaps behaviorism owed less to its famous high priest, John B. Watson, than it did to the work of the many obscure clinical and applied psychologists who labored in private and public agencies to solve immediate problems. It may well have been Watson's function to provide a voice from within the academic establishment for that larger community.[35]

In the 1910s and 1920s, psychologists offered their services to government, private industry, the mass media, commerce and advertising, and other institutions. As Loren Baritz's classic study of the uses of social science in industry suggests, much remains to be explored. The roles and influences of psychologists in these areas have barely been sketched. And while we may know something about the interest of the more prestigious institutions and research groups in applying psychology, we know almost nothing about the "lesser" academic institutions, such as the state universities of the Midwest, not to mention child-saving institutions, social welfare charities, and a host of other institutions that increasingly employed psychologists and other professionals after 1910.[36] The most famous example of applied psychology of the era, of course, was the program of mental measurement of Army recruits during World War I. The results

[33] Joel H. Spring, *Education and the Rise of the Corporate State* (Boston: Beacon, 1972); and Spring, "Psychologists and the War: The Meaning of Intelligence in the Alpha and Beta Tests," *History of Education Quarterly*, 1972, *12*:3–14.

[34] Matthew Hale, Jr., *Human Science and Social Order: Hugo Münsterberg and the Origins of Applied Psychology* (Philadelphia: Temple Univ. Press, 1980); and James B. Gilbert, *Work Without Salvation: America's Intellectuals and Industrial Alienation, 1880–1910* (Baltimore: Johns Hopkins Univ. Press, 1977).

[35] John M. O'Donnell, *The Origins of Behaviorism: American Psychology, 1870–1920* (New York: New York University Press, forthcoming); and David Cohen, *J. B. Watson: The Founder of Behaviorism: A Biography* (London: Routledge & Kegan Paul, 1979) (not the definitive biography of its subject, to say the least).

[36] On psychology in industry, see Loren Baritz, *The Servants of Power: A History of the Use of Social Science in American Industry* (1960; Westport, Conn.: Greenwood, 1974); see also Donald S. Napoli, *Architects of Adjustment: The History of the Psychological Profession in the United States* (Port Washington, N.Y.: Kennikat, 1981). Kuklick, *The Rise of American Philosophy* (cit. n. 3), a stunning achievement, and Daniel J. Bjork, *The Compromised Scientist: William James in the Development of American Psychology* (New York: Columbia Univ. Press, 1983), a book with far narrower concerns, focus on famous people at prestigious institutions.

of the Army tests "proved," to the testers' satisfaction, the superiority of some "races" over others and helped sanction the cause of immigration restriction. The Army program was a spectacular example of social technics and planning. Psychology also seemed to have a powerful message to laypeople. As Dominick Cavallo has argued in his interesting study of the urban recreation and play-ground movement, the new psychology seemed to the movement's organizers and leaders a social technology that had the potential to train city children to be good "team players," by which they meant training them to adapt to society's rules.[37] Yet a review of the corpus of secondary work on American psychology shows how much remains to be investigated. Psychology was indeed a hot item in culture and society, far beyond the ivory towers and silos of academe. It is worth exploring more fully.

American sociologists discovered that when they tried to create their discipline there were already messages aplenty about sociology in culture and society. From the academics' point of view, furthermore, these messages were, if not wrong, certainly not theirs, which was the main point. The academics even had to retrieve the word *sociology* from the various popular groups interested in society and "sociology" and make it theirs. Many historians and sociologists have dated the origins of American sociology to Lester Frank Ward and his ponderous writings. This seems highly doubtful. John C. Burnham pointed out that the academics had little use for Ward, and I have suggested that there was little relationship intellectually or professionally between Ward and early twentieth-century American sociologists.[38] Academic sociologists' most serious competitors in the early twentieth century were the so-called "practical" sociologists, that is, those interested in social work, charities, corrections, and child welfare. As economists withdrew from teaching "social problems" courses in the 1890s, and as social workers became professionalized in the 1910s, this left the academics in charge of sociology, although it was hardly clear what sociology was. Certainly there was no shared sense of sociology as a research enterprise before the 1920s, and inventing a consensus on what constituted the discipline seemed impossible.[39]

---

[37] Daniel J. Kevles, "Testing the Army's Intelligence: Psychologists and the Military in World War I," *J. Amer. Hist.*, 1968, *55*:565–581; and Dominick Cavallo, *Muscles and Morals: Organized Playgrounds and Urban Reform, 1880–1920* (Philadelphia, Univ. Pennsylvania Press, 1981). Still the fullest account of the Army tests remains Robert M. Yerkes, ed., *Psychological Examining in the United States Army* (Memoirs of the National Academy of Sciences, 15) (Washington, D.C.: National Academy of Sciences, 1921). See also Cravens, *The Triumph of Evolution*, pp. 224–265, for a discussion of the mental testing controversy.

[38] John C. Burnham, *Lester Frank Ward in American Thought* (Washington, D.C.: Public Affairs Press, 1956); and Cravens, *The Triumph of Evolution*, pp. 136–137; those putting Ward at the origins of American sociology include Henry Steele Commager, *The American Mind: An Interpretation of American Thought and Character Since the 1880s* (New Haven: Yale Univ. Press, 1950), pp. 204–210; Harry Elmer Barnes, ed., *An Introduction to the History of Sociology* (Chicago: Univ. Chicago Press, 1948), pp. 173–190; and Clifford H. Scott, *Lester Frank Ward* (Boston: Twayne, 1976).

[39] Cravens, *The Triumph of Evolution*, pp. 121–153. For useful practitioner histories of sociology as discipline and profession see, e.g., Roscoe C. and Gisela J. Hinkle, *The Development of Modern Sociology* (New York: Random House, 1954); Anthony Oberschall, ed., *The Establishment of Empirical Sociology: Studies in Continuity, Discontinuity, and Institutionalization* (New York: Harper & Row, 1972); Luther Lee Bernard, "The Teaching of Sociology in the United States," *American Journal of Sociology*, 1909, *15*:164–213; and Robert E. L. Faris, *Chicago Sociology, 1920–1932* (San Francisco, Calif.: Chandler, 1967). For valuable raw data on departments of sociology, see "History of Sociology Departments," Box 4, files 1–8, Luther Lee Bernard Papers, University Archives, Pennsylvania State University.

For the most part, the literature on American sociology focuses on individual sociologists and their ideas. Rarely does one find accounts that attempt to depict the discipline and profession as a whole, let alone its roles and interrelations in culture and society. Some accounts represent a New Left consciousness, assuming that there is a truly positivistic science of society that "would have" confirmed the true insights of left ideology about society. Others are more concerned with issues within the discipline itself, such as building or correcting contemporary theories by examination of older ones.[40] The prevailing view among historians is that the pioneer sociologists were reformers in academic guise who made reform central to their theories of society. Certainly these pioneer sociologists wanted sociology to be socially useful. But the tension between "reform" and "professionalism" is more apparent than real. Sociologists often found themselves in circumstances defined by the academy and academic professionalism. For example, in his lively, well-researched biography of Edward A. Ross, Julius Weinberg demonstrates how easily the irrepressible Ross handled the supposed tension between professionalism and reform. Ross was essentially an academic entrepreneur who used academic and professional institutions and customs to win fame and fortune. For Ross both "reform" and "professionalism" were means to a career.[41]

There have been several studies of the intellectual history of sociology as well. Roscoe C. Hinkle has offered an intensive, almost baroque, taxonomy of the specific disciplinary ideas of those he considers the founders of academic sociology. But its utility is limited by his narrow angle of vision. Only sociological ideas matter. Ellsworth R. Fuhrman has traced the development of a tradition of sociology of knowledge among pioneer academic sociologists, arguing that most accepted a "social-technological" mode of interpretation, that is, they believed that man had emotive, irrational impulses that society had to control. Thus they defined the rather sharp limits of individuality within the context of group identity. William Fine corroborates the importance of group identity from a slightly different perspective in discussing the impact and later transformation of evolutionary models in early sociological thought.[42]

The Chicago School has also received some attention. Fred H. Matthews, in his biography of Robert E. Park, discusses Park's ideas from within the discipline of sociology. He traces Park's ideas to Georg Simmel and William I. Thomas. Much of Matthews's discussion is concerned with Park's shift from "philosophical" to "sociological" levels of argumentation and discourse. J. David Lewis and Richard L. Smith have combined the history of philosophy with the cliometrics of knowledge to find the "roots" of the sociological theory of symbolic interactionism, which Herbert Blumer and other Chicago sociologists eventually elaborated into a major theory of the discipline. In a series of

[40] An example of the former is Herman and Julie R. Schwendinger, *The Sociologists of the Chair: A Radical Analysis of the Formative Years of North American Sociology, 1883–1922* (New York: Basic Books, 1974); an excellent example of the latter is Robert Bierstedt, *American Sociological Theory: A Critical History* (New York: Academic Press, 1981).

[41] Julius Weinberg, *Edward Alsworth Ross and the Sociology of Progressivism* (Madison: State Historical Society of Wisconsin Press, 1972).

[42] Roscoe C. Hinkle, *Founding Theory of American Sociology, 1881–1915* (Boston: Routledge & Kegan Paul, 1980); Ellsworth R. Fuhrman, *The Sociology of Knowledge in America, 1883–1915* (Charlottesville: Univ. Press of Virginia, 1980); and William Fine, *Progressive Evolution and American Sociology, 1890–1920* (Ann Arbor, Mich.: UMI Research Press, 1979).

sometimes negative conclusions, they insist that, contrary to professional folk-lore, George Herbert Mead had very little to do with the development or an-tecedents of symbolic interactionism. In his study of Albion W. Small, Vernon K. Dibble suggests that Small's thought, balanced as it was between a desire for a science of society and for social reform, could not be adapted to the new notions of social science and public policy that swept through culture, society, and the social sciences in the 1920s and beyond. Yet it remains difficult to find the intellectual cohesion within the Chicago School that several of these studies seem to assume.[43] The writing of the history of American sociology is not yet complete. More broadly gauged studies that go beyond single individuals and "disciplinary ideas" are welcome.

The controversy over mental illness and institutions continues in this period. In *Conscience and Convenience: The Asylum and Its Alternatives in Progressive America,* David J. Rothman has returned to the fray to find more black hats. Focusing on the "progressive" ministrations to the problems of crime, delin-quency, and mental illness, Rothman, still a proponent of history as the result of ineluctable social forces, nevertheless is preoccupied with the motives of in-dividuals, not with the impersonal structures and processes, the social forces of the social order—thus the "conscience" and "convenience" of the title. Con-venience, of course, won out. This is not perhaps a helpful insight for someone as interested in "correcting" contemporary social policy and changing "the system" as Rothman is; are social problems the result of "bad" individuals? And his analyses of therapy, rationale, professional ideology, and related phenomena appear both presentist and difficult to corroborate. Thus individuals were *not* thought of by therapists apart from the group to which they "belonged"; "feeble-minded" and "psychopathic" delinquents, for example, were regarded and treated as very different kettles of fish before the 1920s. Gerald N. Grob's *Mental Illness and American Society, 1875–1940* reiterates many of the themes of his earlier work, but, if anything, is an even more impressive piece of careful research, thoughtful interpretation and analysis, and balanced judgment. His findings about the fate of the mentally ill are hardly reassuring. In this work Grob pays far more attention to ideas, therapeutic and others as well, than be-fore. His is not an argument with the past, but an understanding of it. Another contribution of note is Norman Dain's recent biography of that most famous mental patient of the early twentieth century, Clifford Beers. Dain provided a judicious, well-researched account of Beers's life and of the mental hygiene movement in which he played so important a role. In her evolving work on the child guidance movement, an offshoot of mental hygiene, Margo Horn has studied a large number of cases in the Philadelphia Child Guidance Clinic be-tween the world wars and drastically qualified the argument of some social critics and historians that such institutions and their staffs sought to "invade" and "manipulate" the family.[44]

[43] Fred H. Matthews, *Quest for an American Sociology: Robert E. Park and the Chicago School* (Montreal: McGill–Queen's Univ. Press, 1978); J. David Lewis and Richard L. Smith, *American Sociology and Pragmatism: Mead, Chicago Sociology, and Symbolic Interaction* (Chicago: Univ. Chicago Press, 1980); and Vernon K. Dibble, *The Legacy of Albion Small* (Chicago: Univ. Chicago Press, 1975). Faris, *Chicago Sociology, 1920–1932,* has much valuable information on the depart-ment as an institution. Winifred Rausenbush, *Robert E. Park: Biography of a Sociologist* (Durham, N.C.: Duke Univ. Press, 1979) is an account by a former associate.

[44] David J. Rothman, *Conscience and Convenience: The Asylum and Its Alternatives in Progres-*

As historians venture further into the history of mental illness and hygiene, they will find useful a number of studies of psychiatry as discipline, therapy, and profession from the 1870s to the 1920s and beyond. Of particular note are the contributions of John C. Burnham, Nathan Hale, Jr., and Charles Rosenberg, which provide a solid basis for further work on the relations between professionals and the larger society and culture. In his synoptic account of American ideas on human nature, Merle Curti has provided a thorough assessment of psychoanalytic ideas in America. Also useful are several other more specialized studies of mental health and certain practitioners of psychiatry.[45]

Between the 1870s and the 1920s the social sciences became professions and disciplines. The democratic dogma of an earlier age was considerably revised. Professional social science projected its hierarchical and deterministic assumptions into the creation of a science of society. Historians working on the history of the social sciences in this period have yet, as a rule, to transcend the boundaries of disciplinary history and to place the social disciplines and professions in a broader context. At the same time, much work needs to be done on the disciplines and professions themselves.

THE AGE OF PLURALISM

Most historians would agree that after the 1910s, trends of cultural pluralism and cultural relativism swept through American society and culture. Certainly within the context of evolutionary theory itself, the emphasis shifted from the argument that nature "caused" culture to the notion that culture and nature were interrelated and inseparable and that, moreover, evolutionary progress was the result of the interaction of many different kinds of factors. If by cultural relativism one means a belief in the equipotentiality of all groups, with a presumed tolerance for all with regard to social policy, cultural relativism was probably a minority view. Power relations within the larger society and culture did not change substantially with regard to matters of race, religion, national origin, or gender. Cultural pluralism meant a redefinition of power relationships and a new way of describing the relations between the "majority" and the "minorities" in which one conceded complex interrelations and multiple factors. In this new conception of society and culture, each group contributed to the larger national

---

sive America (Boston: Little, Brown, 1980); Gerald N. Grob, *Mental Illness and American Society 1875–1940* (Princeton: Princeton Univ. Press, 1983); Norman Dain, *Clifford Beers: Advocate for the Insane* (Pittsburgh: Univ. Pittsburgh Press, 1980); and Margo Horn, "The Moral Message of Child Guidance, 1925–1945," *Journal of Social History,* Fall 1984, *18*:25–36.

[45] See Rosenberg, *The Trial of the Assassin Guiteau* (cit. n. 3); Burnham, *Psychoanalysis and American Medicine* (cit. n. 3); Nathan G. Hale, Jr., *Freud and the Americans: The Beginnings of Psychoanalysis in the United States, 1876–1917* (New York: Oxford Univ. Press, 1971); Curti, *Human Nature in American Thought* (cit. n. 3), esp. pp. 186–416; and John C. Burnham, *Jelliffe: American Psychoanalyst and Physician, and His Correspondence with Sigmund Freud and C. G. Jung,* ed. William McGuire (Chicago: Univ. Chicago Press, 1983). On late nineteenth-century psychiatry see Barbara Sicherman, "The Paradox of Prudence: Mental Health in the Gilded Age," *J. Amer. Hist.,* 1976, *62*:890–912; Sicherman, "The Uses of a Diagnosis: Doctors, Patients, and Neurasthenia," *Journal of the History of Medicine and Allied Sciences,* 1977, *32*:33–54; Sicherman, *The Quest for Mental Health in America, 1880–1917* (New York: Arno, 1980); Arthur H. Chapman, *Harry Stack Sullivan: His Life and Work* (New York: Putnam, 1976); Helen Swick Perry, *Psychiatrist of America: The Life of Harry Stack Sullivan* (Cambridge, Mass.: Harvard Univ. Press, 1982); and Walter Bromberg, *Psychiatry Between the Wars, 1918–1945* (Westport, Conn.: Greenwood, 1982).

welfare or, more precisely, played its assigned role, on the job, in the kitchen, or at the back of the bus.

Henry D. Shapiro has provided penetrating discussions of the meaning of cultural pluralism as applied to the problems of region and place. In *Appalachia On Our Mind* he examines changing conceptions of Appalachia in American thought between the 1870s and the 1920s. He argues that the early twentieth-century reconceptualization of Appalachia as not "other" to American normality but as instead a distinct but interrelated region provides us with a realistic, if not cheering, understanding of what regionalism and, therefore, pluralism signified. In other words, different was different, not necessarily equal in value or worth. More recently, Shapiro has offered a searching analysis of the role of place in the identification and explanation of groups in society. He points out that although post-1920 social science theory may have abandoned naturalistic racism, in its stead arose a new model. *Place* now became the key to understanding groups. While this might have been expressed partly in terms of culture, in effect each place had its own culture, whether referred to as a "region," a "community," a "ghetto," a "suburb," or the like.[46]

The new sense of interrelatedness of things and the plurality of social reality influenced the social sciences. In the modified doctrine of professionalism, it was appropriate to recognize the valid claims of other professionals in that larger search for the ultimate truths of human existence and behavior. Most social scientists won their various institutional and disciplinary battles. They eagerly established diplomatic relations, if not precisely a foreign policy, with their public constituencies, but, even more importantly, with the new foundations. The new foundations made possible interdisciplinary, cooperative social science. In some instances they mandated it.

An excellent example of the age's penchant for interdisciplinary cooperation was the new discipline of child development. Known before the 1920s as "child welfare," in the 1920s the new science and profession of child development was invented by philanthropic intervention. The Laura Spelman Rockefeller Memorial, created in 1918, had a vague mandate to assist women and children. The appointment of a young economist and gadfly of left politics, Lawrence K. Frank, to oversee the Memorial's interests in women and children led after 1925 to multimillion-dollar investments in several child research institutions, the most notable of which was the Iowa Child Welfare Research Station (founded in 1917). All sciences of the child, ranging from anthropometry to early childhood education, found a place in the new discipline and profession. Child developmentalists embraced the fundamental assumption of natural and social science of that age, that biology and culture were interrelated and must be taken into account in the explanation of child nature and behavior. Some historical work done by practitioners is useful; thus Robert R. Sears's overview discusses ideas and professional history accurately and perceptively. This is a rich field. The developmentalists have launched an archival preservation program for historians.[47]

---

[46] Henry D. Shapiro, *Appalachia on Our Mind: The Southern Mountains and Mountaineers in the American Consciousness* (Chapel Hill: Univ. North Carolina Press, 1978); and Shapiro, "The Place of Culture and the Problem of Identity," in *Appalachia and America: Autonomy and Regional Dependence,* ed. Allen Batteau (Lexington: Univ. Kentucky Press, 1983), pp. 111–141.

[47] See, e.g., Robert R. Sears, *Your Ancients Revisited: A History of Child Development* (Chicago:

There are other discussions of the new view of culture and nature in this period. As Edward A. Purcell, Jr., has perceptively argued, a major substantive issue from the earlier era that still needed resolution was the meaning and utility of scientific naturalism, or, put another way, the meaning of group identity for the individual. Purcell discusses developments in philosophy, jurisprudence, and social science in the several decades following 1910. In the 1920s social scientists and lawyers invented a new science of society. Without perhaps realizing it, they helped undermine such traditional supports of "democratic theory" as the essential rationality of man and the existence of established law. Caught in the 1930s between their commitments to scientific naturalism and democratic theory, they created a new corpus of democratic thought. This new body of thought ultimately equated American democracy with the status quo. As I have noted elsewhere, a concrete manifestation of this discussion was the controversy within the natural and social sciences over the role of heredity and environment in the making of human intelligence, conduct, and morals. The controversy led to the theory of the interrelatedness of nature and culture and to the kind of model-building that Shapiro and Purcell have noted in slightly different contexts. For the professionals involved, the controversy's resolution had the happy result of recognizing the different but legitimate academic and professional turfs of the various claimants in the controversy. And the controversy itself signified the shift from the age of hierarchy to that of pluralism.[48]

Further signs of the new age can be seen in the new "regionalist" geography, based to a large extent at the University of Chicago. This field raises all sorts of fascinating questions, such as its relation with anthropology, sociology, and economics. Physical anthropology finally was revived, slowly, in the 1930s and 1940s. The appointment of E. A. Hooten at Harvard was important, for this allowed the creation of a school of physical anthropology. If Hooten himself wanted to hear nothing of culture, in time the discipline he recreated recognized a multiplicity of factors. And, notwithstanding the flawed historical account Derek Freeman has written of Margaret Mead—really an account of the Boasian anthropologists—Boasian cultural anthropology (and anthropologists) always assumed the interrelatedness of culture and nature. Yet these matters deserve more intensive study.[49]

Univ. Chicago Press, 1975); Milton J. E. Senn, *Insights on the Child Development Movement in the United States* (Monographs of the Society for Research in Child Development, Serial No. 161, Vol. 40, Nos. 3–4) (Chicago: Univ. Chicago Press, 1975); Steven Schlossman, "Philanthropy and the Gospel of Child Development," *History of Education Quarterly,* 1981, *21*:275–299; Elizabeth M. R. Lomax, *Science and Patterns of Child Care* (San Francisco: Freeman, 1978); Hamilton Cravens, "Child-Saving in the Age of Professionalism, 1915–1930," in *History of Childhood in America,* ed. Joseph M. Hawes and N. Ray Hiner (Westport, Conn.: Greenwood, 1985); and Cravens, "The Wandering I.Q.: Mental Testing and American Culture," *Human Development,* forthcoming. See also Henry L. Minton, "The Iowa Child Welfare Research Station and the 1940 Debate on Intelligence: Carrying on the Legacy of a Concerned Mother," *J. Hist. Behav. Sci.,* 1984, *20*:160–176. For a guide to the papers of many key individuals and institutions in child development, see Committee on Preservation of Historical Source Materials, Society for Research in Child Development, "History of Child Development: Primary Source Materials: First Compilation of Abstracts," *Child Development Abstracts and Bibliography,* 1984, *58*:123–141; for Milton J. E. Senn's oral history interviews with professionals in this field, see Michael M. Sokal and Patrice A. Rafail, comps., *A Guide to Manuscripts Collections in the History of Psychology and Related Areas* (Millwood, N.Y.: Kraus, 1982), pp. 177–180.

[48] Purcell, *The Crisis of Democratic Theory* (cit. n. 2); and Cravens, *The Triumph of Evolution.*

[49] See, e.g., Robert E. Dickinson, *Regional Concept: The Anglo-American Leaders* (London: Rout-

There is much evidence, furthermore, that social scientists fit within the constructs of the age, or, at least, that individuals could not easily modify them. Much has been said about how the brilliant refugees from Hitlerite barbarism in the 1930s "upgraded" American science, social science, and the arts. If by this is meant that a goodly proportion continued to do work of high quality in America as in Europe, or that many were able to sustain themselves here, this is probably valid. And in some fields (e.g., atomic physics) the European migration may have been instrumental in furthering knowledge of the subject. Within the social sciences, however, most likely the refugees adapted to the American idiom, as the brilliant topological and child psychologist, Kurt Lewin did successively at Cornell, Iowa, and MIT, or they remained exciting and perhaps slightly exotic presences, as did Franz Neumann or Kurt Koffka, whose work was absorbed and diffused after their deaths. Indeed, this can be said even of Lewin.[50]

Of all of the social technologies of the early twentieth century, perhaps eugenics had the most determined advocates. From the excellent work of Mark H. Haller and Kenneth Ludmerer, we know that the old-time eugenics movement declined in the 1920s and that a new eugenics movement arose that took culture and nature seriously. Thanks to Daniel J. Kevles's intensively detailed book on eugenics, we now can see how the issues of culture and nature were carried on in discussions of human heredity to the present day.[51]

It was in the 1920s and 1930s that social scientists came to sit among the powerful, most often as advisers to policymakers. As Barry Karl and Stephen J. Diner have pointed out, social scientists saw in advising the leaders a new kind of professionalism more useful for the profession and for public policy than populistic appeals to voters.[52] And what were the relationships between power

ledge & Kegan Paul, 1976); Brian W. Blouet, ed., *The Origins of Academic Geography in the United States* (Hamden, Conn.: Archon, 1981); Frank Spencer, ed., *A History of American Physical Anthropology, 1930–1980* (New York: Academic Press, 1982), which contains many informative papers, with huge bibliographies on the history of the field and "progress" in its various specialties in the period 1930–1980; Derek Freeman, *Margaret Mead and Samoa: The Making and Unmaking of an Anthropological Myth* (Cambridge, Mass.: Harvard Univ. Press, 1983); and Cravens, *The Triumph of Evolution*, pp. 157–190. Freeman's anthropological assertions have been seriously challenged, as well as his depiction of Mead.

[50] Much of the literature on the migration of the European intellectuals assumes one of two positions, both ill suited to investigation by historical methods: either that American culture was "matured" by the migration, or that the refugees "succeeded." Among the more useful discussions of the problem are Laura Fermi, *Illustrious Immigrants: The Intellectual Migration From Europe, 1930–1941* (Univ. Chicago Press, 1968; 2nd ed., 1971); Donald Fleming and Bernard Bailyn, eds., *The Intellectual Migration: Europe and America, 1930–1960* (Cambridge, Mass.: Harvard Univ. Press, 1969); H. Stuart Hughes, *The Sea Change: The Migration of Social Thought, 1930–1965* (New York: Harper & Row, 1975); Michael M. Sokal, "The Gestalt Psychologists in Behaviorist America," *American Historical Review*, 1984, 89:1240–1263; and Jarrell C. Jackman and Carla M. Borden, eds., *The Muses Flee Hitler: Cultural Transfer and Adaptation in the United States, 1930–1945* (Washington, D.C.: Smithsonian Institution Press, 1983).

[51] Mark H. Haller, *Eugenics: Hereditarian Attitudes in American Thought, 1870–1930* (New Brunswick, N.J.: Rutgers Univ. Press, 1963; paper ed., 1984); Kenneth Ludmerer, *Genetics and American Society: A Historical Appraisal* (Baltimore: Johns Hopkins Univ. Press, 1972); Kevles, *In the Name of Eugenics* (cit. n. 2); and Garland E. Allen, "The Misuse of Biological Hierarchies: The American Eugenics Movement, 1900–1940," *History and Philosophy of the Life Sciences*, 1984, 5:105–128.

[52] Among the most helpful accounts of social scientists as advisers to policymakers are Richard S. Kirkendall, *Social Scientists and Farm Politics in the Age of Roosevelt* (1966; Ames: Iowa State Univ. Press, 1982); Elliot A. Rosen, *Hoover, Roosevelt, and the Brains Trust* (New York: Columbia

Kurt Lewin (1890–1947) in about 1936, a refugee social psychologist who adapted to the American idiom. Courtesy of the University of Iowa Archives.

and pluralism? Several commentators have insisted that by and large American social science has not been able to understand the nonwhite, and in particular the black, experience in America. Certainly this would be an interesting question to explore for those interested in power relations in science, as would the careers and contributions of nonwhite social scientists.[53] In a related vein, Rosalind Rosenberg has broken new ground in the history of social science by focusing on both female scholars and the "roots" of a feminist intellectual tradition in social science. She portrays the difficulties that a series of talented and determined women faced in attempting to develop a body of feminist, or at least woman-oriented, social scientific investigations and theories. Margaret Rossi-

Univ. Press, 1977); Barry D. Karl, *Executive Reorganization and Reform in the New Deal* (Cambridge, Mass.: Harvard Univ. Press, 1963); Karl, *Charles E. Merriam and the Study of Politics* (Chicago: Univ. Chicago Press, 1974); and Stephen J. Diner, *A City and Its Universities: Public Policy in Chicago, 1892–1919* (Chapel Hill: Univ. North Carolina Press, 1980).

[53] See, e.g., Eleanor Engram, *Science, Myth, Reality: The Black Family in One Half Century of Research* (Westport, Conn.: Greenwood, 1982); Robert V. Guthrie, *Even the Rat was White* (New York: Harper & Row, 1976); Dale R. Vlasek, "E. Franklin Frazier and the Problem of Assimilation," in *Ideas in America's Cultures: From Republic to Mass Society*, ed. Hamilton Cravens (Ames: Iowa State Univ. Press, 1982), pp. 141–155; and Stanford M. Lyman, *The Black American in Sociological Thought: New Perspectives on Black America* (New York: Putnam, 1972), pp. 171–183 *et passim*.

ter's work on women scientists suggests a revealing picture of partly opened doors slammed shut; the creation of a tradition, or even a network in the masculine world of science, was an extraordinarily difficult task. A recent biography of Ruth Benedict demonstrates how hard it was for a woman scientist to develop the kinds of professional networks that male scientists take for granted and that mean so much for the elaboration of ideas and careers. Perhaps the choices for women in social science were encapsulated in the contrasting careers of Ruth Benedict, who remained quite private, and Margaret Mead, who rapidly became one of the world's best-known anthropologists.[54]

Historians interested in the social sciences have ample opportunities for further work in the period since the 1920s. Vast areas remain untouched, including physical anthropology, home economics, urban planning, economics, geography, child development, and the various agricultural social sciences. Nor do we possess sufficient information for an elementary chronicle of anthropology, psychology, and sociology. Plenty of opportunity exists for work on a single discipline, several disciplines, and broader studies. A final point remains speculative, that is, whether the history of the social sciences has been continuous since the 1920s. There is evidence on both sides of the question. One argument would emphasize developments within the disciplines, whereas the opposing view would focus on changes in culture and society.[55]

CONCLUSION

The history of the social sciences has now begun to take shape as a field of scholarly investigation. The secondary literature is vast; nevertheless, in many instances the available information is sparse. Source materials now exist in plentitude, and the hope is that in time more materials will be preserved. All manner of work can and should be done, including biographies, histories of societies, institutions, particular theories and methods, research projects, studies of "popular" and "professional" understandings of social science, and the interrelationships between social science and public policy. Nor does this exhaust the list of possibilities. Perhaps the most important recommendation is that the field be defined as broadly as possible. The social sciences, or social technologies, embrace much of the human experience in modern times.

[54] Rosenberg, *Beyond Separate Spheres* (cit. n. 2); Margaret Rossiter, *Women Scientists in America: Strategies and Struggles to 1940* (Baltimore: Johns Hopkins Univ. Press, 1982); Judith Schachter Modell, *Ruth Benedict: Patterns of a Life* (Philadelphia: Univ. Pennsylvania Press, 1983); Jane Howard, *Margaret Mead: A Life* (New York: Simon & Schuster, 1984); and Mary Catherine Bateson, *With a Daughter's Eye: A Memoir of Margaret Mead and Gregory Bateson* (New York: William Morrow, 1984). The last two works portray Mead less as a thinker than as a promoter and public representative of anthropology. Other work on Mead's life includes Margaret Mead, *Blackberry Winter: My Earlier Years* (New York: William Morrow, 1972); "In Memoriam: Margaret Mead (1901–1978)," *American Anthropologist*, 1980, 82:261–373; Edward Rice, *Margaret Mead: A Portrait* (New York: Harper & Row, 1979); and Robert Cassidy, *Margaret Mead: A Voice for the Century* (New York: Universe, 1982).

[55] See Peter Clecak, *America's Quest for the Ideal Self: Dissent and Fulfillment in the 60's and 70's* (New York: Oxford Univ. Press, 1983) which argues that the 1960s and 1970s were, in a cultural and social sense, quite similar, and united by a sense of individualism, not a larger whole.

Native Knowledge in the Americas

By Clara Sue Kidwell*

T HE WAYS IN WHICH THE NATIVE PEOPLES of the New World lived in and adapted to the environments around them have generally been categorized as religion or magic rather than science, and studies of native cultures have been the province of anthropologists rather than historians. Historians of science have acknowledged native practices but have generally relegated them to the realm of technology. G. Stresser-Péan maintains: "Though it would be misleading to call the religious systems and technical achievement of these people 'scientific,' even the most primitive among them made practical contributions that no historian of science can ignore."[1]

Religion and science are quite different ways of thought. However, useful comparisons can be made between the ways that native peoples and the western Europeans who encountered them beginning in 1492 understood the natural world. There are similarities in these different intellectual traditions that constitute a basis for a discussion of science.

Let us first discuss briefly the differences. The characterization of native Americans' thought as religious generally presupposes that they believed in a transcendent power underlying physical reality, a power not understood in rational terms but one which arouses feelings of awe and sometimes fear. The defining characteristic of religion, as the term is used here, is the sense that the forces of nature are manifestations of transcendent power that has will and volition. And indeed native American societies have so understood the nature of the world, for they ascribe self-movement, will, and choice to physical forces in nature. These attributes, then, remove nature from the realm of pure rationality. Science, on the other hand, presupposes a nature of physical forces acting according to laws. Those forces have no personal aspect but are susceptible to rational understanding because their behavior is lawful, rather than willful, and they can be understood by logical thinking.

The difference between the Europeans and the native peoples they encountered lies in their differing assumptions about the nature of the physical world. To be sure, the natural world of 1492 was alive with Aristotelian concepts of natural place, transmutation of elements under the influence of philosophers' stones, and mysterious forces of magnetism, to name but a few. However,

---

* Native American Studies, University of California, Berkeley, California 94720.

[1] G. Stresser-Péan, "Science in Pre-Columbian America," in *History of Science*, ed. René Taton, trans. A. J. Pomerans, Vol. I: *Ancient and Medieval Science* (New York: Basic Books, 1963), p. 293.

Europeans increasingly regarded these forces as purely mechanical, while native people in the New World continued to view them as personal attributes of spiritual beings.

To give New World cultures a place in the history of science, we must be aware of the ways in which European and native thought diverged over time. European science was increasingly based in concepts of lawful behavior of natural forces. Native American beliefs were based on the willful behavior of those forces. The one presupposed rational ways of understanding those laws; the other sought interaction with the forces of nature through dreams, visions, and ceremonies—through intuitive and personal ways of comprehending and controlling those forces.

The common factors that characterized European and native activities in 1492, and that continue as accepted tenets of scientific activity, are those of observation of a body of physical phenomena existing apart from human beings (this assertion in itself is a statement of belief), the desire to control those phenomena and the forces behind them, and the attempt to exercise that control. The character of these attempts reflects both the common ground and the differences between these two cultures. Native cultures, with their beliefs in the personal will and volition of those forces, sought control by establishing personal relationships with them through ritual; an important aspect of this for many groups was the careful observation of natural phenomena. European scientific culture sought control through rational perception of laws of nature and the ability to predict the outcome of events, an ability also based on careful observation of physical phenomena.

The crucial point at which native and European concepts of science differed was in the role of experimentation. According to the Western view, the lawful behavior of nature will allow similar actions to lead to similar results, and thus the predictive power of science is enhanced. The native concept posits that the personal relationships of groups or individuals with spirit forces is necessary to assure desired results, and thus ceremony and ritual take the place of experiment.

Having explored these similarities and differences, we can define native American science as the activities of the native peoples of the New World in observing physical phenomena and attempting to explain and control them. This brief summary cannot take in the complexities and variations of cultural practices throughout North America at the time of European contact, and it cannot offer a definitive description of the nature of European science at that time. However, this definition of native science forms a basis for evaluating writings about native American science in the history of science.

PROBLEMS IN THE STUDY OF NATIVE SCIENCE

Many problems in dealing with science in the native cultures of the New World arise from the nature of the available sources. Since native people did not have written language, except for some pictographic or hieroglyphic forms in Mesoamerica, most studies of these cultures fall outside the province of history. Even where indigenous records are available, problems exist. For example, studies of Mayan codices have been concerned as much with problems of deciphering the writing as with their scientific and historical content. Floyd Lounsbury acknowledges the problems of interpretation in his studies of Mayan mathematics and calendrical systems; however, his contributions to an understanding of Mayan science are extremely significant.[2]

If historians of science rely on native sources, they face distinct problems of interpretation, but if they rely on sources written about native people by European observers, they face problems stemming from European contact itself. It is possible that some native practices and beliefs might have been the result of European contact rather strictly indigenous. Lynn Ceci's article "Fish Fertilizer: A Native American Practice?" raises the issue of primacy. She maintains from historical evidence that the practice of fertilizing corn hills with fish was not an indigenous practice but one learned from Europeans. Critics of her thesis argue that planting in hills was not a European custom and that although Old World cultivators might have used fish as fertilizer, it was unlikely that the custom would have been taken over by Indians whose planting methods were otherwise so different.[3]

This kind of discussion focuses on the historian's natural subject—change; it does not so much enlighten us about the nature of Indian agriculture as it raises the issue of the result of European contact on native practices. The historian without indigenous records cannot say what is native knowledge and what has been the result of European contact. More often than not, though, contact has led to the loss of native knowledge rather than its advance.

Ceci's study, by focusing on innovation, uses historical records to ascribe primacy to a discovery, thus implying that earlier invention or discovery in some way establishes a scientific achievement. This implicit assumption of the primacy of European science may also motivate accounts that describe native systems of knowledge as they existed at the time of contact but judge them as scientific by modern standards. Bernard Ortiz de Montellano in "Empirical Aztec Medicine" points out that many of the plants used for medicine have chemically active ingredients with physical effects consistent with the desired cure, but many have no such ingredients and were nonetheless used. Ortiz de Montellano indicates that the empirical basis of Aztec medicine entailed more than the purely physical effects that might be expected from herbs. However, he seems to feel compelled to justify the efficacy of Aztec medicine on the basis of chemistry rather than culture.[4]

Ortiz de Montellano is a scientist rather than a historian, and he brings the assumptions of modern science to the study of past cultures. The sources for his study, however, are seventeenth-century descriptions of Aztec culture. The historian finds problems with accounts such as these because they are ethnographic rather than historical. They are synchronic rather than diachronic: they provide a picture of a culture at one point in time rather than providing a sense of change or development in native systems of knowledge. Nonetheless, they

[2] Floyd G. Lounsbury, "Maya Numeration, Computation and Calendrical Astronomy," in *Dictionary of Scientific Biography,* ed. Charles C. Gillispie, 16 vols. (New York: Scribners, 1970–80), Vol. XV, pp. 759–818.

[3] Lynn Ceci, "Fish Fertilizer: A Native North American Practice?" *Science,* 1975, *188*:26–30; see letters to the editor and Ceci's response in *Science,* 1975, *189*:945–948.

[4] Bernard Ortiz de Montellano, "Empirical Aztec Medicine," *Science,* 1975, *188*:215–220.

may illuminate the world views of native cultures. And, incidentally, they may illuminate the world views of those who study native cultures.

The classic dichotomy between primitive and modern ways of thinking dominated the thought of anthropologists and sociologists for many years and shaped attitudes toward native peoples and their patterns of thought. Around the turn of the century, Emile Durkheim and Lucien Lévy-Bruhl formulated their theories of human culture and society along the dichotomy of primitive and modern, as Robin Horton has pointed out. In the mid-twentieth century, in *The Primitive World and Its Transformation,* Robert Redfield, an anthropologist, classified scientific thought as a characteristic of modern civilization and argued that primitive people did not make the crucial distinction between themselves and their physical surroundings that was necessary for science. Hence natives personify and respond emotionally to their environments rather than viewing them in an objective way. Writing in 1932, however, E. A. Burtt, a philosopher of science, attempted to deal with the dichotomy of primitive and civilized and to meld the scientific and religious viewpoints. His statement in *The Metaphysical Foundations of Modern Physical Science* has relevance to the definition of native science: "Possibly the world of external facts is much more fertile and plastic than we have ventured to suppose; it may be that all these cosmologies and many more analyses and classifications are genuine ways of arranging what nature offers to our understanding, and that the main condition determining our selection between them is something in us rather than something in the external world."[5]

Following Burtt's lead, studies of native science must not only deal with the results of native activities but should acknowledge as well the world views and understandings of native people concerning their relationships to the natural world. Most work on native science has been concerned only with the results of native observational efforts that are similar to those produced by Western science. However, as historians of science have begun to realize that science is not a thing sui generis and have begun to study the social context within which scientific method is used, they have attempted to investigate not only the observational results of scientific endeavors but the assumptions about the nature of the world that underlie them.

One criterion by which to evaluate writings about native American science, then, is the extent to which they are both descriptive of native practices of observation and interpretative of their cultural context. One way of approaching the subject is through ethnoscience.

ETHNOSCIENCE AND NATIVE SCIENCE

Ethnoscience is a method in anthropology that examines the boundaries of categories in systems of classification. It attempts to use precise definitions of sets of characteristics that define classes. Because ethnoscience depends upon analysis of native observations and definition of physical phenomena, it has been

---

[5] Robin Horton, "Lévy-Bruhl, Durkheim, and the Scientific Revolution," in *Modes of Thought: Essays on Thinking in Western and Non-Western Societies,* ed. Robin Horton and Ruth Finnegan (London: Faber & Faber, 1973); Robert Redfield, *The Primitive World and Its Transformation* (Ithaca: Cornell Univ. Press, 1953); Edwin Arthur Burtt, *The Metaphysical Foundation of Modern Physical Science* (2d ed.; London: Routledge & Kegan Paul, 1932) p. 224.

used to study native systems of knowledge in a scientific way. If we assume that the classification systems of native people must bear some relationship both to an objective reality and to their own cognitive structures for seeing the world, then ethnoscience is a useful tool to examine both the objective results of classification systems and the underlying cognitive structures that produce them.

William Sturtevant has defined ethnoscience as a method for the study of "the system of knowledge and cognition typical of a given culture," with the caveat that the term *ethnoscience* should not be taken to mean that studies of folk classifications and folk taxonomies are science while other forms of ethnography are not. Ethnoscientific studies are based on linguistic analysis, and Sturtevant sets out principles for the relationships of linguistic categories that will allow the student of a culture to examine native belief systems. The categories' proximity to or distance from those of Western science provides some insight into the cognitive structures of a world view.[6]

Sturtevant's essay raises a basic question for students of native science. Systems of categorization may indeed be an indication of the underlying preconceptions about the nature of the physical world, and they may be recognized by the outside observer of a culture on the basis of their congruence with the observer's own systems of categorization. The systems may be based on similarities and differences in an objectively real world. In that regard, the observer categorizes as "science" the explanations in the native world that match his or her own experiences and definition of science. However, the reasons why each culture adopts a particular explanation may be quite different.

Ethnoscience in its strictest anthropological sense has focused on eliciting information from native informants and discovering the native principles that determine the kind of congruences that must be present to constitute a category. In "An Ethnoscience Investigation of Ojibwa Ontology and World View," Mary Black finds, for example, that the Ojibwa (Chippewa) have higher respect for spiritual beings in their world than they do for animals. The reasons for the rankings (elicited by Black's question "Is *x* more respected than *y?*") have to do with concepts of power in the Ojibwa world view—with which beings have the ability to control the activities of other beings.[7]

Black's study is an excellent example of the method of ethnoscience. Although it would not necessarily be recognized as an exploration of native American "science," it gives insight into the way in which native people understand the physical world around them and is thus indeed an investigation of science in its broadest definition. Based on the premise that how native people categorize all the phenomena of the physical environment is important, it explores the world view of the Chippewa people through phenomena recognized by the Chippewa within a certain framework, that of control of forces of the environment. The beings who constitute the Chippewa world are manifest as objective phenomena, but their actions are understood in terms very different from those of Western science.

---

[6] William C. Sturtevant, "Studies in Ethnoscience," in *Culture and Cognition: Rules, Maps, and Plans,* ed. J. P. Spradley (San Francisco: Chandler, 1972), pp. 129–167, quoting p. 130.

[7] Mary B. Black, "Ojibwa Power Belief System," in *The Anthropology of Power: Ethnographic Studies from Asia, Oceania, and the New World,* ed. Raymond D. Fogelson and Richard N. Adams (New York: Academic Press, 1977).

Ethnoscience as a methodological approach is important because language is a key to understanding a world view, if one may substitute that term for cognition or cognitive studies. However, it is a technical approach that the more general observer cannot always undertake. It depends on the ability of the researcher to impose order upon the data collected.

Eugene Hunn distinguishes ethnoscience from "folk" science. He describes the first as a methodological and critical stance within cultural anthropology and the second as a domain for analysis of content within the understanding of informant and ethnographer. The ethnoscience view is etic: the observer looks at the culture from an outside perspective and attempts to apply certain general rules of knowledge to what is seen. The folk science view is emic: the observer attempts to understand the native way of describing and categorizing experience.[8]

The distinction between ethnoscience and "folk science" is not clear-cut, however. Brent Berlin, Dennis Breedlove, and Peter Raven describe their study of Tzeltal plant classification as botanical ethnography, or "that area of study that attempts to illuminate in a culturally revealing fashion prescientific man's interaction with and relationship to the plant world." Their approach is a linguistic one, and their study can be called ethnoscientific. It is also a study of the Tzeltal world view. Although one might argue with the assertion that native "interaction with and relationship to the plant world" are prescientific, Berlin and his colleagues have made a significant contribution to an understanding of native science.[9]

Not only do native categories provide an entrée into what may be perceived as universal cognitive processes, but they are also valid in and of themselves within the culture of which they are a product. They reveal the ways in which native people organize, understand, and control their worlds. In systems of classification, these ways are often based not on discrete characteristics so much as on a gestalt of the whole animal. This way of looking at the world is significantly different from Western science.

ARCHAEOASTRONOMY AND ETHNOASTRONOMY

Native classification systems deal with objective natural phenomena, and the results of those systems can be compared with Western scientific systems of classification. This comparative approach is used in other areas of native activity as well. Since astronomy is a science based on observation rather than experiment, its methods and analyses should correspond well in Western and native systems of thought. Archaeologists have presented orientation of structures as objective evidence for the observational powers of native people, and hieroglyphic records from the Maya provide systematic documentation. Archaeoastronomy is a field in which the results of the activities of native people are apparent, although the causes may not be so clear-cut; those results lend themselves to comparison with European systems of observation.

[8] Eugene S. Hunn, *Tzeltal Folk Zoology: The Classification of Discontinuities in Nature* (New York: Academic Press, 1977), pp. 3–5.
[9] Brent Berlin, Dennis E. Breedlove, and Peter H. Raven, *Principles of Tzeltal Plant Classification: An Introduction to the Botanical Ethnography of a Mayan-Speaking People of Highland Chiapas* (New York: Academic Press, 1974), p. xv.

An excellent introduction to the subject of archaeoastronomy is Anthony Aveni's *Skywatchers of Ancient Mexico*. His second chapter, "Astronomy with the Naked Eye," explains for the nonastronomer the movements of celestial bodies as they appear to the earthbound observer. Moreover, the study of archaeoastronomy in Mesoamerica has the advantage of a body of recorded evidence, and Aveni introduces the complexities of the Mayan calendar system and the ways in which astronomical data were recorded in codices. The extensive literature on Mayan astronomy and mathematics is based primarily on two codices, housed in Paris and Dresden, that survived the destruction of sources of native knowledge that occurred after the European conquest of the New World. J. E. S. Thompson has studied the Dresden codex, which contains mathematical notations describing the movements of the planet Venus and a table of lunar eclipse predictions. On the basis of information in the Paris codex, Gregory Severin postulates the existence of a Mayan ecliptic and zodiac.[10]

A major focus for scholars has been the calendar system of the Maya. Thompson has been a leader in the attempt to correlate the Mayan and Julian calendars.[11] The problem of correlation has not been resolved, but the work on it is a good example of the conjunction of Western scientific activity, which resulted in the Julian calendar system; native scientific observation, which resulted in the 365-day cycle of the Mayans known as the vague year and the 260-day sacred calendar; and the discipline of history, by which scholars have attempted to correlate the two systems through some objectively verifiable historical events.

The origin of the 260-day calendar, the *tzolkin,* has been the subject of speculation, for it had no relation to the agricultural year. Vincent Malmstrom argues that it originated from the 260-day interval between the times at which the sun passes through the zenith at the latitude of fifteen degrees north, where the phenomenon was first observed.[12] The *tzolkin* regulated religious activities and the highly deterministic astrological system of the Maya. Besides requiring observational skills, it had profound cultural significance.

The Mayan solar year, on the other hand, was reckoned through observation and divided into eighteen months of twenty days with an added five-day period. Much of the Mayan calendar system was devoted to noting and using the conjunctions between the solar and the ceremonial calendar, which occurred once every fifty-two years, a time marked with special ceremonial observances in both Mayan and Aztec culture. E. C. Krupp has described the Aztec ceremony of the "Binding of the Years," which marked the end of the fifty-two-year cycle.[13] Although claims have been made for the impressive accuracy of the

[10] Anthony F. Aveni, *Skywatchers of Ancient Mexico* (Austin: Univ. Texas Press, 1980); John Eric S. Thompson. *A Commentary on the Dresden Codex: A Maya Hieroglyphic Book* (Memoirs of the American Philosophical Society, 93) (Philadelphia: APS, 1972); Gregory M. Severin, *The Paris Codex: Decoding an Astronomical Ephemeris* (Transactions of the American Philosophical Society, 71, 5) (Philadelphia: APS, 1981).

[11] J. Eric S. Thompson, *A Correlation of the Mayan and European Calendars* (Field Museum of National History, Pub. 241, Anthropological Series, XVII, 1) (Chicago: Field Museum of Natural History, 1927); Thompson, *Maya Chronology: The Correlation Question* (Carnegie Institution of Washington, Pub. 456, Contribution 14) (Washington, D.C., 1935).

[12] Vincent H. Malmstrom, "Origin of the Mesoamerican 260-Day Calendar," *Science,* 1973, *181*:939–941.

[13] E. C. Krupp, "The 'Binding of the Years,' the Pleiades and the Nadir Sun," *Archaeoastronomy: The Bulletin of the Center for Archaeoastronomy,* 1982, 5:9–13.

Mayans in determining the length of the solar year, it is obvious that whatever the basis for their observations, their overriding concern was the prediction of the continuation of their world, which was based on the elaborate cosmology within which they worked.

In addition to codices, archaeoastronomy is based on the observed results of native peoples' activities, that is, the orientation of structures, which provide concrete evidence of native behavior. Interest in studies of orientation began with examinations of Stonehenge and other megalithic sites in Europe. Throughout the Americas, structures of varying complexity have been identified as having specific physical orientations to points on the horizon that mark rising or setting points of the sun at its solstices or the first risings of the year of particularly bright stars at sunrise or at sunset (their heliacal risings). There are extensive ruins of early cultures in South and Mesoamerica and less extensive but no less interesting sites in North America.

Aveni has described the difference between astronomy in North America and astronomy in the tropics as one of orientation. In the temperate latitudes, the emphasis is upon observation of events along the horizon, particularly solstices and the rising and setting of certain stars. In the tropics, the orientation is toward observation of events overhead, since the movement of the stars is directly overhead. However, the difference does not hold clearly, as the series of papers in *Ethnoastronomy and Archaeoastronomy in the American Tropics,* a symposium sponsored by the New York Academy of Science in 1981, demonstrates.[14]

Studies of orientation include one by Aveni, Sharon Gibbs (a historian of science), and Horst Hartung (an architect), who have analyzed the structure of the Caracol Tower at Chichén Itzá and have concluded that its doors and windows and its base platform could be aligned with key rising points of the sun, Venus, and certain stars. They support their hypothesis with early accounts of Mayan ceremonies that corroborate the notion that the Maya actually used such orientations. Orientation on a much larger scale was examined by C. Chiu and Philip Morrison in their study of the offset street grid at Teotihuacan in Mexico. They proposed that the streets were oriented to buildings in a way that brought them into astronomical alignment with celestial phenomena overhead. In Peru, R. T. Zuidma has studied the extensive system of *ceques,* forty-one directional lines centered on Coricancha, a temple of the sun in the city of Cuzco, and marked by various natural phenomena near the horizon. The lines were first recorded by Benabe Cobo in 1653, and his historical account provides the basis for Zuidma's study of archaeological sites and natural phenomena. The lines and celestial phenomena played a role in establishing the Inca calendar.[15]

Studies in Mesoamerican cultures, the Maya and Aztec particularly, have the

[14] Anthony F. Aveni, "Introduction," *Ethnoastronomy and Archaeoastronomy in the American Tropics,* ed. Anthony F. Aveni and Gary Urton (Annals of the New York Academy of Sciences, 385) (New York: New York Academy of Sciences, 1982); Owen Gingerich, "Summary: Archaeoastronomy in the Tropics," *ibid.*

[15] Anthony Aveni, Sharon L. Gibbs, and Horst Hartung, "The Caracol Tower at Chichén Itzá: An Ancient Astronomical Observatory?" *Science,* 1975, *188*:977–985; B. C. Chiu and Philip Morrison, "Astronomical Origin of the Offset Street Grid at Teotihuacan," *Archaeoastronomy,* 1980, *2*:55–64; R. T. Zuidema, "Catachillay: The Role of the Pleiades and of the Southern Cross and α and β Centauri in the Calendar of the Incas," in *Ethnastronomy and Archaeoastronomy,* ed. Aveni and Urton.

advantage of permanent records kept by natives. For North America there are no such sources, and theories of origin must be based solely on orientation of prehistoric sites, with corroboration in historical sources being minimal at best.

One of the most studied but controversial orientation sites in North America is found at Fajada Butte in New Mexico, where a trio of stone slabs, resting against a rock face and supported by a narrow ledge, cause a narrow beam of light, a "sun dagger," to fall, on the day of the summer solstice, across the center of a spiral cut into the cliff. Anna Sofaer and others have been studying the phenomenon for several years, trying to determine whether the placement of the stones is deliberate or accidental. Jonathan Reyman has studied another site in which human effort is obvious: Pueblo Bonito in Chaco Canyon, located in eastern New Mexico. Using observational and measurement techniques similar to those used by Aveni, Gibbs, and Hartung, Reyman determined that window openings at Pueblo Bonito could have been used to observe the winter solstice sunrise.[16]

One of the most dramatic sites in North America is the medicine wheel located high on the side of the Big Horn Mountains in Wyoming. The presence of stones laid out in what are called medicine wheels has inspired the idea that native people were aware of celestial phenomena such as solstices and the heliacal rising of certain stars. John Eddy has investigated the astronomical alignment of the Big Horn medicine wheel and concluded that the builders of the wheel (a simple circle of rocks with a central point, twenty-eight spokes, and five stone cairns around the outer rim) aligned their structure with heliacal rising points of certain stars and the solstice points of the sun. Eddy tested his hypothesis by sighting the rising of the summer solstice sun over the point of alignment of two of the stone cairns.[17] The remoteness of the site and the fact that the ancient people who built it were not agriculturalists but hunters raise questions of interpretation. Any attempt at explanation of the wheel's origins remains speculative.

Following Eddy's lead, Thomas and Alice Kehoe examined a number of sites in Saskatchewan where patterns of boulders indicate human construction efforts. They found at two sites significant evidence of orientation along a north-south axis that would indicate the sun's rising at the summer solstice. They also learned from Blackfoot Indians that there were calendar men who marked the days for certain ceremonies, and that sun dance lodges had a north-south orientation like that of the stone constructions, although the Indians denied there was a connection. They concluded that the stones could indeed be a kind of calendrical system.[18]

The Pleiades have been an important point of reference for astronomical systems in many cultures. The Aztecs watched for the passage of the Pleiades through the zenith as the marker of the end of one cycle in their calendar and the beginning of a new cycle. They celebrated the occasion (which occurred

---

[16] Anna Sofaer, Volker Zinser, and Rolf M. Sinclair, "A Unique Solar Marking Construct," *Science,* 1979, *206*:283–291; Jonathan E. Reyman, "Astronomy, Architecture and Adaptation at Pueblo Bonito," *Science,* 1976, *193*:957–962.

[17] John A. Eddy, "Astronomical Alignment of the Big Horn Medicine Wheel," *Science,* 1974, *184*:1035–1043.

[18] Thomas F. Kehoe and Alice B. Kehoe, "Stones, Solstices, and Sun Dance Structures," *Plains Anthropologist,* 1977, *22*:85–95.

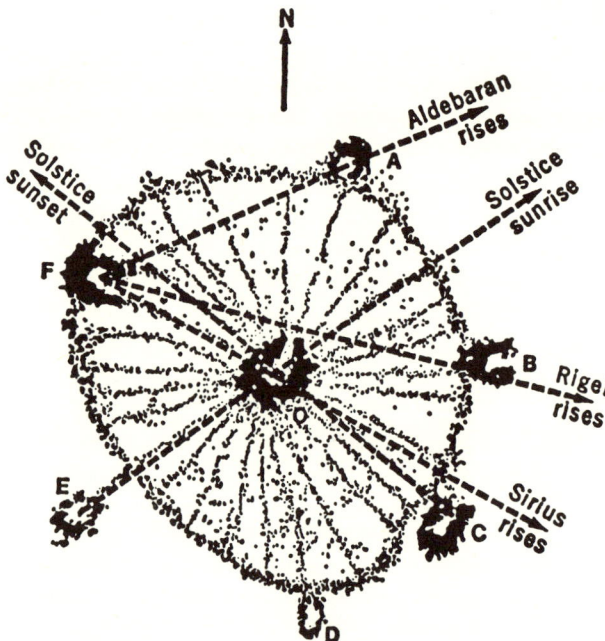

Diagram of Big Horn Medicine Wheel with superimposed arrows indicating observed astronomical alignments. From John A. Eddy, "Astronomical Alignment of the Big Horn Medicine Wheel," Science, 1974, 184 (4141):1040. By permission of the publisher. Copyright 1974 by the AAAs.

once every fifty-two years) by kindling a new fire in the breast of a sacrificial victim. The passage of the stars indicated that indeed another cycle had begun, and the world would continue. Although the explanation does not fit the scientific model, the observational techniques that it inspired led to an understanding of certain objective phenomena.

The Pleiades were an important marker for a number of North American tribes who practiced agriculture, because their appearance and disappearance coincided with the beginning and end of planting seasons. Lynn Ceci describes the phenomena in "Watchers of the Pleiades: Ethnoastronomy Among Native Cultivators in Northeastern North America." The Pleiades also figure in a highly speculative but fascinating attempt to understand the astronomical beliefs of a North American Indian tribe, Von Del Chamberlain's study of the Pawnee.

Chamberlain has used a body of ethnographic information and oral tradition and one permanent record, a star chart inscribed on a finely tanned hide. The ethnographic accounts, both from native Pawnee and nonnative sources, emphasize the role of the stars in the daily lives of the tribe. The Skidi band of Pawnee located their villages in conformity with patterns of the stars. Ceremonies invoked the morning and evening stars to bring their favor on the people. And the Pleiades were used to mark planting seasons. The star chart, with crosses painted on the hide, reproduces recognizable star clusters—the Big and Little Dippers, the Pleiades, and others. Chamberlain speculates that the Pawnee built and oriented certain of their earth lodges with the opening toward the east and the smoke hole opening overhead in such a way that they could observe solstices through the doorway and the passage of certain stars overhead. He must speculate, because there are no longer Pawnee earth lodges built in the traditional way for everyday use, only written descriptions. He asserts the possibility, not the actuality, that the Pawnee were consciously aware of the orientations he describes. The star chart and the rich body of tradition he cites, however, are good evidence for his arguments.[19]

Stephen McCluskey's studies of Hopi astronomy demonstrate the persistence of observations along the horizon in a contemporary Indian culture. He has had the advantage of observing at first hand the correlation of astronomical observation and ceremonial systems, a connection that can only be a matter of speculation for such things as medicine wheels. His field work with the Hopi has allowed him to note the activities of the sun watchers, men who go to certain places to observe the movement of the sun along the horizon as it nears its solstice points. The solstices are still used to mark the time for the beginning of certain ceremonies in the Hopi yearly cycle. The role of the sun watchers in various Pueblos has been documented. McCluskey has demonstrated the correlation of their activities with both the seasonal cycles of frost that determine the agricultural year and the activities of their ceremonial year.[20]

A rather esoteric preoccupation in archaeoastronomy (and yet one that occupies a noticeable part of the literature) is with the possibility that Indians in North America recorded in some way the great supernova of 1054. The existence of records contemporary with the event in China and not in Europe has led to the speculation that some of the rock art of North American Indians, particularly the conjunction of crescents and circles, can be interpreted as a record (and thus independent corroboration) of the supernova.[21]

If archaeoastronomy deals with the objective results of native observation, ethnoastronomy deals with the cultural premises behind the observations. The system of thought that inspired Mayan and Aztec astronomical observations was based on religious beliefs rather than on observation for its own sake. Aveni makes this point in a review article on Mayan archaeoastronomy. David Kelley has argued that the astronomical system of the Maya reflects their religious system. He maintains that the identification of gods with celestial bodies in various codices and their relationships with each other indicates a knowledge of the mechanics of eclipses. He also acknowledges that the thesis is a controversial one. Johanna Broda has explored the influence of culture on Mayan science in "Astronomy, *Cosmovision,* and Ideology in Pre-Hispanic Mesoamerica." She relates the Mayan and Aztec calendar systems, which are very similar, to the structure of their societies, which were very hierarchical. She maintains that the fatalism apparent in the astrological system based upon the calendar (and perhaps the basis for the calendar) is an expression of the highly structured and authoritarian social system. Gary Urton has drawn similar inferences from the perception of contemporary Quechua Indians concerning the heavens. They see animal formations in the dark areas of the Milky Way, which can be called constellations, since they move regularly overhead. Urton finds in this concept

[19] Lynn Ceci, "Watchers of the Pleiades: Ethnoastronomy among Native Cultivators in Northeastern North America," *Ethnohistory,* 1978, *25*:301–317; Von Del Chamberlain, *When Stars Came Down to Earth: Cosmology of the Skidi Pawnee Indians of North America* (Los Altos, Calif.: Ballena; College Park, Md.: Center for Archaeoastronomy, 1982).

[20] Stephen C. McCluskey, "Historical Archaeoastronomy: The Hopi Example," in *Archaeoastronomy in the New World: American Primitive Astronomy,* ed. A. F. Aveni (Cambridge: Cambridge Univ. Press, 1982).

[21] John C. Brandt and Ray A. Williamson, "The 1054 Supernova and Native American Rock Art," *Archaeoastronomy,* 1979, *1*:1–38; Seymour H. Koenig, "Stars, Crescents, and Supernovae in Southwestern Indian Art," *ibid.,* pp. 39–50.

evidence of a system of classification that organizes the physical world of the
Quechua, celestial phenomena, animals, and, by extension, other aspects of the
world as well.[22]

A native account of astronomical phenomena in a North American Indian tribe
is the basis for Berard Haile's study "Starlore among the Navaho." Haile
worked extensively with the Navajo during the early 1900s, and his knowledge
of Navajo culture was extraordinary, as was his working relationship with native
people. His small volume on starlore is based on what he learned in his examination of the extensive body of knowledge encompassed in Navajo chants,
elaborate ceremonies based on mythology and aimed at curing illness. Navajo
starlore describes the way in which the heavens came to be as they are. This
is mythology, and yet it is an intriguing example of a native system of explanation that expresses reality in Navajo thought. It seems that some of the stars
had been carefully placed on the floor of the hogan of creation by Black God,
one of the Navajo deities, and thus constituted constellations (some of which
correspond approximately to Western ones). However, Coyote, a trickster and
practical joker in Navajo cosmology, came along to pick up the deerskin containing the stars and blew them into the sky, thus accounting for the seemingly
random distribution of the stars in the heavens. The story indicates that the
Navajo were keen observers of the heavens, that they picked out certain patterns of stars against a background of seeming randomness, and that they had
a system of explanation for certain distinctive groupings of stars, for example,
the Pleiades, against the randomness of other stars.[23]

Studies of archaeoastronomy and ethnoastronomy indicate that the native
people of the New World were keen observers of the natural world. Observation
of the movements of the stars, the sun, and the moon, and knowledge of their
predictive power was common among native people in North America. If the
nature of the heavenly bodies was a personal one, the results of their actions
were predictable. The Hopi knew that when the Sun returned to his home in
the south and rested there briefly, their ceremonies would cause him to leave
his home and begin his journey to the north, bringing warmth and promoting
the growth of crops. They also knew that the Sun moved at times more quickly
and at times more slowly through the sky relative to the lunar months, by which
they timed most of their ceremonies, and they could predict the conjunction of
solar and lunar events for ceremonial purposes, as McCluskey has pointed out.[24]

Aveni and Urton, in their introduction to *Ethnoastronomy and Archaeoastronomy in the American Tropics,* describe the purpose of the conference whose
proceedings it contains as an exploration

> in a comparative perspective, [of] the traditions of thought and logic whereby
> American Indian cultures in the tropics organize cycles and phenomena perceived

[22] Anthony F. Aveni, "Archaeoastronomy in the Maya Region: A Review of the Past Decade,"
Archaeoastronomy, 1981, *3*:1–16; David H. Kelley, "Astronomical Identities of Mesoamerican
Gods," *ibid.,* 1980, *2*:1–54; Johanna Broda, "Astronomy, *Cosmovision,* and Ideology in Pre-Hispanic
Mesoamerica," in *Ethnoastronomy and Archaeoastronomy,* ed. Aveni and Urton (cit. n. 4); Gary
Urton, "Animals and Astronomy in the Quechua Universe," *Proceedings of the American Philo-
sophical Society,* 1981, *125*:110–127.

[23] Berard Haile, *Starlore Among the Navaho* (Santa Fe, N.M.: Museum of Navajo Ceremonial
Art, 1947).

[24] McCluskey, "Historical Archaeoastronomy" (cit. n. 20) p. 53.

in their terrestrial and celestial environments. If these systems of knowledge are found to be similar to the system of those cultures located in northern temperate latitudes, then we will learn something of the cognitive unity of mankind. If they prove to be dissimilar, then we will be reassured of the human capacity for change and adaptation to diversity.[25]

The first possibility, to learn of cognitive unity, is the aim of the anthropological view. The second, to discover change and adaptation, is closer to the interests of the historian. Although the evidence is certainly not definitive in favor of either view, archaeoastronomy and ethnoastronomy provide intriguing ways of examining not only the physical results of scientific activity but also some thing of the cultural systems that produce that activity.

ETHNOBOTANY

The discovery of the Americas opened up to Europeans a vast new world of plant and animal life (among which native people were generally included). The intellectual tradition of natural science studies in Europe readily subsumed the exploration of the New World. Although the works of early observers treated the Indians as subjects rather than scientists, they include much important information about Indian uses of the environment, particularly plants. They have been extensively used in ethnohistorical studies of Indian tribes.

Spanish explorers provided accounts of Mesoamerican natural lore. Philip II of Spain sent his personal physician, Francisco Hernandez, to collect information in the new realms. Hernandez produced a major compendium of descriptions of plants, including Aztec knowledge of their forms and uses. The work did not reach the public until much later. Hernandez's sixteen manuscript volumes were deposited in the Escurial in Spain and destroyed by fire in 1671. An abbreviated version was published in Italy in 1628 by the Accademia dei Lincei (with a second edition in 1651). Fra Bernardino Sahagún, a Franciscan priest, compiled a vast amount of material from native Aztecs, but his *Historia general de las cosas de Nueva España* was intended to serve not as a treatise on natural history but as a guide to native superstitions so that they could be countered by the Catholic Church. His monumental work was not available to scholars until a partial version was published in 1831; several editions have appeared since.[26]

In North America, interest in native flora and fauna led to works such as John Josselyn's *New England's Rarities Discovered,* which recorded Indian uses of plants; Le Page du Pratz's extensive description of the flora and fauna of the lower Mississippi Valley region, in which Indians were described among the plants and fishes and animals; and Dumont de Montigny's description of the wonders of French Louisiana. The English naturalist William Bartram was probably the most systematic and scientific observer of North American flora and fauna (including Indians).[27]

---

[25] Aveni and Urton, intro. to *Ethnoastronomy and Archaeoastronomy,* p. 7.

[26] Francisco Hernandez, *Rerum medicarum Novae Hispaniae thesaurus, seu Plantarum animalium mineralium Mexicanorum historia* (Rome, 1628); Bernardino Sahagún, *Historia general de las cosas de Nueva España,* 5 vols. (Mexico: Editorial Pedro Robredo, 1938); and Sahagún, *General History of the Things of New Spain,* trans. Charles E. Dibble and Arthur J. O. Anderson, 13 parts (Santa Fe: Monographs of the School of American Research and the Museum of New Mexico, 1950–1965).

[27] John Josselyn, *New-Englands Rarities Discovered: In Birds, Beasts, Fishes, Serpents, and*

The long tradition of observing Indians and the natural environment carried over into studies in anthropology. Observations of Indians and their knowledge of plants and animals in their environments were popular in the first third of the twentieth century. A number of these studies contained the prefix "ethno" with some appended scientific category. Thus one finds J. P. Harrington's *Ethnogeography of the Tewa Indians,* Leland C. Wyman and Flora L. Bailey's *Navaho Indian Ethnoentomology,* Edward F. Castetter and Ruth Underhill's *Ethnobiology of the Papago Indians,* Junius Henderson and John P. Harrington's *Ethnozoology of the Tewa Indians,* and, among an array of ethnobotanies, W. W. Robbins, J. P. Harrington, and B. Freire Marreco's *Ethnobotany of the Tewa Indians* and Huron H. Smith's "Ethnobotany of Menomini Indians."[28]

These studies follow a general pattern—a brief ethnographic description of the tribe followed either by a narrative discussion of uses of plants or animals or by a list of their names (most often the Latin classification, but also the English, or sometimes native, name). There is occasional mention of some kind of native classification system, for instance, Smith's simple statement that the Ojibwa had one.[29] These studies generally do little more than fit native knowledge into Linnaean categories and describe uses of plants and animals.

Considerably more can be learned about classification systems when ethnographers and botanists collaborate. James Teit's field notes on the Thompson River Indians of Canada were edited by Elsie Viault Steedman, a botanist, and the introductory material includes a discussion of linguistic terms and how they indicate general terms for uses and similarities of form. Leland Wyman, a physician, and Stuart Harris, a botanist, worked together on the ethnobotany of the Navajo Indians. Although they correlated native knowledge of plants with botanical classification systems to develop Navajo species and genera, they also acknowledged the existence of the Navajo system that categorizes primarily by male and female, terms that designate not physical sexual characteristics but qualities associated by the Navajo with maleness and femaleness. One of the most extensive studies of native plant uses was done by Frances Densmore

---

Plants of that Country; Together with the Physical and Chyrurgical Remedies wherewith the Native Constantly use to Cure their Distempers, Wounds, and Sores . . . (London, 1672); Antoine S. Le Page du Pratz, *Histoire de la Louisiana, Contenant la Découverte de ce vaste Pays, sa Description géographique; un Voyage dans les terres; l'Histoire Naturelle; les Moeurs, Coutumes & Religion des Naturels, avec leurs Origines* . . . , 3 vols. (Paris, 1758); Jean François Benjamin Dumont de Montigny, *Mémoires Historiques sur la Louisiane,* . . . *le climat la nature & les productions de ce pays; l'origine & la Religion des Sauvages qui l'habitent leurs moeurs & leurs coutumes &c.* 2 vols. (Paris, 1753); William Bartram, *Travels Through North & South Carolina, Georgia, East & West Florida, the Cherokee Country, the extensive Territories of the Muscogulges, or Creek Confederacy, and the Country of the Chactaws* . . . (Philadelphia, 1791).

[28] John Peabody Harrington, *The Ethnogeography of the Tewa Indians* (Twenty-ninth Annual Report of the Bureau of American Ethnology to the Secretary of the Smithsonian Institution, 1907–1908) (Washington, D.C.: GPO, 1909); Leland C. Wyman and Flora L. Bailey, *Navaho Indian Ethnoentomology* (Univ. New Mexico Publications in Anthropology, 12) (Albuquerque: Univ. New Mexico Press, 1964); Edward F. Castetter and Ruth Underhill, *The Ethnobiology of the Papago Indians* (Univ. New Mexico Bulletin, Biological Series, 4) (Albuquerque, N.M., 1935); Junius Henderson and John Peabody Harrington, *Ethnozoology of the Tewa Indians* (Bureau of American Ethnology Bulletin 56) (Washington, D.C.: GPO, 1914); Wilfred William Robbins, John Peabody Harrington, and Barbara Freire-Marreco, *Ethnobotany of the Tewa Indians* (Bureau of American Ethnology Bulletin 55) (Washington, D.C.: GPO, 1916); Huron H. Smith, "Ethnobotany of the Menomini Indians," *Bulletin of the Public Museum of Milwaukee,* 1923, 4:1–174.

[29] Huron H. Smith, *Ethnobotany of the Ojibwa Indians* (Milwaukee: Public Museum of the City of Milwaukee, published by order of the trustees, 1932).

among the Chippewa (Ojibwa) of northern Minnesota. Although Densmore was an ethnomusicologist, she enlisted the aid of botanists and chemists to determine the classifications and chemical components of plants. She provided both native and scientific name, although she made no attempt herself to correlate native plant names with the Linnaean system. More recent studies by Nancy Turner have taken a new approach in ethnobotany. Turner and others have been interested in the food value of plants and have tested nutrients by scientific experiment. They have shown the extent to which native foods were more than adequate to the dietary needs of native people on the Northwest coast.[30]

Descriptions of native plant uses have been heavily oriented toward uses for curing. The work of Wyman and Harris, mentioned just above, is a good example. Ethnobotany has contributed much to knowledge about native medical practices. Daniel Moerman, in *American Medical Ethnobotany: A Reference Dictionary,* has made the attempt to describe systematically uses of plants for medicine in North American native tribes. Drawing on the content of twenty-one ethnobotanical studies and using a computer, he has correlated symptoms or conditions and the plants used to treat them. He points out the chemical components of plants that would cause reactions in the body. Moerman's bibliography is an excellent source of material on ethnobotany. Bernardo Ortiz de Montellano's description of "Empirical Aztec Medicine", already discussed, is a good source for Aztec ethnobotany.[31]

However, studies of native American medical practices other than herbal remedies have tended toward the incredulous, claiming that they were based mainly on trickery, or the too credulous, claiming that American Indians were repositories of knowledge of marvelous cures which have unfortunately been lost as Indian cultures have changed. Although history of medicine is a distinct undertaking in the history of science and not totally within the scope of this essay, several sources in the extensive literature on native American health are worthy of mention. Virgil J. Vogel's *American Indian Medicine* is a survey drawn from historical and ethnographic sources. Although it errs on the side of credulity, it brings together in one place a wide range of material. The appendix lists plants used by native tribes and indicates which have been included in the *United States Pharmacopia,* the standard reference work on drugs; it is a useful reference source, as is the bibliography. In the end, though, Vogel is a compiler, not an analyzer of native practices.[32]

[30] James A. Teit, "Ethnobotany of the Thompson Indians of British Columbia . . . , based on field notes by James A. Teit," ed. Elsie Viault Steedman, *Forty-fifth Annual Report of the Bureau of American Ethnology to the Secretary of the Smithsonian Institution, 1927–1928* (Washington, D.C.: GPO, 1930), pp. 441–552; Leland C. Wyman and Stuart K. Harris, *Navajo Indian Medical Ethnobotany* (Univ. New Mexico Bulletin, Anthropological Series, 3.5) (Albuquerque, 1941); Wyman and Harris, *The Ethnobotany of the Kayenta Navaho: An Analysis of the John and Louisa Wetherill Ethnobotanical Collection* (Albuquerque: Univ. New Mexico Press, 1951); Frances Densmore, "Uses of Plants by the Chippewa Indians," in *Forty-fourth Annual Report of the Bureau of American Ethnology to the Secretary of the Smithsonian Institution 1926–1927* (Washington, D.C.: GPO, 1928); Nancy J. Turner, "A Gift for the Taking: The Untapped Potential of Some Food Plants of North American Native Peoples," *Canadian Journal of Botany,* 1981, 59:2331–2357; Turner and Harriet V. Kuhnlein, "Two Important 'Root' Foods of the Northwest Coast Indians: Springbank Clover (*Trifolium wormskioldii*) and Pacific Silverweed (*Potentilla anserina* ssp. *pacifica*)," *Economic Botany,* 1982, 36:411–432; Turner, "Economic Importance of Black Tree Lichen (*Bryorica fremontii*) to the Indians of Western North America," *Econ. Bot.,* 1977, 31:461–470.

[31] Daniel R. Moerman, *American Medical Ethnobotany: A Reference Dictionary* (New York: Garland, 1977); Ortiz de Montellano, "Empirical Aztec Medicine" (cit. n. 4).

[32] Virgil J. Vogel, *American Indian Medicine* (Norman: Univ. Oklahoma Press, 1970).

More useful are works that attempt to analyze these practices. Erwin Ackerknecht's collection of essays in *Medicine and Ethnology* is important because of its analytic nature. Ackerknecht is interested in native medicine worldwide, and he uses American Indian examples in addressing important questions, such as the role of shamans in curing. His approach is firmly based in Western psychological practices. He is, however, well aware of the nature of native beliefs and appreciates both the emic and etic views in regard to native medicine. Francisco Guerra's work on Aztec medicine employs a more strictly historical approach. Guerra bases his study on the material collected by Sahagún. His intent is to provide a cultural context for Aztec curing practices.[33]

An example of the melding of historical and Western medical approaches—one might say an ethnohistorical study of disease and curing—is an article by Charles Hudson, Ronald Butler, and Dennis Sikes on arthritis. The association of animals and disease, notably deer and arthritis in the southeastern United States, is based in cultural beliefs about the association of angry animal spirits and diseases as punishment. There is evidence for the possibility of transmission of diseases between animals and humans.[34]

Medicine is in many ways an art rather than a science. It is an art at which native people of the Americas were highly adept. Native medicine involved knowledge of the physical effects of natural substances on the human body and was also based on the ways in which people understood the actions of forces in the world around them as they affected their well-being.

AGRICULTURE

An important aspect of health is diet, and the development of stable food supplies through agriculture played an important part in the development of native cultures. Agriculture is another area of human activity in which observation and prediction play basic roles, and thus it represents another common ground between European and native science.

Agriculture figures prominently in the biological exchange between the Old World and the New—the diffusion of animals, foodstuffs, and diseases between continents. That subject has been treated by Alfred Crosby in *The Columbian Exchange: Biological and Cultural Consequences of 1492,* although he gives greater attention to the consequences in Europe than to those in the New World. Among the most important exchanges were those of domesticated crops.[35]

Domestication of crops is one of the most significant agricultural accomplishments ascribed to New World cultures. Richard Yarnell has discussed native uses of plant resources in the upper Great Lakes region, using archaeological techniques to identify plant remains and to indicate their role in the diet of the native inhabitants of the area. Yarnell thinks that the early people in this region

[33] Erwin H. Ackerknecht, *Medicine and Ethnology: Selected Essays,* ed. H. H. Walser and H. M. Koelbing (Bern: Hans Huber, 1971); Francisco Guerra, "Aztec Medicine," *Medical History,* 1966, *10*:315–338.

[34] Charles Hudson, Ronald Butler, and Dennis Sikes, "Arthritis in the Prehistoric Southeastern United States: Biological and Cultural Variables," *American Journal of Physical Anthropology,* 1975, *43*:57–62.

[35] Alfred W. Crosby, Jr., *The Columbian Exchange: Biological and Cultural Consequences of 1492* (Westport, Conn.: Greenwood, 1972).

may have domesticated some plants such as sunflowers. Corn is generally considered the triumph of Indian agriculture. The most extensive work on the development of corn is that of Paul Manglesdorf who deals with the genetic relationships between teosinte, which has been considered the ancestor of corn, tripsacum, another genetic relative, and domesticated corn. Manglesdorf says that teosinte is a hybrid of corn and tripsacum which subsequently crossed with corn to produce modern varieties. He denies that Indian people were consciously practicing breeding techniques to produce new strains of corn, although they were and are careful to preserve certain strains.[36]

Indian agricultural techniques included burning areas to clear fields and to promote the growth of wild plants. Henry T. Lewis has described the practices of the California Indians, who burned off chaparral areas in the foothills of the Sierras to promote better browsing grounds for deer and so improve their own hunting. Indians in the eastern woodlands use fire in a similar manner to control deer hunting and the forest cover, as Calvin Martin recounts. By periodically clearing the understory of forest land, Indians assured that they could move easily in their hunting grounds. More directly representative of agricultural technique is swiddening, the clearing of fields by fire; Mayan use of this technique is discussed below.[37]

Agriculture represents control over or the ability to predict environmental forces—rain, frosts, and supplies of water. Its stability depends ultimately upon control of water, and several native peoples practiced irrigation. Emil Haury has studied the archaeological remains of the extensive irrigation systems that allowed the rise of a sizeable community of Hohokam farmers in Arizona around A.D. 800. Michael Glassow has examined the Colorado Plateau region of the western United States, showing the patterns of field cultivation and water control that made the development of agriculture possible.[38]

Too much water can be as much of a problem as too little. William Denevan has discussed the use of raised planting areas in Mesoamerica, where lowland tropical areas were often flooded. The most notable example of this kind of control was the *chinimpas,* artificial islands created in swampy areas around Lake Tezcoco and in the lake by the Aztecs. The rich bottom mud from swamps and lakes was brought to the surface to create the floating gardens noted by the Spanish. Two other scholars who have focused on the role played by water control in Mayan agriculture have concluded that the Maya were not as dependent on swidden agriculture as earlier studies have led us to believe. B. L. Turner has maintained that lowland reclamation techniques allowed the Maya to practice intensive agriculture in lowland regions and so perhaps to support larger populations than possible with swidden agriculture. Turner thus calls into

[36] Richard A. Yarnell, *Aboriginal Relationships Between Culture and Plant Life in the Upper Great Lakes Region* (Univ. Michigan Anthropological Papers, 23) (Ann Arbor: Univ. Michigan Press, 1964); P. Weatherwax, *Indian Corn in Old America* (New York: Macmillan, 1954); Paul C. Manglesdorf, *Corn: Its Origin, Evolution and Improvement* (Cambridge, Mass.: Belknap Press of Harvard Univ. Press, 1974), pp. 35, 64.

[37] Henry T. Lewis, *Patterns of Indian Burning in California: Ecology and Ethnohistory* (Socorro, N.M.: Ballena Press, 1973); Calvin Martin, "Fire and Forest Structure in the Aboriginal Eastern Forest," *Indian Historian,* 1973, 6:38–42.

[38] Emil W. Haury, *The Hohokam: Desert Farmers and Craftsmen* (Tucson: Univ. Arizona Press, 1976); Michael Glassow, *Prehistoric Agricultural Development in the Northern Southwest: A Study in Changing Patterns of Land Use* (Socorro, N.M.: Ballena Press, 1980).

doubt the theory that increased population with resultant increases of swidden agriculture and concomitant deforestation and decline of agricultural productivity led to the collapse of Classic Maya civilization between 790 and 950. Ray Matheny has investigated agricultural systems in both the highlands and lowlands of traditional Maya territory. He has concluded that there were significant agricultural practices beyond the swidden techniques generally attributed to the Maya, and that they were capable of terracing and water management in highlands and of raised-bed and ridge farming in marshy, low-lying areas.[39]

TECHNOLOGY

Where human activity focuses on control of the environment rather than prediction of changes in it, it is usually called technology rather than science. Although native practices may have been relegated to the realm of technology, they are obviously based on the ability to predict events, as in the case of agriculturalists who used the stars to indicate planting seasons. However, native knowledge was generally more practical than theoretical.

In certain areas technology of a highly sophisticated sort played an important role in Indian cultures. Ethnographic studies of native American cultures have included much information about the material culture of native people, and their technologies have been well presented. The Andean region of South America is particularly renowned for its unexcelled examples of weaving techniques and metalworking. Both have been extensively studied as examples of material culture. However, a recent and intriguing study goes beyond the purely material to a study of Andean values systems as they are exemplified in metallurgy. Heather Lechtman discusses the concern of cultures in the Andean region with form as well as substance. She analyzes techniques of gold plating and the importance of investing external forms with cultural meaning. The plating of shapes with gold reflected cultural values associated with those forms and thus gives additional meaning to the technology involved. An intriguing combination of technology and mathematics is represented by the *quipu,* the knotted cord that was a record-keeping and possibly mnemonic device of the Incas. Marcia and Robert Ascher argue that the *quipu* may be evidence of sophisticated mathematical ability as well as of sophisticated weaving techniques, although their interpretation is problematic.[40]

Form and meaning in native American cultures are more directly related than in contemporary American culture. Symbol, myth, and ritual certainly played more important roles in native cultures than they do in American society today. Mythology and ritual in Indian cultures have been associated with religion, as natural law has been associated with science. Studies such as Lechtman's can

---

[39] William M. Denevan, "Aboriginal Drained-Field Cultivation in the Americas," *Science,* 1970, *169*:647–654; Pedro Armillas, "Gardens on Swamps," *Science,* 1971, *174*:653–661; B. L. Turner, II, "Prehistoric Intensive Agriculture in the Mayan Lowlands," *Science,* 1974, *185*:118–124; Ray T. Matheny, "Maya Lowland Hydraulic Systems," *Science,* 1976, *193*:639–645; Matheny and Deanne L. Gurr, "Ancient Hydraulic Techniques in the Chiapas Highlands," *American Scientist,* 1979, *67*:441–449.

[40] Heather Lechtman, "Andean Value Systems and the Development of Prehistoric Metallurgy," *Technology and Culture,* 1984, *25*:1–36; Marcia Ascher and Robert Ascher, *Code of the Quipu Databook* (Ann Arbor: Univ. Michigan Press, 1978).

reveal the reasons behind technology, and an understanding of Inca culture may finally provide a clue to unraveling the meaning of the *quipu*.

FUTURE DIRECTIONS

Ethnohistory, ethnoscience, and the various ethnobotanies, ethnogeographies, ethnoentomologies, and other "ethno" studies indicate the wide range of disciplines and interests that have been involved in the study of native cultures of the New World. The studies have often been purely descriptive of the ways in which native people perceived, predicted events in, and controlled their environments.

Historians of science have increasingly been interested in the social and cultural context of scientific activity. Anthropologists have always been interested in the cultural context. The problem of combining the diachronic and the synchronic approaches of the two fields has already been mentioned. The possibilities for development, however, are rich. Native cultures in the New World had ways of organizing metaphysical concepts and physical reality into cultural wholes, as Lechtman has demonstrated for metallurgy and Kelley and Urton have demonstrated for astronomy.

Students of native science in the future must look to the native view of the world to understand the activities that they would classify as scientific and to understand that there are many activities that can be classified as scientific within an appropriately broad framework. They can look for elements of control and prediction not only in archaeological remains but also in mythology and ritual. They can realize that the activities of native people can be judged by criteria that do not necessarily reflect only the Greco-Roman, Judeo-Christian, Western scientific world view of modern society.

For those who wish to pursue such studies, the major source of bibliographic information on ethnographic studies of North American tribes is George Murdock's *Ethnographic Bibliography of North America*. It is divided by major cultural areas, subdivided by tribal groups, and includes both books and articles, current to 1975. Still a standard source of citations to ethnographic descriptions, it will provide good background for any serious student of North American tribes who wishes to examine their science. Mark Barrow, Jerry Niswander, and Robert Fortuine have compiled a bibliography of medicine that will also be useful. Richard Ford, whose work in American Indian ethnobotany has been very important, has published an overview of that topic. Stephen McCluskey's article "Archaeoastronomy, Ethnoastronomy and the History of Science" is a cogent discussion of the need for resolving distinctions between "primitive" and "scientific" world views. It also lists major sources in the field.[41]

Since there is no commonly accepted definition of native American science,

---

[41] George Peter Murdock, *Ethnographic Bibliography of North America*, 5 vols. (New Haven, Conn.: Human Relations Area Files, 1975); Mark V. Barrow, Jerry D. Niswander, and Robert Fortuine, *Health and Disease of American Indians North of Mexico: A Bibliography, 1800–1969* (Gainesville: Univ. Florida Press, 1972); Richard I. Ford, "Ethnobotany: Historical Diversity and Synthesis," in *The Nature and Status of Ethnobotany*, ed. Richard I. Ford (Anthropological Papers, Museum of Anthropology, Univ. Michigan, 67) (Ann Arbor, 1978), pp. 33–49; Stephen C. McCluskey, "Archaeoastronomy, Ethnoastronomy, and the History of Science," in *Ethnoastronomy and Archaeoastronomy*, ed. Aveni and Urton (cit. n. 4).

pertinent studies are scattered in a variety of disciplinary journals—*Antiquity* (for archaeological studies), *American Anthropologist, Ethnohistory, Economic Botany,* and *Archaeoastronomy.* There are two journals by the latter name, one an offshoot of the *Journal for the History of Astronomy* and the other the specialized publication of the Center for the Study of Archaeoastronomy at the University of Maryland. There is a journal devoted specifically to ethnobiology (*The Journal of Ethnobiology*).

The various approaches to the study of native American science should continue to produce new information, and a greater appreciation of the many forms of native investigation and explanation of the natural environment will produce exchange of ideas among scholars adopting these approaches. The field needs new directions in research toward understanding cultural systems of thought about the physical world and its explanation. Ultimately, the study of the history of native American science can meld archaeology, anthropology, and history and can enrich the history of science with a way of seeing culture as a system of belief that influences the responses of people to the understanding of the natural phenomena around them.

Science and Technology

By George Wise*

T
HE RELATION OF SCIENCE TO TECHNOLOGY is not the stuff of front page news. But on 7 August 1984 it made the front page of the Science section of the *New York Times*. "Does Genius or Technology Rule Science?" a headline read. The story beneath described a "new school" of historical thought that "lauds technology as an overlooked force in expanding the horizons of scientific knowledge." It attributed the new view to the late historian Derek de Solla Price, who had, in his last lecture and paper, rejected what he described as the "remarkably widespread wrong idea that has afflicted generations of science policy students . . . that science can in some mysterious way be applied to make technology." Instead, he had argued that technology, as embodied in scientific instrumentation, is "autonomous and did not arise from the cognitive core of science, but from other technologies devised for quite different purposes. Much more often than is commonly believed, the experimenter's craft is the force that moves science forward." The *Times* article described a wide range of other historians' reactions to Price's views. William Broad, the reporter, noted that the whole matter was of more than academic interest: at stake, he wrote, was "an ongoing debate on how to spend billions of dollars of federal funds."[1] The purpose of spending the money was to generate technological innovations (that is, inventions that are used). If science drives technology, the money should be spent on science. If technology drives both itself and science, then the money should be spent on technology.

This review will examine two of the main issues raised in that newspaper article. Is technology dependent on science? And, if not, what is the relationship between the two? It will focus on what two groups of people in the United States—science-policy makers and historians—have, since 1945, thought the relationship of science and technology to be. An implicit argument took place in which the policymakers based their policies on a simple but incorrect model, while the historians began to gather the pieces for a new model not yet built.

The oversimplified model favored by the policymakers depicts science and technology as an assembly line. The beginning of the line is an idea in the head of the scientist. At subsequent work stations along that assembly line, operations labeled applied research, invention, development, engineering, and marketing transform that idea into an innovation. A society seeking innovations should, in the assembly-line view, put money into pure science at the front end of the process. In due time, innovation will come out of the other end.

Historians of science and technology have not merely reversed the direction of the assembly line so that technology now generates science. Instead they

* General Electric Research and Development Center Schenectady, New York 12301.

[1] William Broad, "Does Genius or Technology Rule Science?" *New York Times*, 7 Aug. 1984, p. C-1; Derek J. de S. Price, "Of Sealing Wax and String," *Natural History*, Jan. 1984, *93*(1):49–56.

have rejected it, but not yet replaced it. They have created some of the pieces for a new model. But rather than building that new model, they have put forward metaphors depicting science and technology as mirror-image twins, a married couple, a lemon and lemonade, opposing armies, opposing meteorological fronts, or sovereign states. The key idea behind all the metaphors is autonomy. Science and technology are viewed as autonomous with regard to one another, though far from autonomous with regard to economics, politics, and ideologies. But no new model for the way these two autonomous enterprises act on each other has yet emerged.

"Science" will be used in this review primarily to mean knowledge about nature, acquired for its own sake, and secondarily to mean the institutions and people who generate that knowledge. "Technology" will be used primarily to mean knowledge about the man-made world, generated for use, and secondarily the community of people (including engineers, inventors, scientists, and craftsmen) who contribute to this knowledge base. This second definition, if generally accepted, would make the assertion that science provides the knowledge base for technology meaningless. But the definition is not generally accepted. More often, technology is used as a synonym for "tools" or as a synonym for "engineering" and science is used as synonymous with knowledge. This review will regard the knowledge behind the tools, not the tools, as the essence of technology.

Instead of examining the alleged dependence of technology on science, a review of the relation between science and technology might have emphasized the communities where technology and science have been generated, such as nineteenth-century Manchester, or twentieth-century Palo Alto; the interplay of science, technology, and ideology; or the view that distinctions between science and technology are mere reflections of struggles between people or between groups for status or supremacy.[2] The approach presented here, however, avoids diffusing the issue into those more general questions of geography, politics, economics, and ideology. The focus is on a relatively narrow and unintended dialogue that has occurred over the past forty years between the champions of the assembly-line model and the champions of the various autonomy metaphors. That narrow dialogue discloses a major difference in the way two concerned groups of Americans, policymakers and historians, have viewed the relation of science and technology in modern America.

SCIENCE-POLICY MAKERS VERSUS HISTORIANS, 1945–1960

Looking back from the vantage point of 1960, historian John Beer recollected what everybody had known in 1945. It used to be commonly accepted, he wrote,

---

[2] On the study of communities and their role in science and technology, see, e.g., Arnold Thackray, "Natural Knowledge in Cultural Context: the Manchester Model," *American Historical Review*, 1974, *79*:672–710; and Robert Kargon, *Science in Victorian Manchester: Enterprise and Expertise* (Baltimore: Johns Hopkins Univ. Press, 1977). On science as the tool of a business elite, see David F. Noble, *America by Design: Science, Technology, and the Rise of Corporate Capitalism* (New York: Knopf, 1977). On the history of science as a story of groups competing for status and resources, see Robert E. Kohler, "Foreword: the Interaction of Science and Technology in the Industrial Age, "*Technology and Culture*, 1976, *17*:621–623; and the development of the ideas in that brief article in Kohler, *From Medical Chemistry to Biochemistry: The Making of a Biomedical Discipline* (Cambridge: Cambridge Univ. Press, 1982).

that technology was only applied science; that the rate of conversion of science and technology went in direct proportion to the money spent; and that, because the way to set up science-invention assembly lines, as exemplified by giant research laboratories, was now understood, the time between discovery and innovation was rapidly diminishing.[3]

But who had commonly accepted these views? Not Beer himself, who presented them in order to refute them. Not the general public: it largely ignored differences between science and technology. The "everyone" was a small elite of leaders, mainly drawn from academic science departments or deans' offices, who made national science policy. In 1945, that small group wrote a report entitled *Science, the Endless Frontier,* which proclaimed: "New products, new industries, and more jobs require continuous additions to knowledge of the laws of nature. . . . This essential new knowledge can be obtained only through basic scientific research." "Only" is the key word.[4]

Vannevar Bush, the report's principal author, was an MIT engineer with broad experience: he had invented a computer, participated in the creation of the electronics company Raytheon, and headed the major United States military research and development organization during World War II, the Office of Scientific Research and Development. He is sometimes saddled with responsibility for the report's more drastic oversimplifications. But a recent history of the National Science Foundation (NSF) suggests that he permitted them reluctantly. He had hoped that the report would include under the title of research "pioneering efforts of a technical sort," as exemplified by the Wright Brothers. But he found to his annoyance that the panels drawing up the report did not think that "a couple of bicycle mechanics working on a flying machine would . . . be doing research." He hoped briefly that the panels might be enlarged to include members representing "the rugged type of thing that the Wright brothers exemplified," but he did not push his views, and no endorsement of that type of research appeared in the report.[5]

Most members of the science policy elite sided instead with a view expressed by such academic science leaders as James B. Conant. A student of the history of science, Conant was willing to concede that "the cut and try empiricism of practical men" had been important back around 1850. But today, in the mid-twentieth century, "from the labors of those who were interested only in advancing science have come the ideas, the discoveries, the new instruments which have created new industries and transformed old ones." The applied scientist, in Conant's view, inevitably runs into a dead end. "Nine times out of ten," it is the pure scientist who provides the needed knowledge.[6]

Bush's dream of a National Research Foundation supporting the work of modern-day counterparts of everyone from Einstein to the Wright brothers gave way to a National Science Foundation aimed at supporting Einsteins only

---

[3] John J. Beer, "The Historical Relations of Science and Technology" (introduction to papers read at the 7th Annual Meeting of the Society for the History of Technology, 29 Dec. 1964) *Technol. Cult.,* 1965, 6:547–550.

[4] Vannevar Bush, *Science, the Endless Frontier* (Washington, D.C., 1945; rpt. NSF, 1960).

[5] J. Merton England, *A Patron for Pure Science: The National Science Foundation's Formative Years, 1945–57,* (Washington, D.C.: NSF, 1982), p. 14.

[6] James B. Conant, *Science and Common Sense* (New Haven: Yale Univ. Press, 1951), pp. 325–326.

(though in practice, of course, it supported much else, even a little history). The head of that foundation for its first decade, Alan T. Waterman, hewed consistently to the assembly-line ideal. Reissuing *Science, the Endless Frontier* in 1960, Waterman retreated not an inch from its advocacy of pure science. "The general public is still far from a true understanding of the nature of basic research and of the fundamental difference between science and technology." What is that true understanding? Speaking at the 1952 NSF budget hearings, he explained: "Technological advances are made possible only through the application of fundamental knowledge already known." Again, the word "only."[7]

Not much argument was needed to sell that "only" to a generation exposed since the mid 1930s to nylon, radar, synthetic rubbber, the proximity fuse, the atomic bomb, television, and the transistor—all apparently applications of basic scientific discoveries. Surely pure science was paying off. The assembly-line model was sound politics, undertaken not just for the aggrandizement of science, but for the public good.

But was sound politics good history? The histories of science and technology written in this postwar period gave at best limited support to the assembly line view. For example, the first American Ph.D. in history of science, I. Bernard Cohen, was commissioned by the science policy elite to educate the public about the value of science. But his conclusion, as contained in his 1948 book *Science, Servant of Man*, was somewhat less sweeping than those of *Science, the Endless Frontier*. "One inescapable result of studying the history of science," he wrote, "is the conclusion that many practical innovations such as our electric power system, the new weed-killers, radio and radar, nylon, and even advances in the practical art of medicine have come about primarily as the *by products of the search for truth in the scientific laboratory*" (italics in original).[8] No "only" this time; "many" instead.

A pioneering 1957 study of an industrial research laboratory also presented a more cautious view of the power of pure science. Kendall Birr looked at the first half century of the General Electric Research Laboratory and indeed found that "perhaps the outstanding characteristic of the Laboratory . . . was the willingness of both Laboratory and company management to gamble on fundamental research." He judged the gamble a success, but he hedged this conclusion with qualifications. Even in an industrial research laboratory, work was balanced among fundamental research, development, and troubleshooting. And General Electric was unusual, "more a pioneer than a typical example"; its broad technical interests and strong financial status gave it an uncommon capability of exploiting discoveries.[9]

If those sympathetic to the pure-science ideal were cautious in their conclusions, those opposed to it were not. A trio of economists, John Jewkes, David Sawers, and Richard Stillerman, compiled case studies of sixty-one important inventions of the twentieth century, from acrylic fibers to zip fasteners. They concluded that "the theory that technical innovation arises directly out of, and

---

[7] Bush, *Science, the Endless Frontier* (1960), p. xxvi; England, *A Patron for Pure Science*, p. 152.

[8] I. Bernard Cohen, *Science, Servant of Man: A Layman's Primer for the Age of Science* (Boston: Little, Brown, 1948).

[9] Kendall Birr, *Pioneering in Industrial Research: The Story of the General Electric Research Laboratory*, (Washington, D.C.: Public Affairs Press, 1957).

only out of, advance in pure science does not provide a full and faithful story of modern invention."[10]

These examples show that the historians of the period 1945–1960 by no means naively accepted the views of the policymakers of their time. Careful reading by policymakers of this historical literature might have caused them to question the assembly-line model. There is no evidence, however, that those making science policy read that literature, carefully or otherwise.

GROWING DOUBTS AND ALTERNATE DEFENSES, 1960–1975

In the 1960s, some policymakers nevertheless faced growing doubt in Washington that the money spent since 1945 on science had paid off adequately. One senator publicly referred to fusion research, a pet of the scientific community, as a "dead horse." President Lyndon Johnson pointedly told the scientific community that they had done plenty of research; it was time to start applying it. Harry Johnson, an economist asked by the National Academy of Sciences (NAS) to address the economics of pure research, wrote that the justification for pure science presented by the scientists "differs little from the historically earlier insistence on the obligation for society to support the pursuit of religious truth, an obligation recompensed by a similarly unspecified and problematical payoff in the distant future."[11]

One option open to the science policy elite was to abandon the assembly line and switch to a justification of science on cultural grounds alone. But that too had its dangers. "The basic difficulty with the cultural justification for pure science," physicist Harvey Brooks explained in the same NAS study, "is that it does not provide any basis for quantifying the amount of support required."[12] If science could claim credit for driving technology, and through it the economy, then levying an overhead charge for it on the gross national product was justified. If not, then why spend more federal funds on an observatory than on an opera?

Project Hindsight, sponsored by the Department of Defense in the 1960s, sought to answer that question by measuring "the payoff to Defense of its own investments in science and technology." The project team identified the "Events" that led to the development of twenty military systems. It found that "only 0.3 percent of the Events were classified as undirected science." As the team's leaders recognized, this was not a denial of the value of undirected science; it was a denial that undirected science fed directly into invention in the short term (less than twenty years). "It is clear that, on a 50 year or more time scale," the Hindsight team concluded, "undirected science has been of immense value." A subsequent study of civilian innovations, Project TRACES, sponsored by the NSF, confirmed this longer-term impact of science and generally gave results much more supportive of the assembly-line view. A third study team,

---

[10] John Jewkes, David Sawers, and Richard Stillerman, *The Sources of Invention* (London: Macmillan, 1963), p. 7.

[11] Joan Lisa Bromberg, *Fusion: Science, Politics and the Invention of a New Energy Source* (Cambridge, Mass.: MIT Press, 1982), p. 117; Daniel S. Greenberg, "Basic Research: The Political Tides are Shifting," *Science* (1966), *152*:1724–1726; Harry P. Johnson, "Federal Support of Basic Research," in National Academy of Sciences, *Basic Research and National Goals: A Report to the House Committee on Science and Astronautics*, (Washington, D.C.: GPO, 1965), pp. 127–141.

[12] Harvey Brooks, in National Academy of Sciences, *Basic Research and National Goals*, p. 86.

sponsored by the Materials Advisory Board, set out deliberately to fit case studies to the assembly line sequence, but found this to be impossible.[13] With their widely varying conclusions, the exercises tended to deepen the suspicion that the relation between science and technology was more complicated than any simple model such as the assembly line would suggest. Historians of science and technology already doubted that technology was merely applied science. As that view came under increasing attack (though not necessarily because of the attacks), some historians began to explore alternatives. The essence of those alternatives can be summed up in the phrase "technology is knowledge."

In part, this argument emerged from detailed studies of specific relationships between areas of science and technology. Cyril Stanley Smith, a metallurgist turned historian, found that in the history of materials, science lagged rather than led technology, well into the twentieth century. Art, rather than science or economics, drove materials technology: "Almost all inorganic materials and treatments to modify their structure and properties appear first in decorative objects rather than in tools or weapons." Only in very recent times, since about 1950, has science begun answering some simple questions about materials: for example, why you can see through a pane of glass, and why it shatters when dropped, while you can see your reflection in a sheet of metal, and it stays whole and rings when dropped. Even answering these questions has rarely led to new technology. Instead it usually confirms the wisdom of cut-and-try predecessors. That wisdom was not written down, but it was nevertheless a form of knowledge.[14]

Other fields showed similar characteristics. As Lynwood Bryant showed, when inventors tried to apply science to the internal combustion engine, they found surprises. For example, Rudolf Diesel's engine was intended to be science based: an embodiment of the Carnot cycle, an ideal sequence of heating, expanding, cooling and contracting a gas aimed at maximizing efficiency. But the actual invention was shaped by technical realities, rather than a scientific ideal (specifically, Diesel found it necessary for technical reasons to add heat not at constant temperature, as in a Carnot cycle, but at a temperature that first rose, then fell). Again, technology created its own knowledge base, rather than merely applying scientific knowledge.[15]

Similarly, late nineteenth- and early twentieth-century electrical technology proved far more than a simple application of the ideas of Faraday and Maxwell. Thomas P. Hughes and James Brittain have depicted an engineering community sophisticated in its theory as well as daring in its experimentation, epitomized by such figures as Charles P. Steinmetz and Ernst F. W. Alexanderson, who operated parallel to, and in important ways independently of, the scientific community.[16]

[13] Chalmers W. Sherwin, and Raymond S. Isenson, "Project Hindsight," *Science,* 1967, *156*:1571–1577; Illinois Institute of Technology Research Institute, *Technology in Retrospect and Critical Events in Science* (National Science Foundation contract NSF-C535), 2 vols. (Chicago: IIT Research Institute, 1968); Materials Advisory Board, *Report of the Ad Hoc Committee on Principles of the Research-Engineering Interaction* (Washington, D.C.: National Academy of Sciences, 1966).

[14] Cyril Stanley Smith, *A Search for Structure: Selected Essays on Science, Art, and History* (Cambridge, Mass.: MIT Press, 1981).

[15] Lynwood Bryant "The Role of Thermodynamics in the Evolution of Heat Engines," *Technol. Cult.* 1973, *14*:152–165.

[16] James E. Brittain, "C. P. Steinmetz and E. F. Alexanderson: Creative Engineering in a Cor-

Finally, Joseph Schmookler, an economist, studied an important expression of the knowledge base of technology: patents. His graphs of patents issued in a selection of technologies over extended periods of time showed rises and falls that matched very well with the rises and falls of conventional economic indicators. This observation suggested that economic forces, not scientific discovery, drove invention.[17] (Advocates of the science-based assembly line could, however, point out that the fields Schmookler studied—railroads and textiles for example—were hardly the high-technology areas with which the assembly line was most concerned.)

Though the cases were selective and far from comprehensive, some historians felt sufficiently emboldened to try to generalize. Stop asking how science shaped technology, they urged. Start asking, are science and technology separate communities, and if so what is the relation of each to the other? The answers all centered on the autonomy of technology's knowledge base, though expressing it in the form of many different metaphors.

Melvin Kranzberg depicted a married couple, emerging in the twentieth century from a "long and indifferent courtship" into a marriage of convenience, not a love match. The complexity of the problems that arose in the twentieth century forced the scientist to rely more on the technologist for apparatus and information, and the technologist to rely more on the scientist for knowledge and insight. But the two communities remained distinct because their purposes remained distinct. The scientist aims to understand nature; the technologist aims to make useful things. In making useful things he usually does not wait on the scientist for knowledge. "New technology grows mostly out of old technology, not out of science."[18]

Edwin Layton explicitly asserted that technology is knowledge, and then elaborated this insight by looking at one important portion of technology, engineering. In perhaps the most widely quoted metaphor, he portrayed science and engineering as mirror-image twins. Engineering in the late nineteenth century did not become merely dependent on science. Instead, it developed its own explicit knowledge base, professional institutions, and publications practices. These were modeled on parallels in science (hence the twin). But the emphasis placed on elements in the knowledge base was reversed (hence the mirror image): design and hardware occupied the most important roles, with publication serving as a support to professionalism, rather than its essence. The engineer's purpose of creating useful objects, machines, or systems determined the structure and method of acquisition of his knowledge.[19]

Much of Layton's subsequent research has explored a field where technology-as-knowledge is most evident: fluid flow and its applications. Subsequently, Terry Reynolds has traced the roots of water-power technology, and Bruce

---

porate Setting," *Proceedings of the IEEE*, 1976, *64*:1413–1417; Thomas P. Hughes, *Networks of Power: Electrification in Western Society* (Baltimore: Johns Hopkins Univ. Press, 1983).

[17] Jacob Schmookler, *Invention and Economic Growth* (Cambridge, Mass.: Harvard Univ. Press, 1966).

[18] Melvin Kranzberg, "The Disunity of Science-Technology," *American Scientist*, 1968, *56*:21–34; and Kranzberg, "The Unity of Science-Technology," *ibid.*, *55*:48–66.

[19] Edwin Layton, "Mirror Image Twins: The Communities of Science and Technology in 19th Century America," *Technol. Cult.* 1971, *12*:562–580; and Layton, "Technology as Knowledge," *Technol. Cult.*, 1974, *15*:31–41.

Sinclair has described the generation of an engineering-knowledge base for that technology in the mid-nineteenth-century United States as part of his history of the Franklin Institute. Among those carrying the story into the twentieth century, Walter Vincenti has contrasted the analytical methods used by engineers and physicists when each looked at the dynamics of moving fluids.[20]

Another generalizer chose a third and vaguer set of metaphors, featuring fronts of a military or meteorological nature. Derek Price asked, "Is technology historically independent of science?" and answered yes. Science is defined by a body of scientific literature and is most active at the "research front," the leading edge of that body, the new papers appearing today. Technology is not interested in knowledge committed to paper. It is a moving state of the art, drawing occasionally on old science but mainly on itself. In case science-policy makers missed the point of all this, Price spelled it out. "Beware of any claims that particular scientific research is needed for particular technological potentials, and vice versa. Both communities can only be supported for their own separate ends."[21]

By 1972 most historians of science and technology had accepted that technology is knowledge, not merely applied science. This was most graphically demonstrated at a meeting held in that year at the Burndy Library, whose proceedings were published in 1976 in the journal *Technology and Culture*. Ostensibly on the topic "The Relationship of Science and Technology," the meeting was in fact an extended funeral for the old assembly-line or technology-as-applied-science view. Since 1972, historians have left that view behind.

Has the historical consensus convinced the science-policy makers? The signals differ. As recently as 1980, the director of the National Science Foundation reprinted *Science, the Endless Frontier* for a third time and included in his introduction the statements that "basic research . . . creates the fund from which the practical applications of knowledge must be drawn. New products and new processes do not appear full grown. They are founded on new principles and new conceptions, which in turn are painstakingly developed by research in the purest realms of science." He concluded that "those statements are as true today as when they were written thirty-five years ago." And the physicist Leon Lederman, writing in the November 1984 issue of *Scientific American*, gave arguments based on the assembly-line model that might have been taken from *Science, the Endless Frontier*. Though conceding that he had not calculated the direct impact of pure science on technology, he was certain it would be a straightforward matter to show that it is large. Major parts of tomorrow's technology will, in Lederman's view, follow directly from specific discoveries made in experimental elementary particle research or pure theoretical physics.[22]

---

[20] Terry Reynolds, *Stronger Than A Hundred Men: A History of the Vertical Water Wheel* (Baltimore: Johns Hopkins Univ. Press, 1983); Bruce Sinclair, *Philadelphia's Philosopher Mechanics: A History of the Franklin Institute, 1824–1865*, (Baltimore: Johns Hopkins Univ. Press, 1974); Walter G. Vincenti, "Control-Volume Analysis: A Difference in Thinking Between Engineering and Physics," *Technol. Cult.*, 1982, 23:145–174.

[21] Derek J. de S. Price, "Is Technology Historically Independent of Science? A Study in Statistical Historiography," *Technol. Cult.*, 1965, 6:553–568; and Price "A Theoretical Basis for Input-Output Analysis of National R & D Policies," in *Research, Development, and Technological Innovation: Recent Perspectives on Management*, ed. Devendra Sahal (Lexington, Mass., Lexington Books, 1980), pp. 251–260.

[22] Richard C. Atkinson, "Introduction," in Vannevar Bush, *Science, the Endless Frontier* (Wash-

But other important statements made in 1984 indicate that other policymakers are less firmly committed to the assembly line. When the president's science adviser discussed the proposed federal government research and development budget for fiscal year 1985, he still emphasized the "renewed—and considerably strengthened—commitment to federal support for basic research." But in the rationale for that research, training of scientists (those aimed at both basic and applied research careers) got first priority. Challenging intellectual frontiers came next. The statement that basic research "provides new knowledge that drives our economic growth, improves our quality of life, and underlies our national defense" was tossed in as a supporting point. No claim followed that basic research was the only, or even the most important such driver and improver.[23]

An even bigger shift came in an endorsement by the chairman of the National Science Board (the board of directors of the NSF) that the NSF would now treat engineering not as a science discipline, or as an application of science, but as an area with a knowledge base and research needs of its own. That chairman gave no indication of having read the historical literature (or knowing about Vannevar Bush's plea for the Wright brothers). But his reasons echo the reasoning of Layton, Price, and Kranzberg.[24]

FROM METAPHORS TO ALTERNATIVE MODELS

One reason that the battle against the assembly line has not been completely won is that its rivals remain metaphors rather than models, and metaphors are good stimulants to thinking but unreliable tools for answering questions. So far, historians such as Layton, Kranzberg, and Hughes who have put forward the view that technology is more than a stepchild of science have not provided an alternative model that is just as clear as the assembly line and gives a more realistic depiction of the way science and technology influence each other and the society that supports them. They and many other historians have shaped some of the pieces out of which such a model might be made. Those pieces operate on different levels, and are sometimes far more opaque than the ones that go to make up the transparent, if incorrect, assembly-line model. But they seem to point the way toward a future synthesis.

The pieces include one that was well defined by 1972, the concept of technology-as-knowledge, and three others that have emerged since: the presumptive anomaly; the balance of momentum and external pressure; and the role of the research entrepreneur.

Science, Technology, and the Origins of Innovation: The Presumptive Anomaly. In the assembly-line tradition, linear models explain the origins of innovation by a pull from society at one end of the line, or a push from science at the other. But the concept of technology as autonomous suggests that both of these approaches are inadequate.

Edward Constant's study of the invention of the jet engine provides a new

ington, D.C.: NSF, 1980), p. xii; Leon M. Lederman, "The Value of Fundamental Science," *Scientific American* (Nov. 1984), *251*(11):40–47.

[23] George Keyworth, "Four Years of Reagan Science Policy," *Science,* 1984, 224:9–13.

[24] Lewis M. Branscomb, "Engineering and the National Science Foundation," *Science,* 1984, *224*:10.

approach. He describes a community of technologists, those concerned with air-craft engines, making a technological revolution in much the same way that scientists, in one popular view of the history of science, make scientific revolutions. They begin with an old way of looking at things: in this case, that propellers drive airplanes. Within that way of looking at things, a few visionaries see anomalies, or problems that cannot be solved within the old framework. These anomalies do not block current development. They are, instead, "presumptive anomalies": they represent problems that will emerge if the current way of doing things is extended into the future. Here the presumptive anomaly is a limit to the speed of airplanes. A creative inventor able to look beyond today's successes and stare that presumptive anomaly in the face will gain the insight needed to make the jump to the jet engine.[25] Where does science fit in? It is not, as in the assembly-line view, the source of all change; instead, it is a resource to be drawn on by the perceiver of the presumptive anomaly. But not a mere passive resource, for recognition of the possibility of innovation can spur research—particularly engineering research, the acquisition of the knowledge that engineers need to get on with their jobs.

Other studies have taken similar views. Robert Bruce and David Hounshell have looked at the invention of the telephone and have found its origins in a presumptive anomaly. Many inventors in the 1870s envisioned the eventual need for putting many, rather than just one, two or four, telegraph signals onto a single telegraph line. But only two of them, Alexander Graham Bell and Elisha Gray, envisioned the limits of coded pulses as compared to a voice as a message form, and conceived the telephone.[26]

Hugh Aitken, studying the origins of radio, draws on similar ideas in depicting the origins of both the initial "syntony and spark" radio systems and the later "continuous wave" radio systems in terms of the relations of communities of scientists, technologists, businessmen, and government officials. Radio is more than a classic assembly-line invention (Maxwell begat Hertz who begat Marconi): there were important feedbacks from one community to another. Science, in the form of Maxwell's theory of electromagnetism, becomes not the beginning of a process but a resource to be drawn on by people who perceived the ultimate limitations of wired telegraphy. Long before any crisis in communications capacity had actually occurred, they created spark telegraphy. And long before its potential was exhausted, a few other visionaries perceived the need for a new technology based on continuous waves. This change would later draw in science, in the form of the physics behind the ultimately most successful generator of continuous waves—the vacuum tube.[27]

The most prominent embodiment of twentieth-century "high technology," microelectronics, seems superficially to be an ideal case of the assembly line: curiosity-driven discovery in an esoteric scientific field (the quantum theory of solids) spawns an industry. But, as studies by Ernest Braun and Stuart

---

[25] Edward Constant, *Origins of the Turbojet Revolution* (Baltimore: Johns Hopkins Univ. Press, 1980).

[26] Robert V. Bruce, *Bell: Alexander Graham Bell and the Conquest of Solitude* (Boston: Little, Brown, 1973); David Hounshell, "Elisha Gray and the Telephone," *Technol. Cult.*, 1975, *16*:133–161.

[27] Hugh Aitken, *Syntony and Spark: The Origins of Radio* (N.Y.: Wiley Interscience, 1976); and Aitken, *The Continuous Wave* (Princeton, N.J., Princeton Univ. Press, 1985).

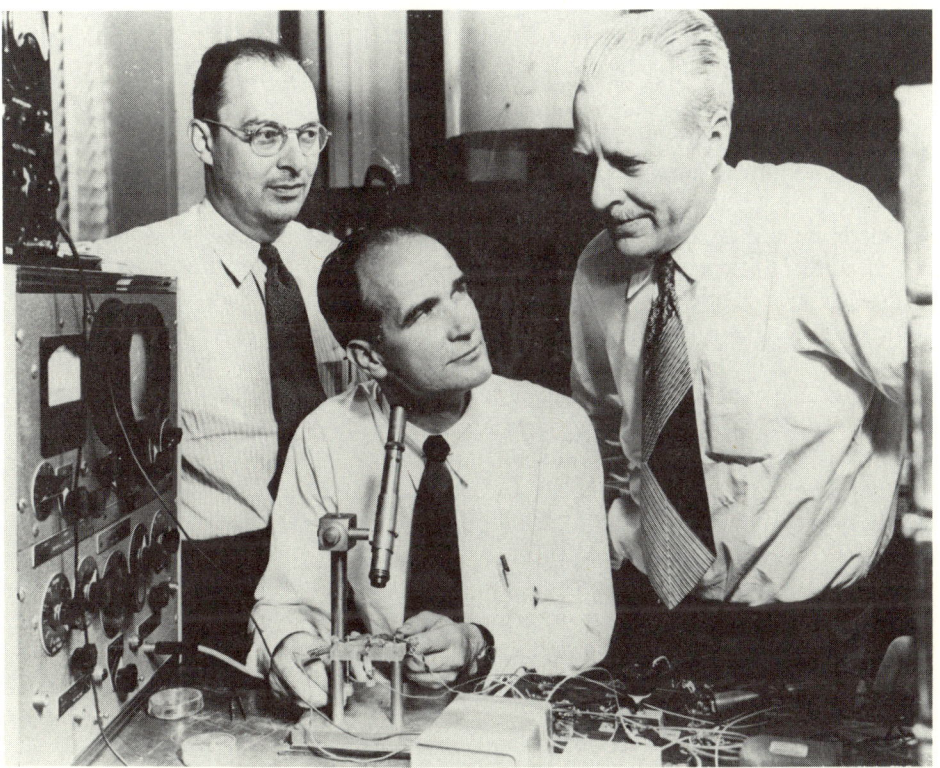

Capitalizing on a presumptive anomaly. Members of the Bell Laboratories team that invented the transistor: William Shockley, seated; John Bardeen, standing left; and Walter Brattain. A T & T Bell Laboratories, Archives/Record Management Services. Printed with permission.

MacDonald, Lillian Hoddeson, and others have shown, technology was involved from the first. Scientific insight may have helped the Bell Laboratories team pick the right area in which to look for a solid-state telephone amplifier. But the fact that they were looking for one at all owed much to the technological insight of a Bell research manager, Mervin Kelly. As early as 1936, more than a decade before the invention of the transistor, he pointed out long-run limitations to vacuum tube switching and the desirability of a solid-state substitute. Other groups, motivated by more purely scientific considerations, looked at the same science but did not find the transistor.[28]

Finally, Joan Bromberg's history of the effort to develop magnetically contained fusion as an energy source shows that some people who see the limits of existing technology may greatly underestimate the time it will take them to provide an alternative. (In the case of fusion, that time may prove to be infinite.) The history of fusion also shows how ambitious innovation efforts proceed under shifting rationales and call forth new types of knowledge. The knowledge needed for invention may not be in the form previously produced by scientists. Fusion

[28] Ernest Braun and Stuart MacDonald, *Revolution in Miniature: The History and Impact of Semiconductor Electronics Re-explored* (2nd ed.; Cambridge: Cambridge Univ. Press, 1982); Lillian Hoddeson, "The Discovery of the Point-Contact Transistor," *Historical Studies in the Physical Sciences,* 1981, *12*:41–76.

research created demands for new knowledge about hydrodynamic instabilities, for example. Much the same thing happened in other fields. The aerodynamics used in inventing better aircraft engines and the electron physics used by vacuum tube inventors was not necessarily the aerodynamics or electron physics already available from the physics laboratory.[29]

Inventions have themselves helped to shape scientific advance. Many examples can be put forward, from the microscope to the laser. But they ought to be accompanied by two warnings: the effects of technology on science vary widely across different scientific fields; and, again and again, from the air pump to the particle accelerator, scientists have conceived and built their own technology, rather than simply learning from the engineers.

The best studies of the influence of technology on science have concerned a field particularly dependent on technology: astronomy. Books by Martin Harwit, Richard Hirsh, and David Edge and Michael Mulkay describe how a series of inventions, from telescopes and diffraction gratings to radio antennas and rockets, opened up new channels for observing the universe, and how surprising messages coming in over these channels have reshaped the thinking of astronomers. In the most general of these books, Harwit concludes that the typical mode of discovery in astronomy is the seizure of a new technology by astronomers, or by outsiders with an interest in astronomy, and its quick exploitation to skim the observational cream off a new field. Technology becomes a resource. Scientists often exploit forefront technology faster than technologists exploit forefront science.[30]

It would be premature to extrapolate this conclusion from astronomy to science as a whole. Astronomy is an observational science, especially dependent on ways of seeing. Other sciences may be more theory-driven and as a result may generate their own technologies more actively. In other cases the results will be mixed, as in high-energy physics, where both theoretical goals and the moving horizon of technological possibilities drive the development of apparatus.[31]

Science, Technology, and the Growth of Systems and Institutions: Momentum, Salients, and Pressures. Once initiated, why does an innovation, a system, a profession, or an institution grow as it does? Why is growth not faster or slower? Why is the form hierarchical or decentralized, anarchic or autocratic? Explanations picturing progressive evolution fueled by science have given way to explanations involving internal momentum, uneven growth, and the opposition of external forces.

The outstanding recent study of growth and form, Thomas P. Hughes's *Networks of Power*, takes just this approach. It traces the growth and organization

[29] Bromberg, *Fusion* (cit. n. 11), pp. 248–256.

[30] David O. Edge and Michael Mulkay, *Astronomy Transformed: The Emergence of Radio Astronomy in Britain* (New York: Wiley, 1976); Richard Hirsh, *Glimpsing an Invisible Universe: The Emergence of X-Ray Astronomy* (Cambridge: Cambridge Univ. Press, 1983); Martin Harwit, *Cosmic Discovery: The Search, Scope, and Heritage of Astronomy* (New York: Basic Books, 1981).

[31] Arthur Norberg, "Cross-Fertilization of Innovations in Science and Technology: Radio-Frequency Circuits and Particle Accelerators in the 1930s," paper read at the XVth International Congress of the History of Science, Edinburgh, 1977; John L. Heilbron, Robert W. Seidel, and Bruce Wheaton, *Lawrence and his Laboratory: Nuclear Science at Berkeley, 1931–1961* (Berkeley: Lawrence Berkeley Laboratory and Office for History of Science and Technology, Univ. California, 1981).

of electric power systems in terms of first technological and then business communities solving their immediate problems. Inventors with a vision of systems get the jump on inventors who see only components. Those inventors and their business partners transfer technology across oceans; the modern light bulb moves east, the transformer moves west. The systems grow unevenly, until the bypassed areas (or reverse salients, as Hughes calls them)—the difficulty of sending direct current over long distances, or the difficulty of running a large motor on alternating current—become critical problems that help shape the research efforts of engineers. The systems' growth has now attained a momentum that enables them to overcome or compromise with outside political or economic pressures. Finally, with technological maturity, the technological problem of creating systems gives way to the mainly economic problem of creating regional networks.[32]

The scientist is little in evidence. Maxwell's laws may make a system possible, but they have less to do with its growth and form than does a more modest conceptual invention, load factor (a measure of how much of the electricity generating capacity of a system is actually used). The role of science in the growth phase is a supporting one. The main influence of science on electrical technology was a transient one, in education, for physicists, not engineers, were the first to perceive the possibilities of educating electrical engineers. The first generation of true electrical engineers came out of the physics laboratories. Then they rewarded the profession that spawned them by creating a discipline of their own independent of physics.[33]

Other studies of the growth and form of innovations and systems echo these conclusions. Martha Trescott's study of electrochemistry has shown how that industry also developed its own internal knowledge base and technical institutions, rather than relying on direct imports from science. No doubt the work of chemical researchers provided the basic ideas for the electrochemical industry. But the influence of science was, again, supporting and transient rather than central. Similarly, John Servos has traced the way the growth in demand for physical chemists was not a direct but an indirect result of industrial growth. That is, the industries did not at first hire physical chemists with Ph.D.s.; they hired engineering graduates who had taken chemistry courses, and the colleges hired the Ph.D. physical chemists to teach those courses.[34]

More recently, in the evolution of computers, the pacemaker of technology has been not science but practical needs, especially national defense. Scientific needs may have inspired such pioneers as John Atasanoff and John Mauchly, but the most successful innovators were the ones who coped most successfully with nonscientific issues: engineering design, patents, funding, project targets, and marketing. Only when computers became established did a discipline of computer science—a branch of engineering research—begin to emerge.[35]

[32] Hughes, *Networks of Power* (cit. n. 16), pp. 140–174.

[33] Robert Rosenberg, "American Physics and the Origins of Electrical Engineering," *Physics Today*, 1983, 36:48–53.

[34] Martha M. Trescott, *The Rise of the American Electrochemical Industry, 1880–1910: Studies in the American Technological Environment* (Contributions in Economics and Economic History, 38) (Westport, Conn.: Greenwood, 1981); John W. Servos, "The Industrial Relations of Science: Chemical Engineering at MIT, 1900–1939," *Isis*, 1980, 71:531–549.

[35] Kent C. Redmond and Thomas M. Smith, *Project Whirlwind: The History of a Pioneer Computer*, (Bedford, Mass.: Digital, 1980); Nancy Stern, *From Eniac to Univac: An Appraisal of the Eckert-Mauchly Computers* (Bedford, Mass.: Digital, 1981).

Reconciling Technological and Scientific Goals with the Immediate Needs of Patrons: The Role of the Research Entrepreneur. In rejecting science as the pacemaker of technology, historians have begun to suggest that the particular balance of scientific and technological efforts undertaken by an institution may depend more on the needs of patrons than on the direct influence of science and technology on each other. This interpretation puts emphasis on a new role, the "research entrepreneur": an individual dedicated to creating new science or new technology, but realistic enough to recognize that he must strike bargains with people who have very different interests if he hopes to accomplish his goals.

The role is not new. Joseph Henry provides a nineteenth-century example. The publication of the Henry papers, edited by Nathan Reingold and colleagues, and studies based on those papers have as one theme Henry's attempts to reconcile science and technology by viewing science as making technology possible, though by no means as the head of an assembly line. Social conditions call invention forth. The inventive genius is needed as well as the scientific. Henry's role in developing the telegraph and discouraging the electric motor was consistent with what he saw as the scientist's proper social and intellectual role in the creation of new technology. Other nineteenth-century figures made uneasier adjustments. Robert Post depicts the physicist-inventor Charles Page as seeking to vindicate his position as a true scientist, even as he got more deeply enmeshed in questionable government-funded schemes to develop a practical electric motor. And David Hounshell has described how even Thomas Edison identified himself as a "scientific man" when he needed scientists' approval of his electric lighting system, then broke with them when their pure-science ideal (and their occasional kind words about rivals' lighting systems) proved incompatible with his further needs for allegiance and support.[36]

Looking at a later part of the nineteenth century, Charles Rosenberg and Margaret Rossiter have shown how the directors of the federal government's agricultural experiment stations exemplified the research entrepreneur role. The burden of requests for advice and technology from farmers dampened the idealism of station scientists. But successful laboratory directors found a way to compromise their ideals in a way that looked a lot like surrender, but did eventually make possible important research (for example, on hybrid corn).[37]

Research entrepreneurs also created major twentieth-century industrial research laboratories. The successful laboratory directors were team players, not rugged individualists. Stuart Leslie has shown how even as individualistic a leader as General Motors' research director Charles Kettering learned the need to make institutional compromises—especially after an innovative air-cooled automobile engine his team developed failed commercially, as much because of organizational problems within GM as because of technical flaws in the engine.

[36] Arthur P. Mollella, "The Electric Motor, the Telegraph, and Joseph Henry's Theory of Technological Progress," *Proc. IEEE* (1976), *64*:1273–1278; Joseph Henry, *The Papers of Joseph Henry,* ed. Nathan Reingold (Washington, D.C.: Smithsonian Institution Press, 1972–present), 4 vols. to date; Robert Post, *Physics, Patents, and Politics: A Biography of Charles Grafton Page* (New York: Science History, 1976); David Hounshell, "Edison and the Pure Science Ideal in 19th Century America," *Science,* 1980, *207*:612–617.

[37] Charles E. Rosenberg, *No Other Gods: On Science and American Social Thought* (Baltimore: Johns Hopkins Univ. Press, 1976) Chs. 8–12; Margaret Rossiter, "The Organization of the Agricultural Sciences," in *The Organization of Knowledge in Modern America, 1860–1920,* ed. Alexandra Oleson and John Voss (Baltimore: Johns Hopkins Univ. Press, 1979), pp. 211–248.

Leonard Reich's studies of Bell Laboratories provide major new insights into the business pressures shaping industrial research, and his work and other studies of General Electric's laboratory show how that organization's director, Willis R. Whitney, extended the role of research entrepreneur. By first proving the laboratory's value as a defender of established company businesses, he earned for a few of his researchers the right to wander into more remote fields of science and technology.[38]

Clayton Koppes's history of the Jet Propulsion Laboratory shows that in the absence of such a research entrepreneur, scientists and technologists can allow external patrons to set priorities. The laboratory's two patrons, California Institute of Technology and the federal government (the Department of Defense and NASA) agreed, Koppes argues, that the laboratory's primary purpose would be service to the "warfare state," not science or technology. "Lacking a strong institutional ethic, the science and research-engineering communities allowed the organization and agenda of research to be determined disproportionately by military funding."[39] But the laboratory's federal patrons also sponsored a remarkable series of planetary explorations. The JPL's history must be seen as a compromise of scientific and nonscientific goals, not merely a surrender. People interested in space exploration as technology, and space scientists interested in knowledge about the solar system, achieved many of their own goals in the process of supporting the government in achieving its goals.

Two studies of twentieth-century science and technology suggest that the mediating role of research entrepreneurs has been eliminated by capitalist managers who dictate the roles of both science and engineering. David Noble's study of the behavior of important science and engineering educators and organizers argues that any idealistic rhetoric from them is only a smoke screen masking their total surrender to capitalist managerial hierarchies. David Dickson, in a survey of science policy since World War II, gives essentially the same message.[40]

Others have found more impressive the amount of autonomy scientists attained, and how well they have insulated themselves from their patrons. John Servos has described how an apparent victory at MIT of applied, industry-allied chemical engineering over "purer" forms of that discipline and of chemistry proved in fact only temporary. After 1930, the verdict was reversed; science

[38] Hoddeson, "The Discovery of the Point Contact Transistor" (cit. n. 28); Lillian Hoddeson, "The Emergence of Basic Research in the Bell Telephone System, 1875–1915," *Technol. Cult.*, 1981, 22:512–544; John J. Beer and W. David Lewis, "Aspects of the Professionalization of Science," in *The Professions in America*, ed. Kenneth S. Lynn (Boston: Houghton Mifflin, 1965); Beer, "Coal Tar Dye Manufacture and the Origins of the Modern Industrial Research Laboratory," *Isis*, 1958, 49:123–131; Birr, *Pioneering in Industrial Research* (cit. n. 9); Kendall Birr, "Industrial Research Laboratories," in *The Sciences in the American Context: New Perspectives*, ed. Nathan Reingold (Washington, D.C.: Smithsonian Institution Press, 1979); Stuart Leslie, "Charles A. Kettering and the Copper Cooled Engine," *Technol. Cult.*, 1979 20:752–778, Leslie, *Boss Kettering* (New York: Columbia Univ. Press, 1983); Leonard Reich, "Industrial Research and the Pursuit of Corporate Security: The Early Years of Bell Labs," *Business History Review*, 1980, 54:504–529; Reich, "Irving Langmuir: Engineer and Scientist," *Technol. Cult.*, 1983, 24:199–221; George Wise, "A New Role for Professional Scientists in Industry: Industrial Research at General Electric, 1900–1916," *Technol. Cult.*, 1980, 21:408–429; and Wise "Ionists in Industry: Physical Chemistry at General Electric, 1900–1915," *Isis*, 1983, 74:7–21.

[39] Clayton R. Koppes, *JPL and the American Space Program: A History of the Jet Propulsion Laboratory*, (New Haven: Yale, 1982), p. 247.

[40] Noble, *America by Design* (cit. n. 2); David Dickson, *The New Politics of Science* (New York: Pantheon, 1984).

and engineering research goals set by faculty members, rather than by external patrons, got priority. In *The Physicists* Daniel Kevles studies a discipline particularly susceptible to clashes between its pure and applied wings. But Kevles denies that business or technology shaped physics. Indeed, he downplays the role technology played in the growth and organization of the discipline. Such technological episodes as the invention of the transistor are left out of the account altogether. Technology is principally mentioned as something people who lacked understanding confused with science. The major treatment of the interaction of science and technology, the discussion of how the modern interdisciplinary research lab was created, appears in a chapter entitled "The Search for New Patrons." In the twentieth century, Kevles concludes, the physicists did not suffer domination by the capitalists. Instead, during and after World War II, the physicists, the capitalists and the politicians sat down around the table and came up with a compromise that satisfied all of them. They institutionalized the role played by research entrepreneurs within a science policy system under which the government would allow scientists to dictate their own government-supported research programs, and the scientists would help meet the technology needs of the government, particularly in the area of national defense. Under that compromise, the physicists enjoyed, until the 1970s, a remarkable degree of autonomy. A major idea that both reflected this privileged position and provided a justification for it is the rationale for pure research with which we began: the science-technology assembly line.[41]

CONCLUSION

Refuting the assembly-line model stands as a main contribution of the historians to the discussion of the relation of science and technology in modern America. In its place, most historians have asserted the autonomy of technology in relation to science (at the same time as they have been emphasizing that technology itself is *not* autonomous in relation to economics, politics, and international relations). All knowledge is not science; technology is knowledge, too. Science is invented, not revealed, and the tools of technology can help scientists invent it.

Treating science and technology as separate spheres of knowledge, both man-made, appears to fit the historical record better than treating science as revealed knowledge and technology as a collection of artifacts once constructed by trial and error but now constructed by applying science. The presumptive anomaly is a big improvement on tired debates about demand-pull versus technology-push. The balance between a technology's momentum and the realities, both scientific and social, that constrain it, offers a more realistic way of tracing the growth of systems and institutions than does the picture of technology as an irresistible juggernaut. And the figure of the research-entrepreneur is a key to understanding the past century's innovations in organizing research.

These pieces do not yet constitute a model. But the shape of such a model appears to be emerging. It will depict technology as an autonomous body of knowledge enriched but not driven by science. Major innovation emerges when

---

[41] Servos, "Industrial Relations" (cit. n. 34); Daniel J. Kevles, *The Physicists: The History of a Scientific Community in Modern America* (New York: Knopf, 1978).

creative individuals understand market needs, envision the future limits of current ways of meeting those needs, and acquire insight into new ways of overcoming the limits. Once innovation creates a new field of technology, that field generates its own internal logic of momentum, reverse salients, and response to external pressures. Research entrepreneurs find ways of drawing idealistic scientists and engineers into attacks on the field's practical problems. Eventually those ways become frozen into institutions and policies.

This model is not universally accepted. Some of its critics argue that it underestimates the domination of both science and technology by capitalism. Others deny that it makes any sense to distinguish science intellectually from technology. Yet others have even dismissed the relation between the two as a dying issue.[42]

An indication of that issue's vitality, however, is the recent undertaking of several major historical projects that depend heavily on it. Science-based companies such as DuPont and Rohm and Haas have commissioned books about themselves by historians of science and technology (David Hounshell and John K. Smith of the University of Delaware and the Hagley Museum, and Sheldon Hochheiser, an employee of Rohm and Haas, respectively). In these books the compatibility of science and technology with each other and with business will be important issues. The Smithsonian Institution's publication of the Joseph Henry Papers continues, and the initial volumes of the Thomas A. Edison Papers, jointly sponsored by Rutgers University, the state of New Jersey, the National Park Service, and a number of private individuals, foundations, and corporations, head for publication. Both Henry and Edison repeatedly found themselves dealing with actual or potential conflicts or compromises between scientific and technological goals.

Two recent innovations in the way history is done also focus on relating science and technology. One is the jointly funded project on the origins of a particular science or technology. Examples are the Project for the History of Solid State Physics, with many public and private sponsors, carried out by an international team under the auspices of the American Institute of Physics; the Laser History Project, largely sponsored by companies that make lasers; and the Polymer Project, under the auspices of the Center for History of Chemistry. These projects have explicitly set out to deal with the science-technology relation as a major theme.

The second major innovation is the permanent center for the study of the history of a discipline or profession. The Center for History of Physics, operating out of the headquarters of the American Institute of Physics, has demonstrated the value of this approach for purposes ranging from archival services to original research, through the outstanding work of Charles Weiner, Lillian Hoddeson, Spencer Weart, Joan Warnow, and others. Its example has now been followed by the History Center of the Institute of Electrical and Electronic Engineers (IEEE), under the direction of Ronald Klein and operating out of the IEEE's headquarters in New York City; the Center for History of Chemistry, directed by Arnold Thackray and associated with the American Chemical Society, the American Institute of Chemical Engineers, and the University of

---

[42] Thomas P. Hughes, "Emerging Themes in the History of Technology," *Technol. Cult.*, 1979, *20*:697–711.

Pennsylvania; and the Charles Babbage Institute for the History of Information Processing, directed by Arthur Norberg, sponsored by the American Federation of Information Processing Societies and various corporations and individuals involved with information processing, and located at the University of Minnesota. Each of the directors mentioned is a professional historian of science or technology.

The challenge these initiatives present to historians mirrors the major theme of this review. Historical studies have shown that the relations between science and technology need not be those of domination and subordination. Each has maintained its distinctive knowledge base and methods while contributing to the other and to its patrons as well. History must now show that it too can be an autonomous discipline capable of contributing value to other disciplines and to corporate, public, or technical-society patrons without becoming their creature or mouthpiece. In pursuing that goal, the study of the historical relations of science and technology can provide both a theme for investigation and a source of guidance.

Science and War

By Alex Roland*

THE BAD NEWS is that military history has been studied often but not well; the history of science has been studied well but not often. Military histories are as old as the *Iliad* and the Old Testament, but as a genre they are dominated by operational accounts of campaigns and battles and hagiography of the great captains. The history of science and technology tends to be more scholarly and critical, but hardly any was written before this century; most of the best work has been done since World War II. Both fields remain outside the mainstream of American historiography; neither, for example, appears on the list of traditional "Fields of Specialization" in the American Historical Association's *Guide to Departments of History*. In a country "born in an act of violence" and risen to world preeminence largely on the basis of science and technology, this neglect is almost inexplicable.[1]

The good news is that we know more than we realize. While the histories of science and war have been poorly integrated in surveys of American development, and while we do lack compelling syntheses of these topics, the monographic literature is substantial and has been growing significantly in recent years. It forms, in fact, such a huge corpus that I cannot claim to have read all or even most of it. I have, however, read enough to know that it has not yet been exploited. Since it has been more than adequately described and evaluated elsewhere, I will not attempt here simply to reshuffle the materials into a new list.[2]

I propose instead to essay a tentative outline of what a synthesis of the existing literature on science and war in the United States might look like. I will try to cite the best literature and address the main issues within it, leaving the reader to consult prior bibliographic essays in this field for more comprehensive

*Department of History, Duke University, Durham, North Carolina 27706.

[1] After almost thirty years, Walter Millis's *Arms and Men: A Study in American Military History* (New York: Mentor, 1956), remains the only survey of American military history worthy of serious scholarly attention. Its only counterpart in the history of technology was published in the same year—John W. Oliver, *History of American Technology* (New York: Ronald Press, 1956)—but it has not stood up nearly so well. In the history of science, George Daniels, *Science in American Society: A Social History* (New York: Knopf, 1971) comes closest to a full survey, though Dirk Struik, *Yankee Science in the Making* (Boston: Little, Brown, 1948) carries the story well through the Civil War. None of these three, however, comes close to matching Millis's achievement.

[2] See Edward C. Ezell, "Science and Technology in the Nineteenth Century," in *A Guide to the Sources of United States Military History,* ed. Robin Higham (Hamden, Conn.: Archon, 1975), pp. 185–215; Carroll W. Pursell, Jr., "Science and Technology in the Twentieth Century," *ibid.,* pp. 269–291; Ezell, "Science and Technology in the Nineteenth Century," in *A Guide to the Sources of United States Military History: Supplement I,* ed. Robin Higham and Donald S. Mrozek (Hamden, Conn.: Archon, 1981), pp. 44–55; Pursell, "Science and Technology in the Twentieth Century," *ibid.,* pp. 69–71; and Harvey M. Sapolsky, "Science, Technology, and Military Policy," in *Science, Technology and Society: A Cross-Disciplinary Perspective,* ed. Ina Spiegel-Rösing and Derek de Solla Price (London: Sage, 1977), pp. 443–471.

guidance to specialized topics. This, then, is intended to be an interpretive rather than a comprehensive survey.[3] In it I will define science to include technology, but I will not cover medicine and war, which is a significant and large subfield unto itself.[4] The survey will conclude with some observations on themes in the literature and some recommendations for further research.

THE COLONIAL PERIOD, 1604–1775

The British colonies in North America shaped the institutions of the United States. The colonists engaged in Indian fighting and wars of empire, adapting European techniques and technology to the wilderness and learning from the Indians the virtues of skirmishing. The militia systems under which most of the colonists fought required them to provide their own arms. Many armed themselves with rifles, not so much because they were wealthier than their European counterparts as because they did much more hunting, where the rifle was worth the added cost. They were, as Charles Winthrop Sawyer observed, "the greatest weapon-using people of that epoch in the world," and gunmaking was one of their most advanced technologies. M. L. Brown has gone so far as to say that "the American rifle . . . stands alone as the first major technological innovation produced in North America."[5]

Circumstances drew the colonists toward natural science and away from the physical sciences that might have influenced the conduct of war. The research equipment, publications, and collaboration that fostered progress in the physical sciences in Europe were largely absent in the colonies. But flora and fauna abounded, and European scientists welcomed all the information they could get from the New World. Of the seventeen colonists elected fellows of the Royal Society before the Revolution, the greatest majority were naturalists. John Winthrop, Jr., who did original research and manufacturing in chemistry, was one of the few elected from the physical sciences. For similar reasons, the colonies produced little theoretical work and few institutions in any branch of science. Benjamin Franklin and his activities provide the exception that proves all these generalizations.[6]

---

[3] For earlier such endeavors see, e.g., Clarence Lasby's well-written and insightful "Science and the Military," in *Science and Society in the United States*, ed. David D. Van Tassel and Michael G. Hall (Homewood, Ill.: Dorsey, 1966), pp. 251–282; and Harvey Sapolsky's somewhat more narrowly focused "Academic Science and the Military: The Years since the Second World War," in *The Sciences in the American Context: New Perspectives*, ed. Nathan Reingold (Washington, D.C.: Smithsonian Institution Press, 1979), pp. 379–399.

[4] On medicine and war see James O. Breeden, "Military and Naval Medicine," in *Guide to the Sources*, ed. Higham, pp. 317–343; and Breeden, "Military and Naval Medicine," in *Guide to the Sources: Supplement I*, ed. Higham and Mrozek, pp. 79–87. In general, army medicine in the Revolution, the Civil War, and World War II has received considerable attention; army medicine in peacetime and in the other wars needs more research. American naval medicine has been badly neglected, in striking contrast with British historiography.

[5] M. L. Brown, *Firearms in Colonial America: The Impact on History and Technology, 1492–1792* (Washington, D.C.: Smithsonian Institution Press, 1980), p. 264. Sawyer is quoted in Millis, *Arms and Men*, p. 20, and Oliver, *History of American Technology*, p. 93 (without attribution). On the military nature of colonial government, see Stephen Saunders Webb, *The Governors-General: The English Army and the Definition of Empire, 1569–1681* (Chapel Hill: Univ. North Carolina Press, 1979).

[6] Raymond P. Stearns, *Science in the British Colonies of America* (Urbana: Univ. Illinois Press, 1970); and Ralph S. Bates, *Scientific Societies in the United States* (2nd ed., New York: Columbia Univ. Press, 1958).

Colonial technology shared many of the same characteristics as science. It was, for example, derivative, pragmatic, and appropriate to the environment in which it was transplanted. It made up for scarcity of labor by exploitation of rich natural resources, especially wood, a commodity already growing scarce in England. Shipbuilding and gunmaking surpassed all other colonial technologies in scale and sophistication, but they did not bring in their wake the modern industrial development that began in England near the end of the colonial period. The only real patronage for invention and industry in the colonies came from local governments through monopolies, tax breaks, and subsidies. These incentives were not enough. War and rumors of war were required before large-scale industrial development could begin in the United States.[7]

THE AMERICAN REVOLUTION, 1775–1783

The war for independence effected a military as well as a political revolution. Not only did it entrain the age of the democratic revolutions; it also upset the pattern of eighteenth-century European warfare. Against the small, professional, highly trained, and tightly husbanded armies of the Age of Reason it pitted the citizen in arms, making up in numbers and patriotism what he lacked in discipline and equipment. Unlike its Napoleonic successor, the American army could not drive its foe from the field, but neither could it be crushed. As Walter Millis observed, "Washington could readily be defeated, but he could not be destroyed." Herbert Agar put the matter nicely when he noted that "Washington had to keep an army in the field long enough for the British to lose the war."[8]

The materiel of war contributed substantially to the outcome. Recent scholarship has drawn attention to the severe handicap the British faced in trying to maintain lines of communication across the Atlantic.[9] The Americans, in contrast, had more material advantages than the British seem to have appreciated, largely through a combination of importation and resourcefulness in local production. They had to import what cannon they had, for example, but they were able to manufacture many of their own small arms. The famous chain across the Hudson River—and many other smaller barriers to intrusions by the British fleet—was entirely native. With the help of European scientific literature, Benjamin Rush was able to teach the Americans how to isolate saltpeter for gunpowder manufacture. David Bushnell relied on European literature only for the idea of submarines and torpedoes; all the skills and materials necessary to build his *Turtle* were available in his native Connecticut.[10] Qualified engineers were the one commodity that Washington could never find in the colonies, with the possible exception of David Rittenhouse; these he had to import, contributing,

---

[7] See most recently Brooke Hindle, ed., *Material Culture of the Wooden Age* (Tarrytown, N.Y.: Sleepy Hollow Press, 1981).

[8] Millis, *Arms and Men*, p. 31. Herbert Agar, *The Price of Union* (Boston: Houghton Mifflin, 1966 [Sentry ed.]), p. 28.

[9] David Syrett, *Shipping and the American War, 1775–83: A Study of British Transport Organization* (London: Univ. London Press, 1970); and R. Arthur Bowler, *Logistics and the Failure of the British Army in America* (Princeton, N.J.: Princeton Univ. Press, 1972). Cf. Erna Risch, *Supplying Washington's Army, 1775–1783* (Washington, D.C.: U.S. Army Center for Military History, 1981).

[10] On Rush, see Oliver, *History of American Technology* (cit. n. 5) Chs. 8, 9; on Bushnell, Alex Roland, *Underwater Warfare in the Age of Sail* (Bloomington: Indiana Univ. Press, 1978), Ch. 5.

perhaps, to his enthusiasm after the war for establishing a school of military engineering in the United States.[11]

The overall effect of the war on science and technology was mixed. On the positive side it spurred the development of many technologies that had earlier been forbidden to the colonists or ignored by them because of the superiority of British goods. Metallurgy is surely in this category, as are such nonmilitary technologies as textiles and leather.[12] War also witnessed, perhaps even spurred, the institutionalization of science and technology, a pattern to be repeated often in American history. The American Academy of Arts and Sciences was founded in Boston in 1780, the Massachusetts Medical Society the following year, helping to compensate for the disruption of intercourse with comparable British institutions.[13] While the war lasted, however, its effects on science were profoundly disruptive, upsetting the natural history circle, curtailing international communication, and hindering the colleges where many scientists worked. Some scientists were lost to emigration, though others immigrated. On the positive side, residual patriotism and enthusiasm after the war moved some to advocate that the military and political achievements be matched in science and other aspects of cultural development. But perhaps the greatest impact of the war on science was to intensify the American preoccupation with the practical application of science—in war and peace.[14]

THE EARLY REPUBLIC, 1783–1815

The Federalists who controlled the United States in the last quarter of the eighteenth century brought from their experience in the Revolution a military agenda that shaped debate through the earliest years of the Jeffersonians. They wanted a standing army, arsenals and magazines, and a military school. Of these, the contest over a standing army has proven to be by far the most important in American history, as alive in the Cold War as it was in the Revolution. Our deepest national prejudice, one we inherited from England, is a distrust of standing armies, which we associate with tyranny, oppression, and loss of freedom. Taking a leaf from the English Mutiny Acts of 1689, Americans embodied this prejudice in the Constitution, establishing that the army could be funded for no more than a year at a time, while the navy could be funded in multiyear increments. This provision has had consequences for the organization of research and development on weapons and the establishment of a reliable industrial base for weapons production in wartime. For better or for worse, the

---

[11] Silvio A. Bedini, *Thinkers and Tinkers: Early American Men of Science* (New York: Scribners, 1975), Ch. 11; and Brooke Hindle, *The Pursuit of Science in Revolutionary America* (Chapel Hill: Univ. North Carolina Press, 1956), p. 242.

[12] Millis, *Arms and Men*, p. 33; and Oliver, *History of American Technology*, pp. 101–103.

[13] Struik, *Yankee Science* (cit. n. 1), pp. 44, 49. Struik observes (p. 40): "Like the Hollanders of 1575 who founded Leyden University while the Spanish were still overrunning their country, like the later French of 1795 who founded the Ecole Polytechnique when the Republic was menaced from all sides, like the Russians of 1920 who began electrification in the midst of foreign invasions, these Americans of 1780 began the organization of science and the establishing of manufactures 'for the happiness of mankind' at a time when the enemy was burning their towns and ravaging the countryside. They were confident that they were building a new and better world." Cf. Hindle, *Pursuit of Science*, p. 379.

[14] Hindle, *Pursuit of Science*, Ch. 11 and pp. 378–379, 384.

navy—the more technologically dependent service—has had more latitude for development.[15]

At a higher level of abstraction, this debate also entailed another issue of paramount importance to the evolution of science and war in the United States: civilian control of the military, especially as it affects development and procurement. Though the United States remains the only country that leaves procurement of materiel in the hands of the military services—instead of entrusting it, for example, to civilian ministries on the British model—the constitutional provision for civilian control of the services nominally ensured that this would not lead to undue military influence on domestic affairs. The rise of the military-industrial complex in the twentieth century casts doubt on this assumption, even while it raises the specter of a standing army in a form never envisioned by the Founding Fathers.[16]

The second major military issue of the Federalist period was more obviously related to the development of science and technology. Prompted in large measure by the potential war crisis with revolutionary France in the 1790s, Congress created national arsenals for the production of weapons and simultaneously contracted with private suppliers for small arms. Not only did this arrangement employ many bureaucrats and artisans throughout the nineteenth century, but it also has employed a comparable number of historians in recent years, for here were the beginnings of what became the American system of manufacture—arguably the major interest of historians of American technology in the last two decades. Here too were the beginnings of a debate that has continued to the present day about how best to procure military materiel—the advocates of government arsenals pitted against proponents of contracting out. Springfield arsenal is thrown up as a model on one side, Eli Whitney and his now infamous contract of 1798 on the other.[17]

Not until they had fallen from power in political suicide over the issue of command of the standing army during the period of conflict with France known as the Quasi War did the Federalists achieve the third leg of their national military program—a service school. A small school for artillerists and engineers had opened at West Point in the 1790s, but it languished through the remainder of that decade. What the Federalists really wanted was a "University of Mars," as one of them put it, to teach all aspects of the "science" of war, but they

---

[15] For the period through 1802, I rely heavily on Richard H. Kohn's masterful *Eagle and Sword: The Beginnings of the Military Establishment in America, 1783–1802* (New York: Free Press, 1975). Kohn, who rightly draws attention to Fletcher Knebel and Charles W. Bailey's insightful novel *Seven Days in May* (New York: Harper & Row, 1962), might also have noted that the military conspiracy to overthrow the government begins to unravel when the *naval* conspirator backs out. This suits the American belief that navies, perhaps because they are remote from the seat of power, do not pose a threat to civilian government.

[16] Paul A. C. Koistinen, *The Military-Industrial Complex: A Historical Perspective* (New York: Praeger, 1980), p. 36. See also the section "Cold War" below.

[17] The mythology of Eli Whitney is rehearsed in Constance McLaughlin Green, *Eli Whitney and the Birth of American Technology* (Boston: Little, Brown, 1956). Robert Woodbury began a quarter century of debunking with "The Legend of Eli Whitney and Interchangeable Parts," *Technology and Culture*, 1960, 1:235–253. This trend reached a high art with Merritt Roe Smith's *Harpers Ferry Arsenal and the New Technology: The Challenge of Change* (Ithaca, N.Y.: Cornell Univ. Press, 1977). For more on the American system, see the section "Civil War" below. On the debate over contracting and arsenals in the Cold War, see H. L. Nieburg, *In the Name of Science* (Chicago: Quadrangle, 1966).

were too vulnerable on military issues to press for such an institution. Rather it was Jefferson, the great democrat and early foe of a military academy, who finally established a military engineering school at West Point modeled on the French *école polytechnique*—a wartime expedient of revolutionary France intended to serve military purposes. For many years the only engineering school in the United States, West Point was to prove a training ground not only of soldiers but also of civilian engineers who would play a crucial role in the technical development of the United States. Though this may have been Jefferson's intent in creating West Point, his motives remain unclear. As Richard Kohn observes, "the man who had risen to power on the ashes of a Federalist party consumed by militarism created the institution destined in the twentieth century to become the heart and soul of the very kind of military establishment so long feared by his party and his countrymen."[18] That it began as a school of engineering hardly diminishes the irony.

While military technology, engineering, and industry were well begun in the early republic, science fared less well.[19] The closest association between science and war from the Revolution through the War of 1812 was the Lewis and Clark expedition, the first of many nineteenth-century explorations by military men or under military auspices. For much of its first century, the most pressing military need of the young republic would be on its expanding frontier, and there too lay one of its greatest opportunities for extending the bounds of scientific knowledge, especially in geology and the life sciences. It is little wonder that the two fields of science and war moved so often in tandem on the frontier, or that botanists and geologists—not physicists or chemists—were the most significant scientific professionals in the United States at midcentury.[20]

Science and technology contributed little to the American effort in the War of 1812 and profited little by it. Robert Fulton used the occasion to dust off the schemes for steam-propelled warships that he had earlier tried to peddle to Napoleon and the British, but the war ended before his *Demologos* saw action.[21] Most other weaponry employed by the Americans was traditional and derivative, save perhaps for the finely designed and crafted frigates that gave the British such fits and sustained our reputation for shipbuilding and other maritime technology—a reputation enhanced by entirely different craft used on the Great Lakes.[22]

---

[18] Kohn, *Eagle and Sword*, p. 303, where Oliver Wolcott's comment on the "University of Mars" also appears.

[19] A. Hunter Dupree, *Science and the Federal Government: A History of Policies and Activities to 1940* (Cambridge, Mass.: Belknap Press of Harvard Univ. Press, 1957).

[20] On botanists and geologists, see Nathan Reingold, "Science in the Civil War: The Permanent Commission of the Navy Department," *Isis,* 1958, *49*:307–318, esp. p. 317. But note that Jefferson instructed Lewis and Clark to pursue all areas of scientific observation, including astronomy and natural history, and that the expedition made significant contributions in botany, zoology, and ethnology; see Dupree, *Science and the Federal Government*, p. 26.

[21] Roland, *Underwater Warfare in the Age of Sail*, Chs. 7, 8; Wallace S. Hutcheon, *Robert Fulton: Pioneer of Undersea Warfare* (Annapolis, Md.: Naval Institute Press, 1981). See Roland for an explication of Fulton's debts to the scientific and technical communities of Europe and the United States.

[22] Harold and Margaret Sprout, *The Rise of American Naval Power, 1776–1918* (1939, Princeton, N.J.: Princeton Univ. Press, 1967), Ch. 6. The Sprouts note correctly that the success of the American frigates did nothing to disrupt British command of the sea.

NATION BUILDING, 1815–1860

American technological progress in the years between the War of 1812 and the Civil War manifested itself at the London Exposition of 1851. American maritime technology continued to shine, as the yacht *America* overcame tremendous odds to win an international sailing competition around the Isle of Wight, capturing a trophy that was to stay in the United States until 1983. But Americans made their biggest impression at the exposition with the McCormick reaper and the Colt revolver, exemplars of American achievement in the first half of the nineteenth century and harbingers of future technological development.

The McCormick reaper was the embodiment of agricultural technology. The opening of the frontier, to which the army contributed significantly in the nineteenth century, created an opportunity for the young country.[23] But the traditional shortage of labor and an inadequate transportation system left the nation unable to exploit it fully. The development of canals and waterways, and later of railroads, solved the transportation problem and opened up the world's markets to the farmers of the Midwest. The army contributed to this development through the Corps of Engineers, engaging, as they have done so often, in civil engineering for domestic purposes.[24] The second obstacle, labor shortage, was solved by the mechanization of agriculture, made possible by advances in technology and made profitable by the availability of economical routes to market. As so often in the first half of the nineteenth century, this is basically a story of civilian technologies proceeding with peripheral, but significant support from the military. Here the support produced an added consequence, increased farm productivity that freed labor for the industrialization to come.

The Colt revolver, and the large story of which it is a part, constituted an entirely different case. Here the influence of the military was paramount, providing the major stimulus and the major funding for the most important technological development of the nineteenth century in the United States. All of the essential ingredients of the American system of manufactures—machine tools, interchangeable parts, standardization, and mass production—were pioneered in army arsenals or under army contracts.[25] From there they spread to civilian industries and helped to stimulate the growth of industrialization in the United States that made the nation the envy of most countries and the model of many.[26] One of the most intriguing aspects of this development is that perhaps only the military could have supported it. Interchangeability, for example, was a luxury most industries could not afford, but the concept had special appeal to the military because it promised repair and cannibalization of weapons on the battlefield, where cost was not the primary consideration. Similarly, machine tools

---

[23] F. Paul Prucha, *Broadax and Bayonet: The Role of the United States Army in the Development of the Northwest, 1815–1860* (Madison: Univ. Wisconsin Press, 1953); William H. Goetzmann, *Army Exploration in the American West, 1803–1863* (New Haven, Conn.: Yale Univ. Press, 1959).

[24] Forest G. Hill, *Roads, Rails, and Waterways: The Army Engineers and Early Transportation* (Norman: Univ. Oklahoma Press, 1957); Harold L. Nelson, "Military Road for War and Peace, 1791–1836," *Military Affairs*, 1955, *19*:1–4.

[25] Smith, *Harpers Ferry Arsenal* (cit. n. 17).

[26] David Hounshell, *From the American System to Mass Production, 1800–1932* (Baltimore: Johns Hopkins Univ. Press, 1984); Russell I. Fries, "British Response to the American System: The Case of the Small Arms Industry after 1850," *Technol. Cult.*, 1975, *16*:377–403.

and standardization required a large capital investment and considerable research and development, commitments beyond the reach of most private manufacturers. This would not be the last time that the military, for its own reasons, judged it prudent to support a technological development that civilian industries found exorbitantly expensive. Once the technology was developed, economical civilian applications could be spun off.[27]

Merritt Roe Smith carries this argument one step further, maintaining that the techniques developed in the Army Ordnance Bureau in the nineteenth century to run the arsenals and administer contracts with private manufacturers established a model of rational management of technology, a model that was imposed on industry by army procurement regulations and in some cases simply copied by industry on its merits. This well-documented and persuasive argument flirts with the more sweeping assertions of Lewis Mumford, Waldemar Kaempffert, and David Noble that the military has insinuated itself and its values into virtually every aspect of modern technology.[28] Kaempffert has asserted, for example, that "industry learned everything, except invention, from war—organization, discipline, standardization, the coordination of transport and supply, the separation of line and staff, the division of labor (cavalry, infantry, artillery)." In Kaempffert's hands, without adequate evidence, this has a ring of "post hoc, ergo propter hoc" about it, far removed from Smith's closely reasoned and meticulously documented work. And even Smith's more circumscribed claim is challenged by Alfred Chandler's equally scholarly conclusion that "in the United States, the railroad, not government or the military, provided training in modern large-scale administration," though he does concede that "modern factory management . . . had its genesis in the United States in the Springfield Armory."[29]

The record of other military contributions to civilian technologies in this period is mixed. Forest G. Hill makes a strong case for army influence on roads, railroads, rivers, and harbors, and he even notes that national security was one motivation behind internal improvements, but he stops short of claiming that military influences were paramount. Like Smith, he addresses army techniques of administration and management, but again he does not assert that these shaped civilian practice. David Hounshell has documented how armory practice spread through the machine tool industry into other technologies such as sewing machines, woodworking tools, and bicycles, but he also demonstrates that other influences were at work as well.[30] Some significant technologies of the first half of the nineteenth century—steamships, telegraph, and iron, for example—have yet to be tied as closely to the military. Walter Millis's judgment on the years before the Civil War still holds up remarkably well:

[27] The most recent example is David F. Noble, *Forces of Production: A Social History of Machine Tool Automation* (New York: Knopf, 1984).

[28] Merritt Roe Smith, "Military Entrepreneurship," in *Yankee Enterprise: The Rise of the American System of Manufactures,* ed. Otto Mayr and Robert C. Post (Washington, D.C.: Smithsonian Institution Press, 1981), pp. 63–102; cf. Waldemar Kaempffert, "War and Technology," *American Journal of Sociology,* 1941, 46:431–444; Lewis Mumford, *Technics and Civilization* (New York: Harcourt, Brace, 1934); and David F. Noble, *America by Design: Science, Technology, and the Rise of Corporate Capitalism* (New York: Knopf, 1977).

[29] Kaempffert, "War and Technology," p. 443; and Alfred D. Chandler, *The Visible Hand: The Managerial Revolution in American Business* (Cambridge, Mass.: Belknap Press of Harvard Univ. Press, 1977), pp. 205, 75.

[30] Hill, *Roads, Rails, and Waterways* (cit. n. 24); and Hounshell, *From the American System to Mass Production* (cit. n. 26).

All the while the industrial and technological base . . . was building up in the United States as elsewhere. In this country, military requirements were never of primary importance—our industry grew on the civilian demand for rails, locomotives, steamboat and factory power plants, textile machinery—but military requirements (to say nothing of civilian requirements for weapons on the frontier) provided a powerful added stimulus. And there was a strong interaction between civilian and military needs and between the engineers, inventors, and factory managers who responded to both.[31]

The same predominance of civilian over military support can be seen in the institutionalization of science and technology. Science continued to be supported by universities and by private gifts. The major institutions formed during this period were just as peaceful in their origins. The Smithsonian Institution had virtually nothing to do with military considerations. The Franklin Institute was similarly civilian in nature, though more commercially oriented. A large part of its early research was for the military, and it turned to the navy for support of its campaign to establish a standardized screw thread in the United States, but these activities hardly distorted its basically civilian orientation.[32] Military support of scientific and technological research in the period was confined largely to the frontier, where the natural sciences were pursued as quiet adjuncts to larger operations, and to the oceans, where exploration and oceanography received significant support from the navy.[33]

THE CIVIL WAR, 1861–1865

The Crimean War notwithstanding, the American Civil War was the first war to witness the full impact of the Industrial Revolution. That European observers—and, indeed, many American participants—failed to appreciate this does not alter the fact. Technology, and to a lesser extent science, was beginning to influence warfare dramatically; less obviously, war was shaping science and technology.

On land, the rifle and the shovel revolutionized the face of battle. The increased range and accuracy of the new small arms extended the killing zone and placed a higher premium on aimed fire. In response, soldiers unpacked their shovels and dug in with a fervor not seen since the Roman legions went into castrametation. Denis Hart Mahan had been predicting this revolution in warfare to his West Point cadets for decades, but the experience of the Mexican War erased the lesson from their minds.[34] Countless soldiers on both sides of the Civil War fell to a technical revolution that should have been more obvious than it was.

[31] Millis, *Arms and Men*, p. 82.

[32] Bruce Sinclair, *Philadelphia's Philosopher Mechanics: A History of the Franklin Institute, 1824–1865* (Baltimore: Johns Hopkins Univ. Press, 1974); and Sinclair, "At the Turn of a Screw: William Sellers, the Franklin Institute and a Standard American Thread," *Technol. Cult.*, 1969, *10*:20–34.

[33] On the frontier, see Goetzmann, *Army Exploration* (cit. n. 23); and Goetzmann, *Exploration and Empire: The Explorer and the Scientist in the Winning of the American West* (New York: Knopf, 1966). For the navy see Dupree, *Science in the Federal Government*, pp. 56–61; Patricia Johns, *M. F. Maury and Joseph Henry: Scientists of the Civil War* (New York: Hastings, 1961); Frances L. Williams, *Matthew Fontaine Maury: Scientist of the Sea* (New Brunswick, N.J.: Rutgers Univ. Press, 1963); and Roland, *Underwater Warfare* (cit. n. 10), Ch. 11.

[34] Millis, *Arms and Men*, p. 115; Grady McWhiney and Perry D. Jamieson, *Attack and Die: Civil War Military Tactics and the Southern Heritage* (University: Univ. Alabama Press, 1982).

Traditional military conservatism was only partly to blame. It worked its worst in cavalry, another victim of the revolution in small arms. Dragoons were still effective, but Napoleonic cavalry charges had grown suicidal. Because they were losing, the Confederates proved more receptive to technical innovation, but they too had been steeped in Napoleonic lore at West Point and they rode their cherished horses into the maw of small arms fire with even greater abandon than the Northerners. They were better at adopting and even inventing new technologies than at abandoning old ones. Both North and South were captives of an inertia that saw the U.S. army continue to use horses right through World War I.

The other technologies that strongly influenced the war on land were those developed in earlier decades for peaceful purposes: the railroad, the telegraph, and the steamboat on inland waterways. Like most ships, the steamboat can carry guns and butter equally well, and it proved readily adaptable to war on the rivers.[35] The railroad was a different story. It was laid down for commercial purposes, with no thought given to military uses, in contrast to the Prussian development of railroads and the later development of the United States interstate highway system.[36] Fortuitously, the routes adopted for commerce served the North well in the war, and served the South poorly. The telegraph often followed the rail lines, but it could be strung more easily than rails could be laid; it too proved adaptable to the needs of war.

At sea, another technological revolution was effected—the transition from sail to steam and from wooden hulls to ironclads. The famous battle of the *Monitor* and the *Merrimac* climaxed a technical revolution to which the United States had contributed significantly in the 1840s and 1850s, and it launched a continuing revolution that carried well into the twentieth century.[37] A second naval revolution, in underwater warfare, followed a somewhat different course. Hitherto eschewed by polite navies as unchivalrous, mines, torpedoes, and submarines were adopted by the Confederates because they were desperate and lacked resources. When these weapons proved effective, the Union navy abandoned its scruples and employed the same weapons in retaliation. After the war, other navies around the world followed suit.[38] In this way, dubious weapons join the international arsenal.

Throughout the war both capitals were flooded by inventions. Most of these were crackpot schemes by well-meaning but unqualified tinkerers and amateurs.[39] But each government was also the benefactor of the best scientific and engineering talent in the country, both military and civilian. Joseph Henry, for example, the secretary of the Smithsonian Institution, contributed his free time to the war effort, inventing devices himself and working on the inventions of others. Matthew Fontaine Maury, dubbed "Pathfinder of the Seas" for his

---

[35] Louis C. Hunter, *Steamboats on the Western Rivers: An Economic and Technological History* (Cambridge, Mass.: Harvard Univ. Press, 1949).

[36] Hill, *Roads, Rails, and Waterways* (cit. n. 24), Chs. 4, 5. Compare any of the numerous histories of American railroading in the Civil War with Dennis Showalter, *Railroads and Rifles: Soldiers, Technology, and the Unification of Germany* (Hamden, Conn.: Shoe String Press, 1975).

[37] James Phinney Baxter, *The Introduction of the Ironclad Warship* (Cambridge, Mass.: Harvard Univ. Press, 1933).

[38] Roland, *Underwater Warfare* (cit. n. 10), Chs. 11, 12.

[39] Dupree, *Science and the Federal Government*, Ch. 7.

enormously influential *Physical Geography of the Seas* (1854), gave up his superintendency of the Naval Observatory to join the Confederacy, where he established the program in underwater warfare that was to change Union and world opinion on the issue.[40] The list of distinguished engineers is even more impressive. Josiah Gorgas, for example, did as much for Confederate ordnance as M. C. Meigs did for Northern supply. Benjamin F. Isherwood and John A. Dahlgren made pivotal contributions to the technical development of the Union navy. Civilians, such as John Ericsson, Joseph Anderson, and John Roach, were equally important.[41]

While these men of science and engineering were shaping the war, the war was shaping their fields as well—by institutionalization. In one of the recurring patterns of American military experience, war served to produce new institutions that would survive the combat and influence the growth of science and technology in the peacetime to follow. The National Academy of Sciences, created in 1863 to advise the Union government on the application of science and technology to war, contributed little to the war effort but went on in the twentieth century to become an important national institution. The Permanent Commission of the Navy Department, which Nathan Reingold describes as "an abortive National Research Council,"[42] hardly accomplished more and additionally proved less durable, but nonetheless established a precedent for enlisting scientific advice in war. Freed from the opposition of Southern senators and representatives, Congress passed the Morrill Act of 1862, establishing the land-grant colleges and laying the groundwork for critical research and education in the decades ahead. The net effect was to create institutions in the war that would affect the war hardly at all but would influence profoundly the future course of science and technology. The major exception to this general pattern was the American Association for the Advancement of Science, which, after twelve years of existence, had to suspend meetings in 1860 for the duration of the war. When it reconvened in 1866, it was a diminished and relatively less important organization.[43]

Whether the Civil War was a boon to American industrial development, as has been claimed for the Revolutionary War, is another matter. Certainly, agricultural and industrial production increased during the war, and new industries were stimulated and nurtured by the demands for war materiel. Even the South revealed an industrial potential that is still not widely appreciated. Walter Millis has observed that "perhaps the most striking demonstration of the strength of

[40] See Johns, *M. F. Maury and Joseph Henry*; Williams, *Matthew Fontaine Maury* (both cit. n. 33); and Roland, *Underwater Warfare*, Ch. 11.

[41] Frank E. Vandiver, *Ploughshares into Swords: Josiah Gorgas and Confederate Ordnance* (Austin: Univ. Texas Press, 1952); Russell F. Weigley, *Quartermaster General of the Union Army: A Biography of M. C. Meigs* (New York: Columbia Univ. Press, 1959); Edward W. Sloan III, *Benjamin Franklin Isherwood, Naval Engineer: The Years as Engineer in Chief, 1861–1869* (Annapolis, Md.: Naval Institute Press, 1965); Alfred D. Chandler, "Du Pont, Dahlgren, and the Civil War Nitre Shortage," *Military Affairs*, 1949, *13*:142–149; Leonard A. Swann, Jr., *John Roach, Maritime Entrepreneur: The Years as Naval Contractor, 1862–1886* (Annapolis, Md.: Naval Institute Press, 1965); and Charles B. Dew, *Ironmaker to the Confederacy: Joseph R. Anderson and the Tredegar Iron Works* (New Haven, Conn.: Yale Univ. Press, 1966).

[42] Nathan Reingold, "Science in the Civil War" (cit. n. 20), p. 311.

[43] Sally Gregory Kohlstedt, *The Formation of the American Scientific Community: The American Association for the Advancement of Science* (Urbana: Univ. Illinois Press, 1976).

the new technological base lay not in its production from the already established plants, but in its ability to create new ones, especially in the South, where virtually a whole war industry had to be, and was, evoked out of almost nothing." But whether America's industrial revolution in the last third of the nineteenth century may be attributed to the Civil War remains a hotly contested issue.[44]

EMERGENCE AS A WORLD POWER, 1865–1914

In the half century following the Civil War, the United States emerged as a world power because of its economic, industrial, and technological strength. The army demobilized into a languishing constabulary force, modernizing only in the twentieth century. The navy was strong enough to crush the decrepit Spanish fleet but not to establish the "command of the sea" advocated by naval apostle Alfred Thayer Mahan, America's only military theoretician of note. The United States was not yet a scientific power, save perhaps in the earth and biological sciences[45]; in the former, both the army and the navy had made their contributions.

In this period, technology contributed to warfare most clearly in naval developments. Innovations demonstrated in the Civil War were perfected and refined at a dizzying and bankrupting pace, setting off an arms race of the modern sort on the eve of World War I and creating the first modern instance, claims William H. McNeill, of a military-industrial complex.[46] The United States ran hot and cold in this race, not so much for want of scientific, industrial, or technical capacity (though these were often outclassed by the huge British and German arms manufacturers) as for reluctance to pay the bill. Only in submarines and gunnery did we contribute significantly to the evolution of naval technology.[47]

Civilian and military leaders combined in this period to retard the development of more and better arms. The politicans balked at paying more for the military than was necessary to support our imperialistic adventures in Latin America and the Pacific. The soldiers and sailors continued their traditional resistance to innovation. Elting Morison has revealed this latter phenomenon with telling effect

[44] Quoting Millis, *Arms and Men*, p. 103; see also George Fort Milton, "Conversion and the Confederacy," *Technology Review*, 1944, 46:141–142, 156 ff. On the long-term effects, see the seminal article by Thomas C. Cochran, "Did the Civil War Retard Industrialization?" *Mississippi Valley Historical Review*, 1961, 48:197–210, and the body of literature it stimulated, as catalogued in Ezell, "Science and Technology in the Nineteenth Century" (cit. n. 2). More sharply focused on science and technology are the essays by A. Hunter Dupree and Robert V. Bruce, in *Economic Change in the Civil War Era: Proceedings of a Conference on American Economic and Institutional Change, 1860–1873*, eds. David T. Gilchrist and W. David Lewis (Greenville, Del.: Eleutherian Mills–Hagley Foundation, 1964). Much of the debate is carried on in the terms staked out by Werner Sombart, *Krieg und Kapitalismus* (Munich: Duncker & Humblot, 1913); and John U. Nef, *War and Human Progress: An Essay on the Rise of Industrial Civilization* (Cambridge, Mass.: Harvard Univ. Press, 1950). Sombart's position is summarized in Kaempffert, "War and Technology" (cit. n. 28).

[45] Daniel J. Kevles, *The Physicists: The History of a Scientific Community in Modern America* (New York: Knopf, 1979).

[46] William H. McNeill, *The Pursuit of Power: Technology, Armed Force, and Society Since 1000 A.D.* (Chicago: Univ. Chicago Press, 1982), Ch. 8.

[47] On the reluctance, see Sprout and Sprout, *The Rise of American Naval Power* (cit. n. 26), Chs. 11–17; on the contributions, see Richard K. Morris, *John P. Holland, 1841–1914: Inventor of the Modern Submarine* (Annapolis, Md.: Naval Institute Press, 1966); Elting E. Morison, *Admiral Sims and the Modern American Navy* (Boston: Houghton Mifflin, 1942); and Paolo E. Coletta, *Admiral Bradley A. Fiske and the American Navy* (Lawrence: Regents Press of Kansas, 1979).

Naval Technology: Comparison of the old and the new Pennsylvanias *and their guns.*
From Albert A. Hopkins, ed., The Scientific American War Book: The Mechanism and
Technique of Warfare *(New York: Munn & Company, 1916), p. 291.*

in numerous studies of the early modern navy, though Lance Buhl has shown
that forces beyond mere conservatism were driving the bureaucracy.[48] The army
proved just as reluctant in dealing with the machine gun and the airplane. Even
when the military potential of new technologies was quickly recognized, as in
the case of radio, well-placed officers of conservative disposition could retard
their adoption simply by establishing impossible standards of performance.[49]

During this half century, it is often easier to see the impact of war on technology than it is to see the revolution in warfare that technology was about to
effect. The American system of manufactures spread to an ever wider circle of
civilian industries. The steel industry grew not only because orders to build the

[48] Morison, *Admiral Sims*; Elting E. Morison, *Men, Machines, and Modern Times* (Cambridge,
Mass.: MIT Press, 1966); Lance C. Buhl, "Mariners and Machines: Resistance to Technological
Change in the American Navy, 1865–1869," *Journal of American History,* 1974, 61:703–727. See
also Paolo E. Coletta, "The Perils of Invention: Bradley A. Fiske and the Torpedo Plane," *American
Neptune,* 1977, *37*:111–127, later incorporated in Coletta, *Admiral Bradley A. Fiske*. Different per-
spectives on military resistance to change in this period can be gained from the civilian insider, e.g.,
in Benjamin Franklin Cooling's *Benjamin Franklin Tracy: Father of the Modern American Navy*
(New York: Archon, 1973); and the civilian outsider, e.g., in Thomas P. Hughes, *Elmer Sperry:
Inventor and Engineer* (Baltimore: Johns Hopkins Univ. Press, 1971).

[49] See David A. Armstrong, *Bullets and Bureaucrats: The Machine Gun and the United States
Army, 1861–1916* (Westport, Conn.: Greenwood, 1982); Susan Douglas, "Exploring Pathways in the
Ether: The Formative Years of Radio in America" (Ph.D. diss., Brown Univ., 1979); and for re-
vealing insights into the army's early ventures in flight (though a definitive study is lacking), John
F. Shiner, *Foulois and the U.S. Army Air Corps, 1931–1935* (Washington, D.C.: Office of Air Force
History, 1983); and Paul W. Clark, "Major General George O. Squier: Military Scientist" (Ph.D.
diss., Case Western Reserve Univ., 1974), which also treats radios.

new navy increased but also because the arms race was spawning improved techniques and products. Engineers and inventors in uniform produced countless new devices that often had civilian applications.[50]

Similarly, in this period the military gave to science more than it received in return. Albert Michelson went to the Naval Academy in the late 1860s because "there was no other college in the country that offered adequate instruction in physics."[51] While an instructor there, he conducted the first of the experiments that would lead to his international fame and Nobel Prize. Nor was he alone; George O. Squier, for example, took a Ph.D. in physics at Johns Hopkins University while on active duty and went on to a long and distinguished career as an inventor, scientist, and Signal Corps officer.[52] The army continued to sponsor surveys of Western lands after the Civil War, including Clarence King's famous "Geological and Geographical Exploration of the Fortieth Parallel," until this responsibility was finally vested in a civilian Geological Survey in 1879 under Civil War veteran John Wesley Powell. The army pioneered the first national weather service from 1870 to 1890, before handing the task over to the Department of Agriculture.[53]

These significant achievements notwithstanding, the last decades of the nineteenth century were, as Hunter Dupree has argued, a time of growing detachment between the military services and science.[54] Two trends were at work. First, the federal government was beginning to approve civilian agencies like the Weather Bureau and the Geological Survey to take over chores that previously had been handled by the military services only for want of another agency. Second, with the constricted budgets of peacetime, military officers were demanding that the funds they expended show immediate, practical returns on investment. Long-term research could be rationalized in some fields, like ordnance and telephony, but not so in such fields as topography, geology, and astronomy. The eventual result of this shift was to be a new emphasis on physics and chemistry that would accelerate the rise of these disciplines in the twentieth century.

The greatest contribution of science to war in this transitional period may well have been conceptual. Ever since the scientific revolution of the seventeenth century, other fields of human activity have attempted to embrace the "scientific method" or something like it in hopes of discovering the fundamental "laws" at work and learning how to master them. This impulse has spawned social science, political science, and any number of pseudosciences.

---

[50] Clark maintains that George O. Squier was one of the last of a dying breed of nineteenth-century scientists and engineers in uniform, men who successfully combined military careers with professional activity in science, technology, invention, and even business. Surely the rise of professionalism, both in the services and in science and engineering, made it increasingly difficult to serve more than two masters.

[51] Dorothy Michelson Livingston, "Michelson in the Navy; the Navy in Michelson," *U.S. Naval Institute Proceedings,* 1969, 95:72; see also Livingston, *The Master of Light: A Biography of Albert A. Michelson* (New York: Scribners, 1973).

[52] Clark, "George O. Squier."

[53] One of the major issues in the creation of the Geological Survey was whether it would be a military or a civilian agency; Thomas D. Manning, *Government in Science: The U.S. Geological Survey, 1867–1894* (Lexington: Univ. Kentucky Press, 1967), Ch. 2. On the weather service, see Joseph M. Hawes, "The Signal Corps and Its Weather Service, 1870–1890," *Military Affairs,* 1966, 30:68–76; and Donald R. Whitnah, *A History of the United States Weather Bureau* (Urbana: Univ. Illinois Press, 1965).

[54] Dupree, *Science and the Federal Government,* Ch. 9.

Military affairs proved to be no exception. Beginning at least with Sebastien de Vauban (1633–1707), claims have arisen that the traditional art of war is complemented by a science of war.[55] Karl von Clausewitz and Baron Jomini, for example, both studied Napoleon—one saw art; the other, science.[56] In the last hundred years, infatuation with the science of war has reached unprecedented levels. The Prussians launched this latest enthusiasm with their quick and conclusive victories over the Austrians and the French in the late 1860s and early 1870s. Their *Kriegsakademie*, general staff, planning, and superior management techniques seemed to have reduced war to a science at just the time when German science was beginning to emerge as superior both in its achievements and in its educational system. As the nineteenth century gave way to the twentieth, countries all over the world adopted the general staff model for their armies and the graduate research laboratory for their universities. The Johns Hopkins University and the restructuring of the armed forces instituted by Secretary of State Elihu Root drew on a common conceptual base.[57]

A closely related development was to have an equally dramatic impact on the military. Scientific management, or Taylorism, came into vogue around the turn of the century, riding the same cultural horse as the general staff concept and the model of the German university. It was applied quickly and with varying success at army arsenals, at navy yards, and in the plants of military contractors.[58] In the form advocated by Frederick Taylor, scientific management has fallen from grace in a way that the general staff and the research university have not. But it made its contribution nonetheless to the emergence of the modern concept of military men as "managers of violence."

THE FIRST WORLD WAR, 1914–1940

The influence of science and technology on World War I is a source of controversy among historians. Daniel Kevles has challenged one body of conventional wisdom—that this was a chemist's war. He has argued in a number of persuasive publications that the contributions of science in general, and physics in particular, have been overlooked by historians.[59] Paul A. C. Koistinen has argued,

---

[55] Henry Guerlac, "Vauban: The Impact of Science on War," in *Makers of Modern Strategy*, ed. Edward Mead Earle (Princeton: Princeton Univ. Press, 1943), pp. 26–48; and Daniel R. Beaver, "Cultural Change, Technological Development, and the Conduct of War in the 17th Century," in *New Dimensions in Military History*, ed. Russell F. Weigley (San Rafael, Calif.: Presidio Press, 1975), pp. 75–89.

[56] Carl von Clausewitz, *On War*, ed. and trans. Michael Howard and Peter Paret (Princeton: Princeton Univ. Press, 1976); Henri Jomini, *The Art of War*, trans. G. H. Mendell and W. P. Craighill (Philadelphia: Lippincott, 1862).

[57] Walter Millis calls the military side of this the "Managerial Revolution," an attempt to control and exploit the forces unleashed by the democratic and industrial revolutions. In *Arms and Men*, he notes that "it was natural, in that age to find an immediate answer in 'science'; and the scientific and methodical Germans led the way" (p. 123).

[58] Hugh G. H. Aitken, *Taylorism at Watertown Arsenal: Scientific Management in Action* (Cambridge, Mass.: Harvard Univ. Press, 1960); and Holden A. Evans, *One Man's Fight for a Better Navy* (New York: Dodd, Mead, 1940).

[59] Daniel J. Kevles, "Flash and Sound in the AEF: The History of a Technical Service," *Milit. Affairs*, 1969, *33*:374–384; Kevles, "George Ellery Hale, the First World War, and the Advancement of Science in America," *Isis*, 1968, *59*:427–437; and Kevles, *The Physicists*, Chs. 8, 9. See also Kevles, "Testing the Army's Intelligence: Psychologists and the Military in World War I," *Journal of American History*, 1968, *55*:565–581; and Kevles, "Federal Legislation for Engineering Experimental Stations: The Episode of WW I," *Technol. Cult.*, 1971, *12*:182–189.

however, that it was engineers and managers, men who saw themselves as "doers" and not researchers, who contributed most to the outcome of the war. This runs close to, but does not quite overlap, Walter Millis's assertion that World War I introduced the "mechanization of war," in contrast to the "scientific revolution" that would follow in World War II.[60]

Whatever the relative merits of these interpretations, several features of the conflict are clear. First, it was a war of industrial production. The Germans were never really defeated in the field; rather, they ran out of the fodder of war, a fate they came close to inflicting on England with submarine warfare. Second, the machine gun and the submarine were the critical technologies. As seen in the previous section, the United States prepared better for the latter technology than the former, both in developing the craft and then in developing the means to combat it. Kevles maintains, with some effect, that this was where the physicists played a role more important than the chemists, for the submarine was more important than chemical warfare and munitions.[61]

Third, some new technologies proved less effective on the battlefield than they might have because their potential was never fully exploited. Surely this is true of the aircraft, largely because of the lack of a doctrinal base, as shown by I. B. Holley in his classic study *Ideas and Weapons*.[62] Gas warfare and the tank were dramatically effective when first employed in battle, but adequate preparations had not been made to exploit the opportunities they offered. By the time the weapons were ready for exploitation, the surprise had worn off and countermeasures had been instituted. Radio was used to good effect at sea, but it was not yet reliable enough to fulfill its potential. A similar pattern emerged on land, where the stasis of trench warfare made conventional land lines appealing, so long as they could be protected against artillery barrage. In the end the runner, a technology as old as warfare, was often preferred to modern inventions. The motor car still took second place to the horse. The machines on the home front producing bullets and canning beans contributed more to the outcome of the war than did any machines on the battlefield, save perhaps the machine gun.

The influence of World War I on science and technology was more pronounced. The Great War was the first of the total wars, in which the entire resources of the state—or very nearly so—were mobilized for military purposes. Not the least of the resources were science and technology. The result was what William McNeill has called a command economy, the end to which Western states had been gravitating since at least A.D. 1000.[63] Virtually all of national life was bent to war.

In the United States the heart of this enterprise was in Washington, where the second major impact of World War I on science and technology occurred. Continuing precedents established in the American Revolution and the Civil War, the government created scientific and technical institutions that would sur-

[60] Koistinen, *The Military-Industrial Complex* (cit. n. 16), Ch. 2; and Millis, *Arms and Men*, Chs. 4, 5.

[61] Kevles, *The Physicists*, Ch. 9.

[62] I. B. Holley, Jr., *Ideas and Weapons: Exploitation of the Aerial Weapon by the United States During World War I: A Study in the Relationship of Technological Advance, Military Doctrine, and the Development of Weapons* (New Haven, Conn.: Yale Univ. Press, 1953).

[63] McNeill, *The Pursuit of Power* (cit. n. 46).

vive the war and become more or less permanent promoters of scientific and technical development. The first of these was the National Advisory Committee for Aeronautics (NACA), created in 1915 for "the scientific study of the problems of flight with a view to their practical solution." Though not a military institution itself, NACA was formed during the war, because of the war, as part of the Naval Appropriations bill of 1915; it had military members on its main committee and subcommittees, and it was committed to research in support of government aviation programs—all of which were military at the time.[64] A few months later, Secretary of the Navy Josephus Daniels invited Thomas Edison to chair a Naval Consulting Board to provide outside technical advice to the navy. Though the Board proved a disappointment, it helped spawn the Naval Research Laboratory, one of the most successful government research establishments in American history.[65] Disgruntled over the lack of scientists on the Naval Consulting Board, George Ellery Hale promoted the creation of a research arm for the National Academy of Sciences, which had done little science advising to the federal government, either in the Civil War or in the intervening years. This was to be the primary conduit through which the scientific talent of the United States would be enlisted in the war effort.[66] Rounding out the principal battery of agencies established to draw on America's scientific, engineering, and industrial talent was the National Defense Advisory Commission (NDAC), an umbrella organization heavily slanted toward the mobilization of economic and industrial resources in war production. In it, at least one scholar has seen the real origins of the military-industrial complex that came so clearly to the fore after World War II.[67]

Collectively these agencies—and the science and technology they attempted to marshal—contributed little to the course of World War I. Like all prior wars, this one was fought with the weapons in existence at the outset. Scientists and engineers made some real contributions in highly technical fields like aviation, underwater acoustics, and artillery spotting, but problems of producing the technology at hand always outweighed those of developing new technologies. By war's end, the services remained skeptical of the uses of science if not of technology. The war did little to draw the scientist and the soldier into the kind of close collaboration that might have made both more appreciative of each other's potentials, limitations, and needs.

WORLD WAR II, 1941–1945

It was this barrier between science and the military that Vannevar Bush sought to raze as World War II approached, by first tapping scientific talent in the National Defense Research Committee (NDRC), then bringing this talent into closer collaboration with the military users of their ideas through the mechanism of the

---

[64] Alex Roland, *Model Research: The National Advisory Committee for Aeronautics, 1915–1958*, 2 vols. (Washington, D.C.: NASA, 1985).

[65] David Kite Allison, *New Eye for the Navy: The Origin of Radar at the Naval Research Laboratory* (NRL Report 8466) (Washington, D.C.: NRL, 1981); A. Hoyt Taylor, *The First Twenty-Five Years of the Naval Research Laboratory* (Washington, D.C.: Department of the Navy, 1948).

[66] Kevles, *The Physicists*, Ch. 8; Kevles, "George Ellery Hale" (cit. n. 59); and Helen Wright, *Explorer of the Universe: A Biography of George Ellery Hale* (New York: Dutton, 1966).

[67] Koistinen, *The Military-Industrial Complex* (cit. n. 16) Ch. 2.

Office of Scientific Research and Development (OSRD). In doing so he demonstrated the inadequacy of previous institutional arrangements intended to exploit science and technology for war—save the National Advisory Committee for Aeronautics, on which NDRC was modeled.[68] By war's end Bush had evolved an institutional form that he thought was too powerful and too important to be left to the generals.

NDRC and OSRD instituted several critical changes in the relationship between science and war in the United States—changes that turned out to be permanent. First, it drew scientists into warfare at an unprecedented rate. Engineers, inventors, and industrialists had served in large numbers before; now scientists joined them on a comparable scale. Second, the scientists stayed at their home institutions or moved into new ones built for them, such as the Radiation Laboratory at MIT; in general, they did not get into uniform and they did not migrate to government arsenals or industry. Third, they were funded by contract, not to produce a product (as contracts would require in the postwar world) but to conduct research. In essence, the government purchased the scientific method on faith that its end result would be worth the candle. Fourth, information was compartmentalized in the interests of secrecy, a radical departure from standard scientific practice that was more or less accepted as a necessary if unpalatable and often counterproductive concomitant of war. Finally, the soldier and the scientist were drawn into close collaboration, so that the developments in the laboratory would be suited to the requirements of the battlefield. Scientists became advisers at the highest levels of policymaking, while soldiers posed some of the questions addressed in the laboratory.

Though none of these features of scientific research in World War II were entirely unprecedented, the scale on which they were conducted and the rigor with which the process was pursued fomented a revolution in the relation of science to war in the United States. This does not necessarily mean that warfare increased the scale and significance of science in society as a whole; Derek de Solla Price's research indicating the contrary has still not been successfully challenged after a quarter of a century.[69] But science was surely serving a different patron on a scale never before seen in America, raising issues such as secrecy,

---

[68] The National Defense Research Committee was created in June 1940, on the model of the National Advisory Committee for Aeronautics, of which Bush was chairman. One year later, the Office of Scientific Research and Development was created, absorbing NDRC. OSRD was needed to provide for development, to effect coordination with the services, and to provide an institutional umbrella for the Committee on Medical Research, a parallel organization to NDRC. See Irvin Stewart, *Organizing Scientific Research for War* (Boston: Little, Brown, 1948); see also James Phinney Baxter, *Scientists Against Time* (Boston: Little, Brown, 1948); A. Hunter Dupree, "The Great Instauration of 1940: The Organization of Scientific Research for War," in *The Twentieth Century Sciences: Studies in the Biography of Ideas*, ed. Gerald Holton (New York: Norton, 1972), pp. 443–467; and Carroll Pursell, "Science Agencies in World War II: The OSRD and Its Challengers," in *The Sciences in the American Context*, ed. Reingold (cit. n. 3), pp. 359–378.

[69] Derek de Solla Price, *Science Since Babylon* (New Haven: Yale Univ. Press, 1961); and Price, *Little Science, Big Science* (New York: Columbia Univ. Press, 1963). In the latter work Price says that World War II "looms as a huge milepost, but it stands at the side of a straight road of exponential growth" (p. 19). He does allow, however, that "the cost of research on a *per capita* basis and in terms of Gross National Product seems to have remained constant throughout history until about World War II and only since that time has met with the new circumstance of an increase that keeps pace with the growth of scientific manpower" (p. 94). He fails to associate this with the Cold War.

ethics, autonomy, and the conflicts between basic and applied research and between the arsenal and the contract.[70]

Nor did the huge infusion of science into World War II mean that science won the war. True, this was the first war in which the weapons in use at the end were significantly different from those available at the outset, but the new weapons were not decisive. The technology that had the greatest impact on World War II was the internal combustion engine. Employed in tanks, airplanes, motor vehicles, and submarines, it dominated the battlefield. Unlike World War I, this was a war of movement and maneuver. Like World War I, it was a war of industrial production. The other great innovations of the war, such as operations research, radar, the proximity fuze, and jet aircraft, were remarkable achievements of science and technology, but they did not determine the war's outcome.[71]

The one possible exception to this generalization, the atomic bomb, warrants special attention. If it has been a "decisive" weapon—and this remains to be seen—it has been so in a way not normally considered: that is, it may have prevented World War III in the last forty years. It can hardly be credited with winning the war in the Pacific; submarines contributed more to that end.[72] In fact, it is not at all clear that ending the Pacific war was the primary goal of the bombing of Hiroshima and Nagasaki. The men who decided to drop the bomb did so because it was there; surely the government was not about to explain to the taxpayer why it spent $2 billion on a weapon it never used. The official rationale was that it spared the lives, both American and Japanese, that would have been lost in an invasion. But it is difficult to see in retrospect why an invasion was necessary. Japan could have been blockaded and bombed into submission. The real objective of the atomic bombs was more likely the Soviet Union—both to end the Pacific war before Russia became more deeply involved and to make clear what power the United States had at its disposal at war's end. Just because the atomic bombs ended the war does not prove that they caused the end.[73]

---

[70] Cf. the list in Stewart, *Organizing Scientific Research for War*, pp. 325ff.

[71] This interpretation is supported by I. Bernard Cohen, "Science and the Civil War," *Technology Review*, 1946, 48:167; Lincoln R. Thiesmeyer and John E. Burchard, *Combat Scientists* (Boston: Little, Brown, 1947), p. 53; and Millis, *Arms and Men*. Many take the opposite view. Baxter, for example, in *Scientists Against Time*, quotes the German admiral Karl Doenitz as saying in December 1943: "For some months past the enemy has rendered the U-boat ineffective. He has achieved this object, not through superior tactics or strategy, but through his superiority in the field of science." (p. 46) Even Baxter's account, however, does not establish that the submarine could have decided the war. On specific developments, see Edward W. Constant II, *The Origins of the Turbojet Revolution* (Baltimore: Johns Hopkins Univ. Press, 1980); Allison, *New Eye for the Navy* (cit. n. 65); on napalm, see Louis F. Fieser, *The Scientific Method: A Personal Account of Unusual Projects in War and Peace* (New York: Reinhold, 1964); on the proximity fuze, see Ralph Baldwin, *The Deadly Fuze* (San Rafael, Calif.: Presidio Press, 1980). See also the official histories of the divisions within OSRD, such as William Albert Noyes, ed., *Chemistry: A History of the Chemistry Components of the National Defense Research Committee, 1940–1946* (Boston: Little, Brown, 1948); and John E. Burchard, ed., *Rockets, Guns, and Targets: Rockets, Target Information, Erosion Information, and Hypervelocity Guns Developed During World War II by the Office of Scientific Research and Development* (Boston: Little, Brown, 1949).

[72] Clay Blair, *Silent Victory: The U.S. Submarine War Against Japan* (Philadelphia: Lippincott, 1975).

[73] For more traditional views that nonetheless embody the evidence to support this conclusion, see Herbert Feis, *The Atomic Bomb and the End of World War II* (Princeton: Princeton Univ. Press, 1966); Len Giovannitti and Fred Freed, *The Decision to Drop the Bomb* (London: Methuen, 1967);

THE COLD WAR, 1945 TO THE PRESENT

The impression created at home by the atomic bomb, however, was something else again. Combined with the other technical developments of the war, the bombs led many to believe that science had won the war, or at least that it would win the next one. The result was a five-year period in which the national military and political establishments raced headlong in opposite directions. The civilians attempted to dismantle the military establishment, while the services sought to bedeck it with the once and future technology. True to tradition, Congress insisted upon a precipitous demobilization, a policy President Truman compounded by imposing on the military services painfully low budget ceilings. The services responded by reorganizing and scrambling to refight the last war with nuclear weapons. The air force led the way by institutionalizing its own mechanisms for getting scientific and technical advice and research and development, which established patterns subsequently followed by the other services.[74] While Vannevar Bush struggled to make permanent the wartime organization of science he had overseen, the Naval Research Laboratory funded most of the basic science the government chose to support.[75]

The struggle reached a climax in the first six months of 1950. In that fateful period, the United States approved development of the hydrogen bomb, formulated a national security policy (NSC 68) that committed the nation to permanent mobilization in the Cold War, committed American troops to the crusade against communism in Korea, and created the National Science Foundation. These steps meant that the nuclear scientists would be unable to prevent the arms race they had predicted and instigated, that the United States would be saddled with the standing military establishment it had always dreaded, that these decisions would be ratified in blood in Korea (making them virtually indelible in the short run), and that the government would fund a modest amount of basic research through a civilian agency and not through the military services. In many ways it was the arms race that proved most important for science, for it opened the government purse wider than ever before in peacetime and set off the mad scramble for new weaponry that President Eisenhower would come to call the military-industrial complex. In the same speech, Eisenhower also warned against the domination of the scientific and technical elite that he knew would come into increasing power as that complex grew.[76]

---

and Lawrence Freedman, "The Strategy of Hiroshima," *Journal of Strategic Studies*, 1 May 1978, pp. 76–97.

[74] Harvey Sapolsky maintains that "neither the navy nor any of the other services was a convert to science at the end of World War II"; "Academic Science and the Military" (cit. n. 3), p. 381. This was surely not true of the air force. See Thomas A. Sturm, *The USAF Scientific Advisory Board: Its First Twenty Years, 1944–1964* (Washington, D.C.: U.S. Air Force Historical Division Liaison Office, 1967); Nick A. Komons, *Science and the Air Force: A History of the Air Force Office of Scientific Research* (Arlington, Va.: Historical Division, Office of Information, Office of Aerospace Research, 1966); and Bruce L. K. Smith, *The Rand Corporation: Case Study of a Nonprofit Advisory Corporation* (Cambridge, Mass.: Harvard Univ. Press, 1966).

[75] Vannevar Bush, *Science, the Endless Frontier: A Report to the President* (Washington, D.C.: GPO, 1945); Daniel J. Kevles, "Scientists, the Military, and the Control of Postwar Defense Research: The Case of the Research Board for National Security, 1944–1946," *Technol. Cult.*, 1975, *16*:20–47; and Kevles, "The National Science Foundation and the Debate over Postwar Research Policy, 1942–1945," *Isis*, 1977, *68*:5–26.

[76] A good overview is Herbert York and G. A. Greb, "Military Research and Development: A

Before Eisenhower made that warning, one more ingredient was added to the stew: Sputnik. As Walter McDougall argues, Sputnik set off a revolution of its own, one comparable in impact to World War II and the momentous decisions of 1950, for it threw nominally civilian activities like space exploration into the total equation of national strength and security.[77] Less than four years after Sputnik I went up, President Kennedy and the Congress committed the United States to a race to the moon, a $25 billion stunt to demonstrate American scientific and technical superiority.[78] All of this was part of the Cold War, a permanent competition in which science and technology play a leading role, quality seems to matter more than quantity, the lines between civilian and military blur, and war becomes more total than ever before. Now all the resources of the state are thrown into the balance—in peace and in war—and scientific expertise weighs as heavily as any other factor save wealth.

CONCLUSIONS

What conclusions might be drawn from the history outlined above? The most significant inference might well be that the role of the military as patron of science in the United States has changed over the years.[79] From the informal and haphazard utilization and appreciation of science in the Revolutionary War, the military has passed through periods of supporting the earth and life sciences as well as oceanography and astronomy, even while integrating the potentials of science only poorly into the actual conduct of war. This remained true until World War II, when a conviction arose that science and technology would determine the outcome of future wars. Since then the military services have supported the physical sciences on an unprecedented scale. Scientists, like scholars in general, often turn bad money to good purposes, but no amount of rationalization can gainsay the dramatic, though often hidden, ways in which patrons shape the work of their benefactors. Illuminating these subtle influences will be one of the most important contributions of the coming scholarly synthesis. Some aspects of the problem are already apparent.

The moral dilemma of the scientist in the service of war is as old as Leonardo,

---

Postwar History," *Bulletin of the Atomic Scientists,* Jan. 1977, *33*(1):12–27. On the hydrogen bomb decision, see David Alan Rosenberg, "American Atomic Strategy and the Hydrogen Bomb Decision," *J. Am. Hist.,* 1979, 66:62–87. The division of the scientific community over the Hiroshima and Nagasaki bombings, the Baruch Plan, and the hydrogen bomb decision may be traced in Alice Kimball Smith, *A Peril and a Hope: The Scientists' Movement in America* (Chicago: Univ. Chicago Press, 1965); Robert Gilpin, *American Scientists and Nuclear Weapons Policy* (Princeton: Princeton Univ. Press, 1962); Arthur Steiner, "Baptism of the Atomic Scientists," *Bull. Atomic Sci.,* Feb. 1975, *31*(2):12–28; Brian Villa, "A Confusion of Signals: James Franck, the Chicago Scientists and Early Efforts to Stop the Bomb," *ibid.,* Dec. 1975, *31*(10):36–43; and Herbert York, *The Advisors: Oppenheimer, Teller, and the Superbomb* (San Francisco: Freeman, 1976). On NSC 68 and the Korean War, see Samuel Wells, "Sounding the Tocsin: NSC 68 and the Soviet Threat," *International Security,* Fall 1979, 4:116–158. On the NSF see J. Merton England, *A Patron for Pure Science: The National Science Foundation's Formative Years, 1945–1957* (Washington, D.C.: NSF, 1982).

[77] Walter A. McDougall, *The Heavens and the Earth: A Political History of the Space Age* (New York: Basic Books, 1985).

[78] Vernon Van Dyke, *Pride and Power: The Rationale of the Space Program* (Urbana: Univ. Illinois Press, 1964); and John Logsdon, *The Decision to Go to the Moon: Project Apollo and the National Interest* (Cambridge, Mass.: MIT Press, 1970).

[79] On the theme, see Richard Westfall, "Science and Patronage: Galileo and the Telescope," *Isis,* 1985, 76:11–30.

if not Archimedes.[80] In a democracy such as that of the United States, in which a modicum of popular support is necessary to sustain a war, scientists have not until recently been much troubled by the ethics of serving the military. In peace they were not called upon; in war they contributed in the same spirit of patriotism as their fellow citizens. Spared the worst carnage of World War I, American scientists did not begin to worry about their contributions to total war until the atomic bomb made the modern drift of events all too clear.[81] Since that cathartic event, the scientific community has been torn by the growing tension between the allure of lucrative military grants and contracts and the increasingly obvious and deadly consequences to which their research can contribute.[82] Nothing in the foreseeable future suggests that this tension will soon decrease.

The conflict between basic and applied research is also older than the republic; it too emerged early in our national history. The Revolution had reinforced a pragmatism already rampant in the colonies. While the Founding Fathers were not hostile to science, neither were they anxious to spend government funds on activities with no practical return. Still, the "common defense" is a broad term, as vague as our current "national security." Under the former banner the government funneled support for science through the military services throughout the nineteenth century. Sometimes the military payoff was obvious, as in the arsenals. In other cases, such as exploration and weather analysis, the armed services had the only infrastructure and the only constitutional mandate available. After the Civil War, when the government began to support science under the provisions of the general welfare clause, basic research found other sources of support and the military returned for a while to a narrower construction of practical application. In recent decades, the distinction between basic and applied research has blurred, and arguments have been advanced that basic research provides the indispensable base on which applications are built. Though this reasoning persuades few outside the choir, it is true that the military services attach fewer strings to their research funds than other federal agencies.[83]

The institutionalization of science and technology for purposes of war is a multifaceted problem still wrapped in controversy. The relative merits of government arsenals and contracting with private firms remain unclear after almost

---

[80] See Roland, *Underwater Warfare in the Age of Sail* (cit. n. 10), passim; Bernard Brodie, "Defense and Technology," *Technol. Rev.*, 1941, *43*:109; and Monte D. Wright and Lawrence J. Paszek, eds., *Science, Technology, and Society: Proceedings of the Third Military Symposium*, 8–9 May 1969 (Washington, D.C.: Department of the Air Force, 1971).

[81] Smith, *A Peril and a Hope* (cit. n. 76).

[82] R. W. Reid, *Tongues of Conscience: Weapons Research and the Scientists' Dilemma* (New York: Walker, 1969). See also Leonard A. Cole, *Politics and the Restraint of Science* (Totowa, N.J.: Rowman & Allanheld, 1983), which argues persuasively that this problem is not peculiar to military research; and Carol Gruber, *Mars and Minerva: World War I and the Uses of the Higher Learning in America* (Baton Rouge: Louisiana State Univ. Press, 1975), which is equally persuasive in establishing that the problem is not peculiar to science. To see the dilemma in context, see Fieser, *The Scientific Method* (cit. n. 71); and Robert Harris and Jeremy Paxman, *A Higher Form of Killing: The Secret Story of Chemical and Biological Warfare* (New York: Hill & Wang, 1982).

[83] The classic studies are C. W. Sherwin and R. S. Isenson, *First Interim Report on Project Hindsight: Summary* (Washington, D.C.: Office of Director of Defense Research and Engineering, 1966); and Illinois Institute of Technology Research Institute, *Technology in Retrospect and Critical Events in Science (TRACES)*, 2 vols. (Chicago: IIT Research Institute, 1968). Compare these with *R & D Contributions to Aviation Progress (RADCAP): Joint DOD–NASA–DOT Study* (Washington, D.C.: DOD, NASA, DOT, 1972); and W. Henry Lambright, *Governing Science and Technology* (New York: Oxford Univ. Press, 1976), pp. 118–119.

two centuries of experience. The think tank, a relatively modern variant on this dichotomy, exists in both military and private forms; its main distinction from the arsenal or other contractors is that it seldom has facilities for production or laboratory research.[84] How to buy, maintain, and operate research equipment is a related problem with sharply contending advocates, as is the dilemma of maintaining a scientific infrastructure without putting the scientific community on permanent retainer. All of these problems of institutionalization hover on the fringes of the military-industrial complex, insulated from the worst controversies of that political morass by the fact that scientific research usually remains aloof from production and does not operate for profit. When scientists cross those thresholds, as they do from time to time, the issues of science and war usually disappear in a fog of political rhetoric.[85]

Secrecy is as old as warfare; compartmentalization is its modern version. The former precludes scientists from publishing research results with military implications; the latter, which came into prominence in World War II, precludes the scientist from exchanging views with his colleagues unless he can establish a priori that they need to know his ideas to prosecute their own work.[86] Both restraints are anathema to the scientist and disruptive of the advancement of knowledge as the scientist understands the process. The existing literature boasts countless instances where these policies have retarded or even precluded important military technologies; more work is needed on where the line might practically be drawn.[87]

Warfare has politicized science in the Cold War, influencing not only the agenda of science but the method by which science proceeds.[88] David Rittenhouse might suspend his peacetime pursuits to help the colonists find better ways of producing gunpowder, but this lasted only for the duration of the war. Joseph Henry placed his duties at the Smithsonian Institution first; in his spare time he evaluated inventions and advised the government on the technology of the Civil War. In the Cold War, however, science is permanently mobilized. Scientists sit on advisory panels, assess military needs, evaluate enemy capabilities, and finally advise the government on what can and should be done. No amount of detachment and objectivity in the first three tasks can make the last one anything but a political act, as scientists learned in the hydrogen bomb decision, the test-ban controversy of the late 1950s, and the Vietnam War. In some of these cases, scientific research agendas are shaped to prove or disprove one

[84] See Paul Dickson, *Think Tanks* (New York: Atheneum, 1971). The National Aeronautics and Space Administration was free to select either arsenals or contracting when it was created in 1958; it opted instead for a combination of both; see Arnold S. Levine, *Managing NASA in the Apollo Era* (NASA SP-4102) (Washington, D.C.: NASA, 1982). On the pitfalls NASA hoped to avoid, see Nieburg, *In the Name of Science* (cit. n. 17).

[85] Nieburg, *In the Name of Science*. The best introduction to the vast literature on the military-industrial complex is Steven Rosen, ed., *Testing the Theory of the Military-Industrial Complex* (Lexington, Mass.: Lexington Books, 1973).

[86] The classic rationale for this policy is given in Leslie Groves, *Now it Can Be Told: The Story of the Manhattan Project* (New York: Harper, 1962), pp. 167–169.

[87] See, e.g., John Sloop, *Liquid Hydrogen as a Propulsion Fuel, 1945–1959* (NASA SP-4404) (Washington, D.C.: NASA, 1978); and Constant, *Origins of the Turbojet Revolution* (cit. n. 71).

[88] Joseph Rotblat, *Scientists in the Quest for Peace: A History of the Pugwash Conferences* (Cambridge, Mass.: MIT Press, 1972). For the views of two insiders, see James R. Killian, Jr., *Sputnik, Scientists, and Eisenhower: A Memoir of the First Special Assistant for Science and Technology* (Cambridge, Mass.: MIT Press, 1977); and George B. Kistiakowsky, *A Scientist at the White House: The Private Diary of President Eisenhower's Special Assistant for Science and Technology* (Cambridge, Mass.: Harvard Univ. Press, 1976).

political position or another.[89] The dispassionate search for truth that characterizes the scientific endeavor often disappears in the struggle.

The conclusions one might draw about the influence of science on war in American history are at once simpler and less clear. Science and technology have not decided any American wars, with the possible exception of Vietnam, which we may have lost through overreliance on inappropriate technology. But science and technology have been decisive in preventing a war with the Soviet Union, which surely would have come save for nuclear weapons.

By these sweeping and intentionally provocative generalizations I do not mean to suggest that science and technology have been unimportant in America's wars. British logistics in the American Revolution, the railroad and the telegraph in the Civil War, sonar in World War I, radar in World War II, the helicopter in Korea and Vietnam—all had significant impact, but they did not decide the outcome. The reason is not that the technology of war is unimportant; quite the contrary, it grows more important all the time. The reason is that since the Industrial Revolution, during which the United States was born, most nations have come to appreciate the importance of the technology of war and have striven to arm themselves with the best and most modern weapons they can afford. Because the arms bazaar is and has been for many decades an international market, most combatants come to the field of battle comparably equipped. Technologies tend to cancel each other out. In practice, new technologies are indispensable. In theory, they are decisive. In fact, they seldom decide anything.

Vietnam provides the clearest example of the national myopia about the role of technology in warfare. The argument is often made, understandably enough, that smart bombs were decisive in Vietnam, forcing the North Vietnamese back to the bargaining table for the concluding talks that brought a settlement early in 1973.[90] Advocates of this position forget, however, that the United States lost the war. Helicopters, gun ships, electronic sensors, B-52s, saturation bombing, defoliants, infrared gunsights—the whole electronic battlefield[91] was conquered by a determined and ruthless enemy who recognized that politics wins more wars than science. Military historians might profitably spend more time investigating how science and technology—and scientists and technologists—shape warfare as a social institution, and less time trying to decide what was decisive.

NEEDS AND RESOURCES

The historian who attempts the synthesis suggested here will find the existing secondary literature surprisingly rich. The works cited above suggest some of the spadework already done. Several characteristics of this literature deserve special mention. First, it is alive with biographies and autobiographies that rise far above the hagiography and insipid memoirs of earlier generations. Thomas Hughes's model study of Elmer Sperry, for example, provides penetrating

[89] Harold Jacobson and Eric Stein, *Diplomats, Scientists, and Politicians: The United States and the Nuclear Test Ban Negotiations* (Ann Arbor: Univ. Michigan Press, 1966); and Robert A. Divine, *Blowing on the Wind: The Nuclear Test Ban Debate, 1954–1960* (New York: Oxford Univ. Press, 1978).

[90] See, e.g., U. S. G. Sharp, *Strategy for Defeat? Vietnam in Retrospect* (San Rafael, Calif.: Presidio Press, 1978).

[91] Paul Dickson, *The Electronic Battlefield* (Bloomington: Indiana Univ. Press, 1976).

insights into the formative years of the military-industrial complex, without indulging in the opprobrium and moralism that too often attach themselves to that topic. For all its special pleading and self-righteousness, Holden Evans's autobiography provides a revealing glimpse into the operation of naval yards around the turn of the century. Elting Morison's classic study *Admiral Sims and the Modern American Navy* delivers just what the title promises.[92]

Official histories provide a better source of information and analysis than is generally allowed in academic circles. Court history is still written under government auspices, but some good history is being written there as well. Clayton Koppes's prize-winning history of the Jet Propulsion Laboratory was a sponsored study that ended up focusing on the influence of national security considerations on scientific methods. In his study of radar at the Naval Research Laboratory, David Allison has shown the importance of organizational placement of the research facility within a bureaucratic hierarchy dominated by operations. The highly regarded Army official history series on World War II is graced by such significant works as *Buying Aircraft,* by I. B. Holley, and *The Ordnance Department,* by Constance McLaughlin Green and others. Forty years after the event, the two official histories of the Manhattan project are still the best available.[93] Official histories must often be taken with a grain of salt, but they are usually a better resource than is generally known or admitted.[94]

The military context has produced important studies of the nature of technological change, but not all of these have yet been well incorporated into the literature. One thinks immediately of Merritt Roe Smith's *Harpers Ferry Arsenal and the New Technology* and Edward Constant's *The Origins of the Turbojet Revolution.* Studies of other military technologies have been comparably illuminating, such as James Phinney Baxter's classic study, *The Introduction of the Ironclad Warship,* I. B. Holley's *Ideas and Weapons,* and Robert Coulam's *The Illusion of Choice.*[95] There is already enough data here to allow us to make

[92] Hughes, *Elmer Sperry* (cit. n. 48); Evans, *One Man's Fight* (cit. n. 58); Morison, *Admiral Sims* (cit. n. 47). See also, e.g., Harold G. Bowen, *Ships, Machinery, and Mossbacks: The Autobiography of a Naval Engineer* (Princeton: Princeton Univ. Press, 1956); and Arthur Holley Compton, *Atomic Quest: A Personal Narrative* (New York: Oxford Univ. Press, 1956). Cf. the disappointing William Frederick Durand, *Adventures: In the Navy, in Education, Science, Engineering, and in War—A Life Story* (New York: American Society of Mechanical Engineers, 1953).

[93] Clayton R. Koppes, *JPL and the American Space Program: A History of the Jet Propulsion Laboratory* (New Haven, Conn.: Yale Univ. Press, 1982); Allison, *New Eye for the Navy* (cit. n. 71); I. B. Holley, Jr., *Buying Aircraft: Materiel Procurement for the Army Air Forces* (Washington, D.C.: Office of the Chief of Military History, Department of the Army, 1964); Constance McLaughlin Green, Harry C. Thomson, and Peter C. Roots, *The Ordnance Department: Planning Munitions for War* (Washington, D.C.: Department of the Army, 1955); and, on the Manhattan project, Richard G. Hewlett and Oscar E. Anderson, Jr., *The New World, 1939–1946* (University Park: Pennsylvania State Univ., 1962); and Henry DeW. Smyth, *Atomic Energy for Military Purposes: The Official Report on the Development of the Atomic Bomb Under the Auspices of the United States Government, 1940–1945* (Princeton: Princeton Univ. Press, 1945).

[94] The series The United States Army in World War II, begun under the general editorship of Kent Roberts Greenfield, enjoys a much better reputation abroad than it does in the United States. Though the quality of the volumes varies, it is generally remarkably high. The same may be said for the series of histories published by Little, Brown in the late 1940s on the activities of the OSRD during World War II (see Thiesmeyer and Burchard, *Combat Scientists,* Noyes, ed., *Chemistry,* and Burchard, ed., *Rockets,* in n. 48 above).

[95] Smith, *Harpers Ferry* (cit. n. 17); Constant, *Origins of the Turbojet Revolution* (cit. n. 71); Baxter, *Introduction of the Ironclad Warship* (cit. n. 37); Holley, *Ideas and Weapons* (cit. n. 62); and Robert Coulam, *Illusions of Choice: Problems in the Development of the F-111 Fighter-Bomber* (Cambridge, Mass.: Public Policy Program, John F. Kennedy School of Government, Harvard University, 1973).

some generalizations about the differences between military and civilian development and to test W. Henry Lambright's suggestion that these differences
may not be as important as the differences between big and little science and
between basic and applied research.[96]

Other issues of considerable importance have been tentatively addressed, but
cry out for more study. Some work on the scientific and engineering professions
has noted the impact of war, but our understanding is still embryonic.[97] The
effect of science and war on the military profession has received scarcely any
serious study. Military conservatism in adopting new weapons, what Bernard
Brodie called "the traditional reluctance of the military professions to be killed
by anything but traditional weapons,"[98] needs far more investigation than it has
received, as useful as the existing studies are. Particularly wanting is critical
investigation of the antitechnology stereotype and of how and why it has been
so completely reversed in the period since World War II. Civilian applications
of military technology run throughout American history, from early exploration
and collection of weather data to modern aviation and space travel, yet little
attention has been paid to the mechanisms by which this technology transfer
takes place.[99]

A number of important topics with significant implications for the histories of
science, technology, and war in the United States have hardly been touched.
Americans have had a penchant in their history to replace men with machines
on the battlefield, to trade treasure for lives. Most advances in military technology have come from professional, civilian scientists and engineers, not from
military officers and not from inspired amateurs, for reasons that are not entirely
clear.[100] The impact of secrecy and compartmentalization needs more dispassionate study than it has received to date. Objective evaluations are also needed
of the relative impact of basic and applied research, to test the conclusions of
the self-serving studies already extant. More research is needed on the origins
of our current methods of organizing scientific advice to the government; the
military might well be the model for this important dimension of modern science.[101]

This list could easily be extended. The main point is that a significant corpus
of secondary literature exists on science and war in the United States, but many
important questions remain unanswered or in doubt. To fill this gap historians
of science and technology will have to overcome the traditional academic bias
against military topics, and military historians will have to overcome their penchant for operational history. Surely there are enough distinguished entrants in
this field to make it a respectable and inviting arena for any scholar.

---

[96] Lambright, *Governing Science and Technology* (cit. n. 83), Ch. 8.

[97] Smith, *A Peril and a Hope* (cit. n. 76); Daniel H. Calhoun, *The American Civil Engineer: Origins
and Conflict* (Cambridge, Mass.: MIT Press, 1960); Sinclair, *Philadelphia's Philosopher Mechanics*
(cit. n. 39).

[98] Brodie, "Defense and Technology" (cit. n. 80), p. 108.

[99] The modern enthusiasm for new military technology is captured in Herbert York, *Race to
Oblivion: A Participant's View of the Arms Race* (New York: Simon & Schuster, 1970). The best
overview of civilian applications is Steven R. Rivkin, *Technology Unbound: Transferring Scientific
and Engineering Resources from Defense to Civilian Purposes* (New York: Pergamon, 1968).

[100] Kaempffert, "War and Technology" (cit. n. 28), p. 441; Reingold, "Science in the Civil War"
(cit. n. 20), p. 318.

[101] On the last issue, compare the post–World War II mechanisms with those described by Koistinen in *The Military-Industrial Complex* (cit. n. 16), Ch. 2.

Science and Public Policy since World War II

By Margaret W. Rossiter*

THE PERIOD SINCE WORLD WAR II has long been considered the most important in the history of American science. Even those early skeptics who said that little science of consequence had taken place in America customarily exempted the post-1933 (or post-1940 or -1945) period. The rapid increase in the number of scientists in the United States in the postwar years and the large number of Nobel Prizes going to Americans (whether immigrants or native born) during 1945–1970 testify to the large quantity and high ("world class") quality of American science in this period. One reason was the sudden collapse of European science in the 1930s; another was the new and highly visible stress on science and technology in the United States in the postwar era, as both became the basis of international competition and even national survival.

Despite the importance of the science done in the United States in recent decades and its centrality to certain portions of American public policy, the whole subject remains badly understudied. For example, previous anthologies or collective volumes on the historiography of American science have dealt with the post–World War II era only rarely if at all, and then simply to stress the later activities of Vannevar Bush, J. B. Conant, and the several presidential advisers. More material with greater breadth and diversity has appeared, however, in recent years, especially since 1979 (to judge from the publication dates in many footnotes here). The situation is in fact changing so rapidly that it is already both possible and necessary for a preliminary mapping of what has been done, what can be done, and what ought to be done. For this endeavor, the whole overwhelming subject needs to be subdivided into manageable units and sources need to be identified and suggested for use.

The eight topics discussed here certainly cannot pretend to cover the entire subject; quite the contrary, each was designed more like an exploratory probe into relatively unknown territory—a separate, finite, containable foray that would constitute a preliminary investigation of one portion of the whole. Ideally, each topic would have some unity and development of its own, but since each would overlap or adjoin at least some of the others, all the topics together would give a comprehensible overview. The topics are material and personnel shortages and surpluses around 1950; federal aid to nonmilitary (especially "basic") research; loyalty oaths and security checks; the rise of the behavioral sciences; science education from the Cold War to creationism; antinuclear protests and the limited test ban treaty, 1954–1963; Sputnik and the space program; and

* American Academy of Arts and Sciences, Cambridge, Massachusetts 02138.
This project was supported by a grant from the National Science Foundation, SES 83-20601.

health and safety issues, including the environmental movement in the 1950s and 1960s. The end point is roughly 1970. A ninth major topic, military research and development, was omitted to avoid overlap with another essay in this book.

SURPLUSES AND SHORTAGES

Because the late 1940s and early 1950s were very turbulent times in the United States diplomatically and politically, it seems wise to sidestep these areas for the moment and begin instead with some assessment of the country's economic and technological status around 1950. This is appropriate and feasible, because such issues as surpluses and shortages were a major preoccupation at the time. Curiously, the country was in an almost contradictory position. On the one hand, it had tremendous strengths and assets in a war-ravaged world; on the other, its economy was vulnerable on several counts. The United States was blessed as perhaps no other country with vast surpluses of such agricultural commodities as cotton and corn and with an abundance of such essential energy supplies as hydroelectric power (it was then only minimally dependent on imports of petroleum from Saudi Arabia). Yet at the same time the nation's leaders had a continuing fear, stemming from World War II experiences, of serious shortages in essential areas, particularly shortages of certain "critical" commodities such as minerals (e.g., tungsten, manganese, or tin) and of certain "critical" personnel such as highly trained scientists and engineers. Without sufficient supplies of either, the nation could be in considerable danger.

When, in June 1950, North Korea invaded the south, these fears of impending shortages took on a starkly military connotation. By 1951 President Harry Truman had declared that the nation was in a "state of emergency" and that it should be prepared for "economic and technological warfare for the foreseeable future." He reinstated the military draft and placed most of the civilian economy under various price, wage, and other controls. Thus, less than six years after the end of World War II, the United States was again being mobilized for a large war of indefinite duration.

This series of events offers many topics for research. Already a few secondary works have appeared that mine trade journals and the National Archives and Records Service (including the Harry S Truman and Dwight D. Eisenhower presidential libraries) in order to trace the shifts and lurches in industries and various parts of the federal government when emergency planners reacted to strategic external events, complained about insufficient data, and tried to shape the nation's economic and domestic response.

Perhaps the greatest amount of work in this area has been done on agricultural surpluses. The major book so far is Trudy Huskamp Peterson's *Agricultural Exports, Farm Income, and the Eisenhower Administration* (1979), which describes the complex economic problems of trying to dispose of ever-mounting surpluses. Also helpful are a recent article by Douglas Bowers in *Agricultural History,* in which he traces the great need for and subsequent ineffectiveness of the Research and Marketing Act of 1946, and a USDA report Bowers wrote with Vivian Wiser. Narrower in focus but pursuing the scientific and technological roots of the surplus are Gale E. Peterson's "The Discovery and Development of 2,4-D," the major herbicide of the era (also known as "Weedone" and later used in Vietnam as part of "Agent Orange"), and Gilbert C. Fite's "Mechanization

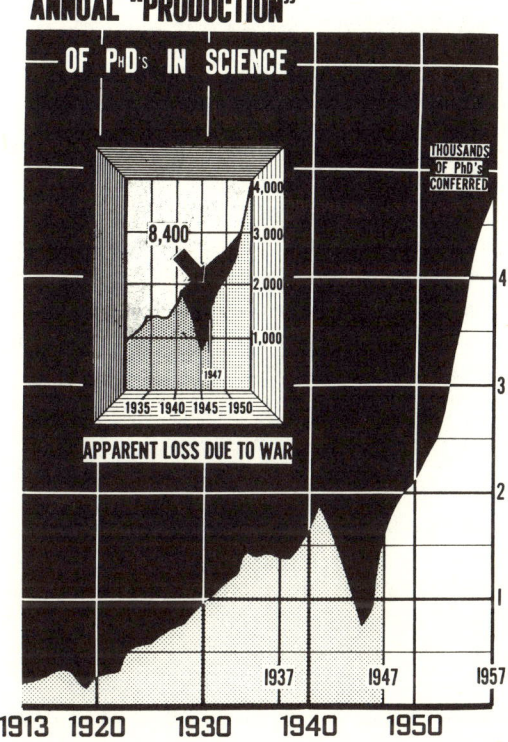

ANNUAL "PRODUCTION" OF PhDs IN SCIENCE

From John R. Steelman, Science and Public Policy: A Report to the President, *Volume 1:* A Program for the Nation *(Washington, D.C.: GPO, The President's Scientific Research Board, 1947), p. 17.*

of Cotton Production Since World War II," both of which appeared in *Agricultural History.*[1]

In the area of shortages, a recent book by Alfred E. Eckes, Jr., *The United States and the Global Struggle for Minerals* (1979), discusses Cold War minerals policy, including the "Paley Report" of 1952, which recommended depletion allowances and a great expansion of the national stockpiling program. The problems of personnel shortages during the Korean emergency are discussed in James M. Gerhardt's *The Draft and Public Policy: Issues in Military Manpower Procurement, 1945–1970* (1971). Gerhardt describes the scientists' attempts to get deferments for certain college students and exemptions for those in "critical" occupations. Also somewhat related to the Cold War stress on economic preparedness was the need for an efficient domestic transportation network, examined in Mark Rose's *Interstate: Express Highway Politics, 1941–1956* (1979). It too mines the National Archives and Records Service, including its several presidential libraries, to tell the essential tale of various interest groups battling for a major federal program flush with lavish public works monies.[2]

[1] Trudy Huskamp Peterson, *Agricultural Exports, Farm Income, and the Eisenhower Administration* (Lincoln: Univ. Nebraska Press, 1979); Douglas E. Bowers, "The Research and Marketing Act of 1946 and Its Effects on Agricultural Marketing Research," *Agricultural History,* 1982, *56*:249–263; Vivian Wiser and Douglas E. Bowers, *Marketing Research and Its Coordination in USDA* (USDA Agricultural Economic Report No. 475) (Washington, D.C.: USDA Economic Research Service, 1981); Gale E. Peterson, "The Discovery and Development of 2,4-D," *Agricultural History,* 1967, *41*:243–253; Gilbert C. Fite, "Mechanization of Cotton Production Since World War II," *Agricultural History,* 1980, *54*:190–207.

[2] Alfred E. Eckes, Jr., *The United States and the Global Struggle for Minerals* (Austin: Univ.

The whole area seems to offer many possibilities for worthwhile monographs. There were numerous commissions and offices set up to plan America's postwar economic and military problems, such as the National Security Resources Board, the Office of Defense Mobilization, and the Council of Economic Advisers.[3] Almost any of these groups and commissions would merit study, and many of their participants deserve articles if not full biographies. In addition, certain major pieces of legislation, such as the Defense Production Act of 1950 (an act whose further amendment underlies much recent talk of "reindustrialization") have apparently never been studied. The role of philanthropic foundations in this postwar economic and technological planning also remains to be explored. To date it can be grasped only indirectly, through studies that individual foundations have commissioned or financed. The full extent of their collective influence can at present only be imagined.

FEDERAL AID TO NONMILITARY RESEARCH

Despite certain debates in both Britain and the United States in the late 1940s about whether truly "basic" research could be planned as projects and budgeted in advance, by the 1950s federal aid to such research was a highly developed activity. Although, compared with other topics in this essay, this subject has been rather well studied, major gaps still remain. The story is usually told in terms of the administrative history of the funding agencies, such as the National Institutes of Health (NIH), the Office of Naval Research (ONR), the Atomic Energy Commission (AEC), and the National Science Foundation (NSF). Much less is known about the impact or effect on scientists or science of any of the money spent.

Although certain precedents were set by the "research contracts" devised during World War II by the Office of Scientific Research and Development (and since studied by A. Hunter Dupree and Carol Gruber), it was the NIH that pioneered in both federal fellowships and research grants. Its Division of Research Grants was functioning well, with "study sections" for several specialties, by 1947. Unfortunately the several existing histories of the NIH all come to an end just at the beginning of this new era, one that saw the NIH establish a massive external grants program and develop its intramural research program (seven separate institutes by 1951 and a large clinical center by 1953) on the new Bethesda campus. The classical history of NIH to 1950 remains Donald Swain's 1962 article in *Science,* and the chief political analysis is Stephen P. Strickland's *Politics, Science and Dread Disease* (1972) on the very close ties between the lobbyists, congressional committees, and NIH leaders through the first two decades of the postwar era. Histories could be written of each of the institutes (e.g., Cancer, Heart, Eye, Dental Research, Mental Health) that would treat NIH's science and scientists as well as the administrative and congressional angles. The records of the NIH are, however, in such an uncertain state and vary

Texas Press, 1979); James M. Gerhardt, *The Draft and Public Policy: Issues in Military Manpower Procurement, 1945–1970* (Columbus: Ohio State Univ. Press, 1971); Merriam H. Trytten, *Student Deferment in Selective Service* (Minneapolis: Univ. Minnesota Press, 1952); Mark Rose, *Interstate: Express Highway Politics, 1941–1956* (Lawrence: Regents Press of Kansas, 1979).

   [3] See Edward S. Flash's *Economic Advice and Presidential Leadership: The Council of Economic Advisers* (New York: Columbia Univ. Press, 1965).

so much from institute to institute that it is hard to be optimistic about what can be accomplished there.[4]

Historians of science have generally praised the Office of Naval Research for its sponsorship or support of much "basic" research in the late 1940s. Led by Admiral Harold Bowen (a long-time rival of Vannevar Bush), chief scientist Alan T. Waterman (later of NSF), and in applied mathematics by Mina Rees (on leave from Hunter College), it supported some early computer work, crystallographic research, and studies of underwater acoustics, including even some forms of animal behavior, as in a study of communication among dolphins and other marine animals. The full extent of ONR's impact is not yet known, and a major study of this agency remains to be written. Political scientist Harvey Sapolsky has written an admirable piece on Bowen and ONR in a recent anthology edited by Nathan Reingold, and Carol Gruber has touched on it in a recent paper on the origins of research contracts. She hints that, contrary to the prevailing belief, the funds given were not all that disinterested. Despite knowing the Navy's interests, the academic researchers took the money.[5]

The Atomic Energy Commission, which came into being in late 1946, inherited from the Army several wartime atomic laboratories, including Los Alamos, Oak Ridge, and Argonne. The future role of these federal laboratories was in doubt, and their proper peacetime organization and management problematic. The staffs preferred to be managed by universities, however remote, or other outside contractors (such as Union Carbide at Oak Ridge) rather than to become government (i.e., civil service or military) employees; but no one objected to the AEC's maintaining the equipment and facilities directly. Thus a curious kind of hybrid (private-public) organization resulted, spread, and persists to this day. In addition, the research laboratories had certain geographic rivalries. In the East, physicists led by I. I. Rabi of Columbia University, anxious not to be outdone by Ernest Lawrence's Radiation Laboratory at Berkeley, California, went so far as to establish a consortium of nine interested universities called the Associated Universities, Inc., to set up and manage a new laboratory at a former Army camp at Brookhaven on Long Island. This arrangement has apparently worked successfully, perhaps because no one university dominates the scene, unlike a similar effort by several midwestern universities at Argonne in Illinois.

A few studies exist of these evolving postwar arrangements, including Robert Seidel's excellent "Accelerating Science: The Postwar Transformation of the Lawrence Radiation Laboratory" and Allan Needell's "Nuclear Reactors and

[4] A. Hunter Dupree, "The *Great Instauration* of 1940: The Organization of Scientific Research for War," in *Twentieth Century Sciences: Studies in the Biography of Ideas,* ed. Gerald Holton (New York: Norton, 1972), pp. 443–467; Carol Gruber, "Contracts for A-Bombs: Government-University Collaboration in World War II and the Emergent Cold War," paper presented at annual meeting of the Organization of American Historians, Los Angeles, April 1984; Donald Swain, "The Rise of a Research Empire: NIH, 1930–1950," *Science,* 1962, *138*:1233–1237; on the NIH study sections see James Cassedy, "Stimulation of Health Research Science," *Science,* 1964, *145*:897–902; and Stephen P. Strickland, *Politics, Science and Dread Disease: A Short History of United States Medical Research Policy* (Cambridge: Harvard Univ. Press, 1972).

[5] Harvey Sapolsky, "Academic Science and the Military: The Years Since the Second World War," in *The Sciences in the American Context: New Perspectives* ed. Nathan Reingold (Washington, D.C.: Smithsonian Institution Press, 1979), pp. 379–399; Gruber, "Contracts for A-Bombs." See also Dan H. McLachlan, Jr., and Jenny P. Glusker, eds., *Crystallography in North America* (New York: American Crystallographic Association, 1983), pp. 85, 138, 226; Juliet Lamont, "From Mechanics to Mind: The Development of Dolphin Research in the United States from 1945 to 1963" (Undergraduate honors thesis, Harvard Univ., 1984).

the Founding of Brookhaven National Laboratory," both of which appeared in recent issues of *Historical Studies in the Physical Sciences*. Also related is Leonard Greenbaum's *A Special Interest: The Atomic Energy Commission, Argonne National Laboratory, and the Midwestern Universities* (1971), which was commissioned after the many scathing attacks on Argonne by Daniel S. Greenberg in *Science* and in his book *The Politics of Pure Science* (1967). Greenbaum's book reveals how very difficult it had been to get the jealous and warring members of the midwestern group behind Argonne to work together. They both needed leadership but distrusted it, especially any hint of domination from the University of Chicago.[6]

To date the National Science Foundation, established in 1950 in the shadows of the Cold War to promote "basic research" in science, is the best studied of all these federal funding agencies. J. Merton England's recent account to 1957 is an administrative history with chapters or portions on such topics as the social sciences, science education, the International Geophysical Year (IGY), and other relevant topics, since in one way or another NSF was involved in almost every scientific issue after 1950. England is now working on a sequel that will bring the Foundation down to the mid-to-late 1960s. The NSF Director's Central and Subject Files are available through at least the mid 1960s at the Science, Technology and Natural Resources Branch of the National Archives in Washington, D.C.[7]

Yet even when the full administrative histories of these agencies are known, when biographies of the major figures (Thomas Parran, Alan T. Waterman, Lloyd Berkner, Leland Haworth, James Shannon, and Detlev W. Bronk, to name a few) are written, and when histories of the major grantee institutions are ready, certain other questions will remain about the federal funding of research. Surely it force-fed some fields (such as space science, discussed below) faster than others or than was advisable, and with costs and tradeoffs that are unclear. Federal aid also had other very visible limits in the 1950s and 1960s. For example, the NIH, despite being funded with millions and eventually billions of dollars annually, could not because of congressional sensitivities support research on the birth control pill. That was left to the private philanthropist Katherine Dexter McCormick and the Worcester Foundation for Experimental Biology. Nor did the NIH move into the area of polio research, which the National Foundation for Infantile Paralysis had already staked out so well, as is ably described by Saul Benison in an article in Gerald Holton's *Twentieth-Century Sciences: Studies in the Biography of Ideas* (1972). Another even more glaring

[6] Harold Orlans, *Contracting for Atoms* (Washington, D.C.: Brookings Institution, 1967) has long been the classic in this area. It has now been supplemented by Robert Seidel, "Accelerating Science: The Postwar Transformation of the Lawrence Radiation Laboratory," *Historical Studies in the Physical Sciences,* 1983, *13*:375–400; Allan A. Needell, "Nuclear Reactors and the Founding of Brookhaven National Laboratory," *ibid.,* 1984, *14*:93–128; Leonard Greenbaum, *A Special Interest: The Atomic Energy Commission, Argonne National Laboratory and the Midwestern Universities* (Ann Arbor: Univ. Michigan Press, 1971); Daniel S. Greenberg, *The Politics of Pure Science* (New York: New American Library, 1967). See also Anton G. Jachim, *Science Policy Making in the United States and the Batavia Accelerator* (Carbondale, Ill.: Southern Illinois Univ. Press, 1971), Ch. 4; and Richard G. Hewlett and Francis Duncan, *Atomic Shield, 1947–1952,* Vol. II of *A History of the United States Atomic Energy Commission* (University Park: Pennsylvania State Univ. Press, 1969).

[7] J. Merton England, *The Patron of Pure Science: The National Science Foundation's Formative Years* (Washington, D.C.: GPO, 1983).

limit to federal aid to science and scientists in the 1950s was the apparatus of loyalty oaths, security checks, and potential passport revocations to which researchers, especially highly prominent scientists, were subject.[8]

LOYALTY OATHS AND SECURITY CHECKS

If federal aid to both "basic" and military research was one response to the insecurities of the Cold War, another was the concurrent rise of loyalty oaths and security checks. With the aid of the Freedom of Information Act of 1974, several recent studies, most notably David Caute's extensive *The Great Fear: The Anti-Communist Purge Under Truman and Eisenhower* (1978), have begun to delve into various cases and episodes. Caute has chapters on scientists, teachers, and government employees, among other occupations. Although in theory the type of the scientists' employer could make an important legal or administrative difference in what charges could be made or proven and what penalties imposed, in practice most accusations were made and publicized so politically and extralegally that the victim's right to due process was violated— he or she was immediately tainted almost irretrievably, regardless of the outcome of any eventual hearing or other quasi-legal proceeding.[9]

Alan Harper has studied the federal loyalty program that Harry Truman instituted with an executive order in 1947. This action applied to all federal employees, not just military personnel and not just those handling classified or secret materials. Truman rationalized that little harm would result, since congressmen were barred from seeing actual personnel files, but whether they saw them or not, few accusers felt constrained by mere documentation. Among the most harrassed scientists in the federal government was the physicist Edward U. Condon, director of the National Bureau of Standards. Accusers hounded him for several years, and any hearings, which might have exonerated him, were always postponed. He merits a full biography. Consultants to the government, being temporary employees, could also be victimized, as was J. Robert Oppenheimer in a celebrated case in 1954, when the Atomic Energy Commission revoked his security clearance after years of distinguished service. The only persons to lose their lives, however, were the so-called "atomic spies" Julius and Ethel Rosenberg, who were electrocuted in 1953 for allegedly handing to Soviet agents atomic secrets obtained from a relative who worked at Los Alamos.[10]

[8] Homer Newell, *Beyond the Atmosphere: Early Years of Space Science* (Washington, D.C.: NASA, 1980) discusses the impact of federal money on related sciences. Saul Benison, "A History of Polio Research in the United States: Appraisals and Lessons," in *Twentieth-Century Sciences*, ed. Holton (cit. n. 4), pp. 308–343.

[9] David Caute, *The Great Fear: The Anti-Communist Purge Under Truman and Eisenhower* (New York: Simon & Schuster, 1978). There is more on schoolteachers in Diane Ravitch, *The Troubled Crusade: American Education, 1945–1980* (New York: Basic Books, 1983), Ch. 3.

[10] Alan Harper, *The Politics of the White House and the Communist Issue, 1946–1952* (Westport, Conn.: Greenwood, 1969); the most comprehensive treatment I have seen of the Condon case is Stuart A. Kirsch, "Science and Security: The Case of Dr. Edward U. Condon" (Undergraduate paper, Harvard Univ., 1984), based on the many periodical accounts of the case over the period 1945–1955. (The Condon Papers are at the American Philosophical Society.) The literature on the Oppenheimer case is enormous. The best places to start are Herbert York, *The Advisors: Oppenheimer, Teller, and the Superbomb* (Berkeley/Los Angeles: Univ. California Press, 1976); Philip M. Stern, *The Oppenheimer Case: Security on Trial* (London: Hart-Davis, 1971), and Barton J. Bernstein, "In the Matter of J. Robert Oppenheimer," *Hist. Stud. Phys. Sci.*, 1982, *12*:195–252. For the Rosenberg case, also much written about, the most recent and most scholarly accounts are Rogers

Holders of government grants and contracts represented a new kind of potential subversive in the early 1950s. When several biochemists learned in 1954 that their grants were not being renewed by the U.S. Public Health Service (the parent agency of the NIH) for security reasons, they protested in the *Bulletin of Atomic Scientists* and in *Science*. Before long, the National Academy of Sciences began an investigation, and soon Oveta Culp Hobby, then secretary of health, education and welfare, admitted that about forty such grants had been withheld, including one to the famous chemist Linus Pauling, whose passport to an international meeting on proteins was also denied by the State Department at about this time, as readers of *The Double Helix* will recall. J. Merton England's recent history of the NSF in these years notes how adroitly the National Science Board sidestepped the issue: since NSF made grants only to institutions (and not directly to individuals), it did not consider itself responsible for establishing the political acceptability of an individual principal investigator and left that sticky task up to the institution.[11]

Federal fellowship holders, however, were an entirely different matter. The AEC set the precedent in 1949, when it learned from a radio reporter that one of its first fellows was an avowed Communist and that another had once attended Party meetings. Since the responsibility here could not be passed to some other institution or group, Congress authorized the FBI to investigate potentially subversive fellows. At this point the National Academy of Sciences, whose National Research Council had been administering the program for the AEC, objected to the necessity for security clearances for unclassified work. It requested that the program be sharply reduced in size, and before long it was ended. This whole episode was deplored by Walter Gellhorn, a professor of law, in his classic study *Security, Loyalty, and Science* (1950), a work supported by the Rockefeller Foundation, whose officials were aghast at what was happening. To this day, holders of federal fellowships are required to sign loyalty statements.[12]

Yet universities were often not the strong bastions of academic freedom that they proclaimed to be, as recent research has shown. State universities, including even those with seemingly independent boards of regents, could be quite susceptible to popular hysteria. David Gardner has written of the loyalty oath problems at the University of California in 1949–1951, and more recently Jane Sanders has published a major archive-based study of events at the University of Washington, where several faculty members were either fired or put on probation in the late 1940s. In the face of these difficulties the American Association of University Professors was powerless, as were all other scientific and professional associations, although the AAUP later voted to censure eight universities, and the American Association for the Advancement of Science elected Edward Condon its president in the late 1940s at the height of the battle against him.[13]

M. Anders, "The Rosenberg Case Revisited: The Greenglass Testimony and the Protection of Atomic Secrets," *American Historical Review,* 1978, *83*:388–400; and Ronald Radosh and Joyce Milton, *The Rosenberg File: A Search for the Truth* (New York: Holt, Rinehart & Winston, 1983).

[11] "Loyalty and U.S. Public Health Service Grants," *Bulletin of Atomic Scientists,* May 1955, *11*(5):196–197; John Edsall, "Government and the Freedom of Science," *Science,* 1955, *121*:615–619; and England, *Patron of Pure Science* (cit. n. 7), pp. 329–337.

[12] Walter Gellhorn, *Security, Loyalty, and Science* (Ithaca: Cornell Univ. Press, 1950), pp. 185–202; and Hewlett and Duncan, *The Atomic Shield* (cit. n. 6), pp. 340–342.

[13] David Gardner, *The California Oath Controversy* (Berkeley: Univ. California Press, 1967); and

It is hard to know whether the so-called private universities were any more secure or whether their presidents were just more secretive or hypocritical about what was going on. A few years ago Sigmund Diamond published a major exposé of Yale University's cooperation with federal investigators in exchange for reduced publicity of its possible security cases. More recently he published an essay on J. B. Conant of Harvard and his "arrangement" with the Boston office of the Federal Bureau of Investigation. (Harvard, with its new Russian Research Center, supported by the Carnegie Corporation of New York, and many employees named "Igor," "Boris," and "Olga," attracted special attention.) To its credit and despite intense local pressure, the Massachusetts Institute of Technology kept the famed mathematician and historian of American science and technology Dirk Struik on the faculty (though on administrative leave) from 1951, when he was accused of being a Communist, until 1956, when charges against him were finally dropped. Invaluable here are the Oswald Veblen Papers at the Library of Congress's Manuscript Division, which contain files on several mathematicians who faced security charges in the 1940s and 1950s.[14]

It is perhaps ironic that McCarthyism and its stress on loyalty and conformity to true-blue "Americanism" led to one of the first social science classics of the 1950s, Paul Lazarsfeld and Wagner Thielens's *The Academic Mind: Social Scientists in a Time of Crisis* (1958). That work showed on the basis of a statistical survey that even when not charged personally or directly with security violations, most American social scientists were sufficiently intimidated by the fear of such charges to modify their behavior in the classroom and near campus so as to seem more conformist than they really were.[15] Security issues remain an area in which detailed local studies can shed a lot of light.

THE "BEHAVIORAL SCIENCES" AND DESEGREGATION

The rapid development of the "behavioral sciences" in the 1950s and 1960s is one of the riper areas for scholarly pursuit, since abundant materials are generally available, the topics fresh and interesting, and the researcher's need for security clearance minimal. The Margaret Mead Papers at the Library of Congress Manuscript Division, the Ford Foundation Archives in New York City, and the Talcott Parsons Papers at the Harvard University Archives are open and should in the next decade become the mecca for historians of science that the Rockefeller Foundation Archives has been in the last ten years. (Other collections exist, such as the David Riesman Papers at the Harvard University Archives, but are not yet arranged and ready for use.)

Probably the best starting area for this whole realm, especially for some

Jane Sanders, *Cold War on the Campus: Academic Freedom at the University of Washington, 1946–1964* (Seattle: Univ. Washington Press, 1979).

[14] Sigmund Diamond, "God and the FBI at Yale," *Nation*, 1980, *230*:422–428; Diamond, "More on Buckley and the FBI," *ibid.*, 1980, *231*:202, 206; Diamond, "The Arrangement: The FBI and Harvard University in the McCarthy Period," in *Beyond the Hiss Case: The FBI, Congress, and the Cold War*, ed. Athan G. Theoharis (Philadelphia: Temple Univ. Press, 1982), pp. 341–371; Caute, *The Great Fear*, pp. 410–411 and 551; and *Facts Relevant to the Struik Case*, pamphlet put out by the Struik Defense Committee, headed by George Sarton, 1952, in the Oswald Veblen Papers, Library of Congress Manuscript Division, Box 24.

[15] Paul F. Lazarsfeld and Wagner Thielens, *The Academic Mind: Social Scientists in a Time of Crisis*, with a field report by David Riesman (Glencoe, Ill.: Free Press, 1958).

understanding of the Ford Foundation's great interest in what its officers termed the "behavioral sciences" in the 1950s, are Peter Seybold's two articles on the Ford Foundation's manipulation of support for the field of political science. They appeared in *Science for the People* and in Robert Arnove's most useful, because highly critical, anthology *Philanthropy and Cultural Imperialism: The Foundations at Home and Abroad* (1981). Seybold shows how the Ford Foundation's officers sought to emphasize the field's least radical and least politically threatening subjects (notably voting behavior) in the early 1950s by offering large grants to young scholars willing to move in this direction.[16]

Beyond this work, many biographies and autobiographies already exist. Alfred Marrow on Kurt Lewin, David Sills on Bernard Berelson, Robert Cassidy on Margaret Mead, and Michael Pollack on Paul Lazarsfeld all open up new and interesting territory.[17] Autobiographies, whether brief ones in *Daedalus* or longer ones in *Perspectives in American History* and elsewhere, have appeared by Paul Samuelson, Talcott Parsons, Erik Erikson, Robert K. Merton, Edwin Boring, Jerome Bruner, Margaret Mead, Hortense Powdermaker, Frederica de Laguna, and numerous others. Other equally prominent social scientists, like Clyde Kluckhohn, so far have only lengthy obituaries.[18]

Many social science organizations also merit full histories. Among these would be the Brookings Institution (whose archives are now being readied), the Menninger Foundation, the RAND Corporation, the Russell Sage Foundation, the Population Council, the National Opinion Research Center in Chicago, the Social Science Research Council (which already has a fiftieth anniversary volume by Elbridge Sibley), the Harvard Department of Social Relations, the Center for Advanced Study in the Behavioral Sciences, and the numerous other centers on many campuses, including Henry Kissinger's Center for International Affairs at Harvard University and other "foreign area" training programs. John King Fairbank has described in a recent autobiography, *Chinabound: A Fifty Year Memoir* (1982), how he trained a generation of China scholars at Harvard with (he admits on occasion) considerable financial help from the Ford Foundation. What effect this outside funding had on the social or behavioral sciences

---

[16] Peter Seybold, "The Ford Foundation and Social Control," *Science for the People*, May/June 1982, pp. 28–31 and "The Ford Foundation and the Triumph of Behavioralism in American Political Science," in *Philanthropy and Cultural Imperialism: The Foundations at Home and Abroad*, ed. Robert Arnove (Boston: G. K. Hall, 1980), pp. 269–302.

[17] Alfred J. Marrow, *The Practical Theorist: The Life and Work of Kurt Lewin* (New York: Basic Books, Inc., 1969); David L. Sills, "Bernard Berelson: Behavioral Scientist," *Journal of the History of the Behavioral Sciences*, 1981, *17*:305–311; Robert Cassidy, *Margaret Mead: A Voice for the Century* (New York: Universe Books, 1982); and Michael Pollack, "Paul Lazarsfeld: A Sociointellectual Biography," *Knowledge*, 1980, *2*:157–177.

[18] Paul Samuelson, "Economics in a Golden Age: A Personal Memoir," in *Twentieth-Century Sciences*, ed. Holton (cit. n. 4), pp. 155–170; Talcott Parsons, "On Building Social System Theory: A Personal History," *ibid.*, pp. 99–154; Erik Erikson, "Autobiographic Notes on the Identity Crisis," *ibid.*, pp. 3–32; Robert K. Merton, "The Sociology of Science: An Episodic Memoir," in *The Sociology of Science in Europe*, ed. Merton and Gerry Gaston (Carbondale: Southern Illinois Univ. Press, 1977), also published separately as Merton, *The Sociology of Science: An Episodic Memoir* (Carbondale: Southern Illinois Univ. Press, 1979). Edwin G. Boring, *Psychologist at Large: An Autobiography and Selected Essays* (New York: Basic Books, 1961); Jerome Bruner, *In Search of Mind: Essays in Autobiography* (New York: Harper & Row, 1983); Margaret Mead, *Blackberry Winter: My Earlier Years* (New York: Morrow, 1972); Hortense Powdermaker, *Stranger and Friend: The Way of an Anthropologist* (New York: Norton, 1966); Frederica de Laguna, *Voyage to Greenland: A Personal Initiation into Anthropology* (New York: Norton, 1977); Talcott Parsons and Evon Z. Vogt, "Clyde Kay Maben Kluckhohn, 1905–1960," *American Anthropologist*, 1962, *64*:140–161.

well before there was much federal money in the area also awaits exploration and critical analysis.[19]

One important political backdrop to the rise of federal aid to the social and behavioral sciences in the 1950s and 1960s was the Supreme Court's 1954 decision that separate schools for whites and blacks were no longer legal. Though unanimous, the decision proved not only unpopular in some places but also controversial, because it was the first to cite social science works (including Gunnar Myrdal's *The American Dilemma* of 1944) in its footnotes. For years southern congressmen denounced the social sciences and all court decisions based either on them or on any decisions that cited them. Then in the early 1960s racist ideologues, especially Delta Airlines executive Carleton Putnam, began to change their tactics and to develop their own kind of social science ("scientific racism") so that their "experts," including certain heretofore respectable psychology professors, could testify in court cases in favor of racial segregation. Idus Newby of the University of Hawaii has pursued this fascinating development in *Challenge to the Court: Social Scientists and the Defense of Segregation, 1954–1966* (1967; rev. 1969), which should interest sociologists and philosophers of science as well as historians. Here one can see the beginnings of mirror-image or mock science decking itself out with such appropriate scholarly apparatus as its own journal and board of editors in order to defend its now manifestly "scientific" beliefs. In some ways this group and its *Mankind Quarterly* foreshadowed the "creation science" movement of the 1970s.[20]

Despite such instances of hostility, the federal aid to "basic research" in the social sciences that began in the late 1950s increased greatly in the 1960s. Gene Lyons's *The Uneasy Partnership: Social Sciences and the Federal Government in the Twentieth Century* (1969) recounts this growth as well as earlier events and episodes. (The U.S. Department of Agriculture had long supported applied work in the social sciences, through its Bureau of Agricultural Economics, for example). The partnership remained uneasy, however, because even if the issues of domestic loyalty and segregation had faded by the time significant federal aid came to the social sciences in the late 1950s, other political issues and pressures had arisen in their stead. For example, foreign area research supported by private foundations had been one thing; covert espionage supported by the military or the Central Intelligence Agency under the guise of scholarship was another. As the involvement of the United States abroad increased in the 1960s and such

[19] Elbridge Sibley, *Social Science Research Council: The First Fifty Years* (New York: SSRC, 1974); John King Fairbank, *Chinabound: A Fifty Year Memoir* (New York: Harper & Row, 1982); Robert A. McCaughey, *International Studies and Academic Enterprise: A Chapter in the Enclosure of American Learning* (New York: Columbia Univ. Press, 1984). For an overview of many social science (and other) organizations, see Joseph C. Kiger, ed., *Research Institutions and Learned Societies* (Westport, Conn.: Greenwood, 1982).

[20] Gunnar Myrdal, *The American Dilemma: The Negro Problem and Modern Democracy*, 2 vols. (New York: Harper, 1944); Walter Jackson, "The Making of a Social Science Classic," paper presented at the American Academy of Arts and Sciences, 1982; Idus Newby, *Challenge to the Courts: Social Scientists and the Defense of Segregation, 1954–1966* (Baton Rouge: Louisiana State Univ. Press, 1967; rev. ed., 1969). A full account of the 1954 decision is given in Richard Kluger, *Simple Justice, The History of* Brown vs. Board of Education *and Black America's Struggle for Equality* (New York: Knopf, 1976). Recent assessments of the impact of the Supreme Court's decision include Raymond Wolters, *The Burden of* Brown: *Thirty Years of School Desegregation* (Knoxville: Univ. Tennessee Press, 1984); George R. Metcalf, *From Little Rock to Boston: The History of School Desegregation* (Westport, Conn.: Greenwood, 1983); and Ravitch, *The Troubled Crusade* (cit. n. 9), Chs. 4 and 5.

deceptions came to light, outcry arose. The most highly publicized episode was Project Camelot, a U.S. Army project started in 1964 ostensibly to study "social systems" in Latin America. But it was abandoned a year later when the Chilean government, fearing a coup, denounced the Camelot workers as spies and counterinsurgents and demanded that they be removed. This whole episode is summarized in Irving Horowitz's anthology *The Rise and Fall of Project Camelot: Studies in the Relationship Between Social Science and Practical Politics* (1967). The issue was not limited to Project Camelot, however, and in 1969 Congress passed the Mansfield Amendment to the military budget banning further military aid not only to the social sciences but also to any other "basic" research on campus.[21]

Another politicizing influence in the late 1960s was the Vietnam War. If professors (other than the redoubtable Linus Pauling) had been intimidated into silence in the 1950s, they now took to protesting as publicly as they could (as in petitions and whole pages in the *New York Times*), proving perhaps that they had not, as some persons might have feared, been bought off by their federal research grants. Everett Carll Ladd, Jr., has discussed this phenomenon in *Science* and with Seymour Martin Lipset in a subsequent volume, *The Divided Academy: Professors and Politics* (1975). There are thus abundant issues here not only of interest to historians of science but also of importance to the public.[22]

SCIENCE EDUCATION: FROM COLD WAR TO CREATIONISM

Because of the postwar "baby boom," secondary and higher education would have expanded greatly in the 1950s and 1960s anyhow, but with the coming of the Cold War, the whole process involved much more scientific emphasis and federal interest than might otherwise have been the case. By the late 1960s, however, a rising right-wing criticism was changing the atmosphere considerably, and the proper limits to federal aid to science soon became the central issue. There are a few historical works on these issues, but more studies are needed.

Among the themes of the period should be the increased emphasis on science in the media and the deliberate recruitment of students into training and careers in science and engineering. For example, the traditional forms of science journalism (in magazines such as *Scientific American*) expanded in the late 1950s into the new realm of science television. The series *Continental Classroom*, which featured early morning lectures on physics and chemistry, first appeared on NBC–TV in October 1958; its records are deposited at the State Historical Society of Wisconsin. Supported by the Fund for the Advancement of Education

[21] Gene Lyons, ed., *Social Science and the Federal Government*, published as an entire issue of *Annals of the American Academy of Political and Social Sciences*, 1971, *394*, is concerned about the independence of social science research supported by federal funds. See also Irving Horowitz, *The Rise and Fall of Project Camelot: Studies in the Relationship Between Social Science and Practical Politics* (Cambridge: MIT Press, 1967).

[22] Everett Carll Ladd, Jr., "Professors and Political Petitions," *Science*, 1969, *163*:1425–1430; Everett Carll Ladd, Jr., and Seymour Martin Lipset, *The Divided Academy: Professors and Politics* (New York: McGraw-Hill, 1975); McCaughey, *International Studies* (cit. n. 19), pp. 222–224, 228–234; and Robert McCaughey, "American University Teachers and Opposition to the Vietnam War," *Minerva*, 1976, *14*:306–329.

(a spinoff from the Ford Foundation), this program represented just one part of the FAE's larger goal of bringing educational technology to the classroom. Other nongovernmental science education projects of the 1950s were the National Merit Scholarship Program (initiated by the Coca Cola Company in 1955) and numerous local science fairs and clubs. Many of these local and private efforts received some nationwide coordination in 1956–1958 from the National (renamed the President's after Sputnik) Committee on the Development of Scientists and Engineers. This committee, whose records are at the National Archives, had a national newsletter that endorsed such measures as more recognition, including higher salaries, for science teachers and more appreciation, publicity, and scholarships for bright students.[23]

After Sputnik I was launched in 1957, an event that linked science education even more closely with national defense, the federal government's role in science education became more explicit, though it still carefully avoided any hint or threat of federal control of education. One further step was the creation of new federal fellowship programs for the training of future scientists and college teachers. Barbara Barksdale Clowse's *Brainpower for the Cold War: The Sputnik Crisis and the National Defense Education Act of 1958* (1981) is the chief work in this area. The NDEA, which was administered by the Department of Health, Education, and Welfare, distributed over the years more than $1 billion in federal aid to graduate students in a variety of fields, including the so-called area studies and languages as well as science and mathematics. In addition to this aid, the National Science Foundation had several fellowship and postdoctoral programs for scientists, and some other agencies (such as the U.S. Public Health Service, AEC, and NASA) had special programs to train urgently needed experts in specific fields such as nuclear engineering or astrophysics. Most works written about the impact of such crash training programs are final reports full of statistics of a highly self-congratulatory nature. Their full impact, including the tradeoffs that were made, remains to be assessed.[24]

Another way for the federal government to improve the nation's science education indirectly was to train or improve the existing training of science teachers for the secondary schools. NSF, whose budget for science education tripled in the late 1950s, ran several programs in this area, including summer teachers' institutes on campuses around the nation, as described in Hillier Krieghbaum and Hugh Rawson's *An Investment in Knowledge* (1969). One of the more interesting problems faced by the federal authorities supporting these summer institutes was just how strongly to insist on racial integration in those held in Southern states.[25]

---

[23] Paul Woodring, *Investment in Education: An Historical Appraisal of the Fund for the Advancement of Education* (Boston: Little, Brown, 1970). The FAE, which existed only from 1951 to 1967, also supported early admissions, advanced placement, and M.A.T. (Master of Arts in Teaching) programs for liberal arts college graduates.

[24] Barbara Barksdale Clowse, *Brainpower for the Cold War: The Sputnik Crisis and the National Defense Education Act of 1958* (Westport, Conn.: Greenwood, 1981); England, *A Patron for Pure Science* (cit. n. 7), Ch. 12; Philip N. Powers, "The History of Nuclear Engineering Education," *Journal of Engineering Education*, 1964, *54*:364–370; Glenn Seaborg and Daniel M. Wilkes, *Education and the Atom, An Evaluation of Government's Role in Science Education and Information, Especially as Applied to Nuclear Energy* (New York: McGraw-Hill, 1964).

[25] Hillier Krieghbaum and Hugh Rawson, *An Investment in Knowledge: The First Dozen Years of the National Science Foundation's Summer Institutes Programs to Improve Secondary School*

In addition, NSF supported several now-historic curriculum reform measures in the 1950s through the early 1970s. Four of them have now been discussed in three dissertations and one book, at least two written by former participants: Paul Marsh's dissertation for the Harvard Graduate School of Education on "Project Physics," 1956–1961 (1964); Robert Hayden's dissertation for Iowa State on the "new math" movement (1981); Arnold B. Grobman's book, *The Changing Classroom: The Role of the Biological Sciences Curriculum Study* (1969), and, on the most controversial curriculum project of all, Peter Dow's Harvard Graduate School of Education dissertation on MACOS, the acronym for *Man: A Course of Study* (1979). The latter project stressed the social sciences but took a more "relativistic" position on values than many parents would have liked. It thus incurred much criticism and investigation from Congress, not only for its subject matter but also for its seemingly elaborate subsidies to publishers and distributors. Jerome Bruner has discussed his involvement in the project in his recent autobiography *In Search of Mind* (1983).[26]

Because the MACOS experience of the early 1970s was not an isolated example but symptomatic of a wider curriculum-criticism movement that included the whole issue of "creationism" (which went beyond the matter of indirect federal aid to science education), a stream of articles and books have appeared on the subject, including John Moore's article on creationism in California in *Daedalus*, Dorothy Nelkin's two books on textbook controversies, and most recently Ronald Numbers's article on twentieth-century creationism in *Science*.[27]

ANTINUCLEAR PROTEST AND THE LIMITED TEST BAN TREATY, 1954–1963

The best overview of the essentials of the antinuclear protest movements of the 1950s from a historian of science's point of view is Carolyn Kopp's 1979 article, "The Origins of the American Scientific Debate Over Fallout Hazards," in *Social Studies of Science*. It analyzes the positions taken by various scientists by specialty (the biologists and physicists differed) and by source of grant support

---

Science and Mathematics Teaching, 1954–1965 (New York: New York Univ. Press, 1969). There have also been several government-funded reports on the value of all this science education since the 1950s. See esp. Stanley L. Helgeson et al., *The Status of Pre-College Science, Mathematics, and Social Education, 1955–1975*, 3 vols. (Columbus: Ohio State Univ. Center for Science and Mathematics Education, 1977), which has extensive bibliographies.

[26] Paul E. Marsh, "The Physical Sciences Study Committee: A Case History of Nationwide Curriculum Development, 1956–1961" (Ed.D. diss., Harvard Graduate School of Education, 1964); Robert W. Hayden, "A History of the New Math Movement" (Ph.D. diss., Iowa State Univ., 1981); Arnold B. Grobman, *The Changing Classroom: The Role of the Biological Sciences Curriculum Study* (Garden City, N.Y.: Doubleday, 1969). BSCS also published a historical review of previous biology curricula: Paul DeHart Hurd, *Biological Education in American Secondary Schools, 1890–1960* (Washington, D.C.: American Institute of Biological Sciences, 1961). On MACOS, see Peter Dow, "Innovation's Perils: An Account of the Origins, Development, Implementation, and Public Reaction to *Man: A Course of Study*," (Ed.D. diss., Harvard Univ. Graduate School of Education, 1979). An entire issue of *Social Education*, Oct. 1975, *39*:388–396 is devoted to the MACOS controversy. See also Bruner, *In Search of Mind* (cit. n. 18), Ch. 10.

[27] John A. Moore, "Creationism in California," *Daedalus*, 1974, *103*:173–189; Dorothy Nelkin, *Science Textbook Controversies and the Politics of Equal Time* (Cambridge: MIT Press, 1977); Nelkin, *The Creation Controversy: Science or Scripture in the Schools* (New York: Norton, 1982); and Ronald Numbers, "Creationism in Twentieth-Century America," *Science*, 1982, *218*:538–544. See also the classic study Judith V. Grabiner and Peter D. Miller, "Effects of the Scopes Trial," *Science*, 1974, *185*:832–837.

(AEC grantees and contractors tended to support their agency). Also interesting because it gives more attention to the fluctuating public opinion on fallout and testing is Robert Divine's *Blowing on the Wind: The Nuclear Test Ban Debate, 1954–1960* (1978). It curiously ends midstream, however, with Eisenhower's second term and the unfortunate U-2 incident, rather than pursuing the subject into the Kennedy years, with the Cuban missile crisis and the eventual passage of the Limited Test Ban Treaty in October 1963. It is also unfortunate that Richard Hewlett's two-volume history of the AEC ends about 1952 and does not continue into the later 1950s to cover the "Atoms for Peace" program and the public criticisms of nuclear testing and fallout. It would have been interesting to see how a government historian (or a team of them) would have handled the political issues surrounding testing, especially the difficulties that Herman J. Muller, Linus Pauling, Barry Commoner, and others experienced and thought characterized the treatment given any critics of AEC policy in these years.[28]

Paying more attention to the actual events leading up to the signing of the treaty (which banned testing in the atmosphere but not underground), including the scientists' many roles in the negotiations, are two works by political scientists: Robert Gilpin's *American Scientists and Nuclear Weapons Policy* (1962) and Harold Jacobson and Eric Stein's *Diplomats, Scientists and Politicians: The United States and the Nuclear Test Ban Negotiations* (1966). The authors rely on astute analyses of government documents, contemporary books and articles, interviews, and newspaper accounts.[29]

In recent years two scientists who were major participants in the test ban negotiations have published their own personal memoirs: George B. Kistiakowsky's excellent *A Scientist in the White House: The Private Diary of President Eisenhower's Special Assistant for Science and Technology* (1976), and Glenn Seaborg's detailed *Kennedy, Khrushchev, and the Test Ban* (1981). Unfortunately Kennedy's science adviser Jerome Wiesner has not published a similar memoir, merely a volume of speeches.[30]

The least well known aspect of this whole episode is the experience of the outside critics or reformers. This is strange, since these were the scientists best known to the public at the time of the fallout debates. Of special note is Linus Pauling, whose revoked passport, Nobel Prize, book *No More War!* (1958 and 1962), and petition to the United Nations to ban testing were well publicized. Elof Axel Carlson's recent biography of Herman J. Muller has a chapter on the latter's experiences as an observer rather than actual member of the American delegation to the Geneva talks in August 1955. As one of the nation's few experts on the genetic effects of radiation on humans but also as an avowed former Communist, Muller was closely watched by the AEC. He suspected that his

[28] Carolyn Kopp, "The Origins of the American Scientific Debate Over Fallout Hazards," *Social Studies of Science*, 1979, 9:403–422; Robert A. Divine, *Blowing on the Wind: The Nuclear Test Ban Debate, 1954–1960* (New York: Oxford Univ. Press, 1978).

[29] Robert Gilpin, *American Scientists and Nuclear Weapons Policy* (Princeton: Princeton Univ. Press, 1962); Harold Jacobson and Eric Stein, *Diplomats, Scientists, and Politicians: The United States and the Nuclear Test Ban Negotiations* (Ann Arbor: Univ. Michigan Press, 1966).

[30] George B. Kistiakowsky, *A Scientist in the White House: The Private Diary of President Eisenhower's Special Assistant for Science and Technology* (Cambridge: Harvard Univ. Press, 1976); Glenn Seaborg, *Kennedy, Khrushchev, and the Test Ban* (Berkeley: Univ. California Press, 1981); Jerome Wiesner, *Where Science and Politics Meet* (New York: McGraw-Hill, 1965).

prepared remarks were deliberately approved too late to be delivered at the actual meetings. Much remains to be told about the whole episode.[31]

Recently Paul Boyer of the University of Wisconsin has pursued the topic's
aftermath—the public's rapid loss of interest in nuclear weapons and testing
after the 1963 treaty—in an article in a recent issue of the *Journal of American
History*. He sees much of the former intense interest and anger about fallout
shifting over to the movement against the Vietnam War but does not follow it
into the burgeoning environmental movement.[32]

THE SPACE PROGRAM

Like the antinuclear movement just discussed, the space program was also
closely tied to the military and political requirements of the Cold War. Yet diplomatically it seems to move in the opposite direction, that is, just as President
Kennedy was moving toward a ban on nuclear testing in the atmosphere, he
was proclaiming a space race and urging ever higher budgets for a highly publicized and very expensive program to put a man on the moon by 1970.

The history of space exploration in this country begins with Robert Goddard
and his homemade rockets in the 1910s. It continues with the voluntary migration to the United States in 1930 of Theodor von Kármán of the California Institute of Technology, as Paul Hanle has described in his recent book. The next
stage was the Army's "Project Paperclip," which brought many German V-2
rocket experts, including Wernher von Braun, to the United States after World
War II, an episode that Clarence Lasby has described so well in his book of the
same title.[33]

Also entering into the story are the events of the International Geophysical
Year (IGY), which Lloyd Berkner of Brookhaven first suggested at the height
of the McCarthy era (a circumstance often cited to show how truly international
geophysical research is), and eventually carried out with the cooperation of over
forty nations in thirty months in 1957–1959. (As such, it constituted the largest
scientific project ever organized.) Its scientific results were extraordinary and
covered eleven fields in the oceanic, atmospheric, and earth sciences, including
meteorology, seismology (important in the detection of underground atomic
tests), many areas of geophysics, including the exploration of Antarctica, and
the worldwide sharing of scientific data. Because NSF coordinated America's
participation in the IGY, J. Merton England's history of NSF has a very useful
section on it. The IGY's international director Sydney Chapman of the United
Kingdom has written his own mutedly self-congratulatory volume, and several
journalists and popular science writers have provided accounts, as has physicist
James A. Van Allen of Iowa State University, who that year discovered the

---

[31] Linus Pauling, *No More War!* (1958; enlarged ed., New York: Dodd, Mead, 1962); Elof Axel
Carlson, *Genes, Radiation, and Society: The Life and Work of H. J. Muller* (Ithaca: Cornell Univ.
Press, 1981), Ch. 25.

[32] Paul Boyer, "From Activism to Apathy: The American People and Nuclear Weapons, 1963–
1980," *Journal of American History*, 1984, 70:821–844.

[33] Milton Lehman, *This High Man: The Life of Robert H. Goddard* (New York: Farrar Straus,
1963); Paul Hanle, *Bringing Aerodynamics to America* (Cambridge, Mass.: MIT Press, 1982); Clarence Lasby, *Project Paperclip* (New York: Atheneum Press, 1971).

famous Van Allen belts and initiated the specialty of magnetospheric physics.[34]

Although part of the plans for the IGY included the development of small rockets with which to explore near space, and the Russians had announced their intention to launch some during the "Year," Americans were apparently caught off guard when in October 1957 the Soviets successfully launched their first Sputnik. Realizing that any nation that could put a hundred-pound sphere into earth orbit could also send a bomb to the United States, President Eisenhower quickly appointed James R. Killian of MIT to be his "science adviser." Before long, Senator Lyndon Baines Johnson of Texas helped push through Congress a bill that merged the rocket efforts of the military (whose tests kept failing) with the old civilian National Advisory Committee for Aeronautics (NACA) to form the new civilian National Aeronautics and Space Administration (NASA). Various uninspired studies exist on the formation and administrative history of NASA.[35]

For several years NASA tried unsuccessfully to catch up with the Soviets, but in 1961, shortly after the latter launched Yuri Gagarin into orbit, President Kennedy proclaimed (and Congress approved with hardly any discussion) the new, face-saving, and possibly attainable goal of landing a man on the moon by 1970. The Apollo program was completed successfully in July 1969 but at costs so tremendous as to incur escalating criticism. Scientists thought more data could be obtained by sending instruments rather than men into space, and social critics questioned the need for such an expensive "stunt" when Americans were starving and ghettoes were burning. Some of the more interesting and amusing items written on NASA and the space program were by such critics. By the end of the Apollo program and in a manner reminiscent of the ambivalence over the triumph of the Manhattan Project in World War II, critics were using its very high level of success to raise the level of expectation for other government crash programs, an attitude exemplified in comments of the type: "If we can put a man on the moon, why can't we cure cancer, fix up the slums, desegregate the schools, end pollution . . . ?"[36]

A few scholarly and critical works on aspects of the space program have

---

[34] England, *A Patron for Pure Science*, pp. 297–304; Sydney Chapman, *IGY, Year of Discovery: The Story of the International Geophysical Year* (Ann Arbor: Univ. Michigan Press, 1959); Walter Sullivan, *Assault on the Unknown: The International Geophysical Year* (New York: McGraw-Hill, 1961); J. Tuzo Wilson, *IGY: The Year of the New Moons* (New York: Knopf, 1961); James A. Van Allen, *Scientific Uses of Earth Satellites* (Ann Arbor: Univ. Michigan Press, 1958); Van Allen, *Origins of Magnetospheric Physics* (Washington, D.C.: Smithsonian Institution Press, 1984). There is an extensive collection of Sydney Chapman's papers at the Regional and Polar Collections Department, Rasmuson Library, University of Alaska, Fairbanks.

[35] James R. Killian, *Sputnik, Scientists, and Eisenhower* (Cambridge, Mass.: MIT Press, 1977); Enid Curtis Bok Schoettle, "The Establishment of NASA," in *Knowledge and Power: Essays on Science and Government,* ed. Sanford A. Lakoff (New York: Free Press, 1966), pp. 162–270; Alison Griffith, *The National Aeronautics and Space Act: A Study of the Development of Public Policy,* with intro. by Lyndon Baines Johnson (Washington, D.C.: Public Affairs Press, 1962); Frank W. Anderson, Jr., *Orders of Magnitude, A History of NACA and NASA, 1915–1980* (NASA History Series NASA SP-4403) (Washington, D.C.: NASA, 1981).

[36] Leonard Mandelbaum, "Apollo: How the United States Decided to Go to the Moon," *Science,* 1969, *163*:649–654; John Logsdon, *The Decision to Go to the Moon, Project Apollo and the National Interest* (Cambridge: MIT Press, 1970); Amitai Etzioni, *The Moon Doggle* (Garden City, N.Y.: Doubleday & Co., 1964); William L. Crum, *Lunar Lunacy and Other Commentaries* (Philadelphia: Dorrance & Co., 1965); Erlend A. Kennard and Edmund H. Harvey, Jr., *Mission to the Moon: A Critical Examination of NASA and the Space Program* (New York: William Morrow & Co., 1969).

begun to appear. The final chapters of Clayton Koppes's excellent and archive-based *JPL and the American Space Program* (1982) covers the Jet Propulsion Laboratory's activities on the planetary probes. Homer Newell's *Beyond the Atmosphere: Early Years of Space Science* (1980) shows that a scientist and former NASA administrator—one who had worked hard to maximize the scientific knowledge (mostly physics) gained from the space program—can raise interesting and critical questions, about NASA's relationships with the universities and the scientific community, for example. Ken Hechler's participant history of the House of Representatives Committee on Science and Astronautics (of which he was a long-time member) is useful because of his unusual vantage point. It is, however, more of a chronicle than a critical history. About the only study of such generally underpublicized spinoffs of NASA as the communications and weather satellite programs is Pamela Mack's superb article on Landsat in a recent anthology. But any entrant to NASA history (or space history in general) should definitely start with Tom Wolfe's fictionalized but inestimable *The Right Stuff* (1979), which gives a gripping and memorable introduction to the excitement and mania of the early space program.[37]

Still, much work remains to be done. Biographies need to be written of most of the major participants, including Lloyd Berkner, James Webb, Hugh Odishaw, and Hugh Dryden. For other topics, NASA's own *Bibliography of Space: Books and Articles from Non-Aerospace Journals, 1955–1977* (1979) would be a good starting point. In particular, NASA's advances in and contributions to the life sciences, especially the field of space medicine, merits a useful monograph. More mention should be made in future works on the space program of military pressure or influence on NASA. Though the agency has a "civilian" status, like the similarly "civilian" AEC, many if not most of its employees are exmilitary personnel and its practices and equipment are military-made. Surely the Defense Department (itself another "civilian" agency at the top) has kept a sharp eye on NASA and has tried to steer its projects into directions that would eventually benefit the military. In a recent article Clayton Koppes warns that in the 1980s, with stagnant budgets for NASA and growing ones for the military, the militarization of the space program is already upon us.[38]

SAFETY AND ENVIRONMENTAL ISSUES

By the late 1950s a few of the new, improved agricultural chemicals and pharmaceuticals in widespread daily use were beginning to cause breakdowns in the ecosystem and in human health. As these dangers were publicized, people

[37] Clayton Koppes, *JPL and the American Space Program, A History of the Jet Propulsion Laboratory* (New Haven: Yale Univ. Press, 1982); Homer Newell, *Beyond the Atmosphere: Early Years of Space Science* (Washington, D.C.: NASA, 1980); Ken Hechler, *The Endless Space Frontier: A History of the House Committee on Science and Astronautics, 1959–1978* (San Diego: American Astronautical Society, 1982); Pamela E. Mack, "Space Science for Applications: The History of Landsat," in *Space Science Comes of Age: Perspectives in the History of the Space Sciences*, ed. Paul Hanle and Von Del Chamberlain (Washington, D.C.: Smithsonian Institution Press, 1981), pp. 135–148; Tom Wolfe, *The Right Stuff* (New York: Farrar, Straus & Giroux, 1979).

[38] John J. Looney, *Bibliography of Space, Books and Article from Non-Aerospace Journals, 1955–1977* (Washington, D.C.: NASA History Office, 1979); Clayton Koppes, "The Militarization of the American Space Program: An Historical Perspective," *Virginia Quarterly Review*, 1984, *60*:1–20. See also Walter A. McDougall, "Technocracy and Statecraft in the Space Age—Toward the History of a Saltation," *American Historical Review*, 1982, *87*:1010–1040.

immediately looked to the federal government to protect them. At first there was a series of temporary scares (like that over fallout and strontium 90), followed by some scandals that were blamed on lax procedures or cooptation if not outright bribery of regulators. Then in 1962 a single book sounded the alarm that helped to launch the wider environmental movement of the 1960s. The movement quickly coalesced into such a potent political force that by 1969 there was a major institutional reform, the creation of the new Environmental Protection Agency.

Among the early scares was the cranberry crisis of Thanksgiving 1958, when, because of the recently passed Delaney Act prohibiting the use of carcinogens on food, the secretary of the Department of Health, Education, and Welfare (Arthur Flemming) had to condemn as contaminated certain batches of cranberries sprayed with agricultural chemicals approved (for slightly different use) by the USDA. In the early 1960s another scare turned into a scandal, as the true story behind the heroism of Food and Drug Administration medical officer Frances Kelsey became known. She had stubbornly withstood outside and inside pressure to certify quickly and on inadequate medical evidence a faulty drug, thalidomide, already suspected of implication in an outbreak of deformed babies in Germany. When the story came out, public pressure caused the FDA to be more stringent in its tests and approvals.[39] Also in the early 1960s, after finding evidence that linked smoking and lung cancer, a new Public Health Service advisory committee urged the Federal Trade Commission to require the Surgeon General's warning on all packages of cigarettes. Thus case by case, improved analytical methods were showing statistical linkages between certain chemicals and certain illnesses, and pressure was building on the federal government to take greater steps to limit usage or at least warn consumers.[40]

Of greater lasting import than any of the above incidents was the publication in 1962 of Rachel Carson's epoch-making book *Silent Spring*. As an employee of the federal Fish and Wildlife Service during World War II, when the miraculous properties of DDT were discovered and first published, Carson had long distrusted the claims made for this new wonder drug. Once alerted to the possible effects of DDT residues on bird populations, she began thorough research on the subject. Before long she predicted that continued widespread use of DDT would silence all birdlife and perhaps other animals and even humans as well (whom it might already be harming in as yet undetected ways). Her book caused an uproar not only in agrichemical circles but also among the public, whose consciousness it helped raise, leading to an effective environmental movement. In 1969 President Richard Nixon signed into law the new Environmental Protection Act, which set up an Environmental Protection Agency that controlled the usage of dangerous chemicals like DDT and required "environmental impact statements" of anyone likely to modify current conditions by releasing such chemicals.[41]

---

[39] Kistiakowsky, *Scientist in the White House* (cit. n. 30), Ch. 7; John Lear, "The Unfinished Story of Thalidomide," *Saturday Review*, 1 Sept 1962, pp. 35–40; Will Jonathan, "The Feminine Conscience of FDA: Dr. Frances Oldham Kelsey," *ibid.*, pp. 41–43.

[40] *Smoking and Health: Report of the Advisory Committee to the Surgeon General of the Public Health Service* (Public Health Service Publication, 1103) (Washington, D.C., GPO, 1964); A. Lee Fritschler, *Smoking and Politics: Policymaking and the Federal Bureaucracy* (Englewood Cliffs, N.J.: Prentice-Hall, Inc., 1969; 1975).

[41] Rachel Carson, *Silent Spring* (Boston: Houghton Mifflin, 1962).

Some of these topics have received scholarly attention. Donald Fleming's "Roots of the New Conservation Movement," in *Perspectives in American History* (1972), is perhaps the best general introduction not only to Carson and her thought but also to a host of other environmental thinkers of the 1950s and 1960s. These include Paul Ehrlich, who advocated population control, and Barry Commoner, who turned from the issue of fallout in the atmosphere to stress other needed reforms for air and water pollution, which, he would argue, entailed a major restructuring of the nation's economy. Several books have appeared specifically on DDT and economic entomology, such as Frank Graham's journalistic *Since Silent Spring* (1970); Thomas Dunlap's *DDT: Scientists, Citizens and Public Policy* (1981), which relies not only on federal records but also on some key state-court cases in Wisconsin; and John Perkins's *Insects, Experts, and the Insecticide Crisis* (1982), which uses oral history to document the entomologists' reactions and their shift in the 1970s to "integrated pest management" (biological control methods as well as insecticides). Aside from works written for children, there is no biography of Carson herself except for the personal memoir by her editor, Paul Brooks (*The House of Life: Rachel Carson at Work*, [1972]), though her voluminous papers have been deposited at the Beinecke Rare Book Library at Yale University.[42]

Because so many of these topics involved the federal government in one way or another—in court cases, in congressional hearings and legislation, and in enforcement procedures by executive agencies—there should be an abundance of relevant materials in federal documents and archival repositories as well as in newspapers and scientific and popular journals and magazines. At this point, however, it is not clear how well the National Archives has coped with the period since 1950. As more and more scholars pursue these issues the picture should become clearer. In any case, many of the participants in the environmental movement are still alive, though they have not published as many memoirs and diaries as have the highly prolific presidential science advisers.[43]

CONCLUSION

Clearly, science in one form or another was very much involved in many of the central issues of post–World War II public policy. It penetrated such key areas as diplomacy, economics, health, safety, and education at all levels. In fact, science and politics were so intertwined that periodic shifts in the so-called national agenda loomed large and required layers of science administrators to mediate between them. This inexorable tendency increased, partly because the Cold War and defense planning dominated the economy, partly because large sums were needed for the new "big science," and partly because with this infusion of federal funds new, nonscientific agendas had to be accommodated in

[42] Donald Fleming, "Roots of the New Conservation Movement," *Perspectives in American History,* 1972, 6:7–91; Frank Graham, *Since Silent Spring* (Boston: Houghton Mifflin, 1970); Thomas Dunlap, *DDT: Scientists, Citizens, and Public Policy* (Princeton: Princeton Univ. Press, 1981); John H. Perkins, *Insects, Experts and the Insecticide Crisis* (New York: Plenum, 1982); Paul Brooks, *The House of Life: Rachel Carson at Work* (Boston: Houghton Mifflin, 1972). See also James Whorton, *Before Silent Spring* (Princeton: Princeton Univ. Press, 1974); and Lynton K. Caldwell, *Man and His Environment: Policy and Administration* (New York: Harper & Row, 1975), Chs. 1–4.

[43] Perhaps some type of archival survey of the records of the environmental movement is necessary. The papers of temporary political action groups are in danger of being lost.

areas of emphasis and even of problem choice. This has been a highly cyclical process: vast sums of money lead to rapid expansion for the favored programs of the moment, such as atomic energy, space, science education, and biomedical research, and then, when the crisis passes or the luster wears off, either projects are tapered off and personnel let go, or a frantic search begins for new goals. This dynamic led in the 1970s to numerous reorganizations and ever more frequent shifts in emphasis. Such close ties between science and politics has also led at times to severe security checks, the imposition of loyalty oaths, and even defamation of character.

Because maintaining large budgets usually requires elaborate justifications, which in turn generally generate long trails of self-serving paperwork, historians with the advantages of hindsight and distance may have a key opportunity and role in untangling, interpreting, and at times even debunking the rhetoric that accompanies such standard bureaucratic defense maneuvers. This is a necessary first step to any further understanding of the politics of science in recent decades. Some initial works have been published that help point the way. Dorothy Nelkin, in particular, has had an eye for significant politico-scientific topics, but she has pursued them whiggishly as the past roots of some present crisis or problem. More sympathetic works based on archives are needed, like Clayton Koppes's, Walter Jackson's, and Barbara Barksdale Clowse's contributions, simply to provide accurate accounts of what has occurred. Only when this stage has been reached will we be able to step back and discuss what the major themes of this period have been.

Such books and articles rely, however, on the availability of good records. Good archives, large, rich, and ready for use, beckon in some areas, such as the behavioral sciences. Their absence, inaccessibility, or restricted use remains a possibly critical stumbling block in other areas. The recent final report of the Joint Committee on Archives in Science and Technology, by stressing the need for more finding aids and archival attention and by avoiding issues of security and classified documents, seems to indicate that bulk and disorganization rather than security or privacy are the major stumbling blocks in this area. The reality of this assessment remains to be tested. Meanwhile, the status of the records of federal contractors and such "soft-money" pressure groups as environmental organizations remains unclear and worrisome.[44] But unless the records have been destroyed (as has happened at the National Heart, Lung, and Blood Institute at the NIH), there is hope that the growing interest in recent history among both historians and the participants themselves will lead to a usable collection someday. (Centennials in particular can stimulate a sudden if uncritical interest in an organization's past.) After all, any collection, even one closed or restricted in use for twenty or even fifty years (as at Harvard University) is preferable to outright destruction, which is necessarily irreversible.

Archives are so important, in fact, crucial, that historians of all periods should expect to keep a watchful eye on their status, for without them it is hard to believe that one is doing "history." Journalists, political scientists, sociologists,

---

[44] Clark A. Elliott, ed., *Understanding Progress as Process: Documentation on the History of Post-War Science and Technology in the United States, Final Report of the Joint Committee on Archives of Science and Technology (HSS–SHOT–SAA–ARMA)* (Chicago: Society of American Archivists, 1983).

and statisticians, among others, have worked on data from more recent years, but the price they have paid—apparently willingly—for access to information, confidential or not, is an assurance of anonymity to their sources. Some historians have followed suit, as is evident from the veiled references in footnotes to an unverifiable "personal communication" that one sees on occasion (or from there being no footnote at all).[45] But most historians have been reluctant to provide such "cover," for verifiable documentation is the heart of a historical study. Without archives and accessibility to them, this essential requirement is missing. Although the volumes and articles discussed here reveal that there can be difficulties, even serious problems, in working with recent materials and with participants who are still alive, work on recent periods also offers certain challenges and advantages, including overwhelming amounts of previously untouched material and a broader and more varied readership that for better or worse stretches beyond one's fellow historians.

[45] See Orlans, *Contracting for Atoms* (cit. n. 6), pp. x–xi, for a brief discussion of this tradeoff.

Bibliographies, Reference Works, and Archives

By Clark A. Elliott*

THE SHORT-TERM VITALITY of a scholarly field is judged by the number and quality of original works of investigation and interpretation published within a given period. Further, it is a truism of historical investigation that successive generations reinterpret the meaning of the past. While the explanations change, they continue to be based on the cumulative data and documentation that all historians must consider. The underlying health and long-term prospects for a field of historical study, therefore, must also be judged by the character and quantity of its research guides, bibliographies, biographical directories, and archival collections and related finding aids, as well as by the published secondary literature. The preparation of bibliographies or the carrying out of similar projects may be done by individuals, but these activities ideally are part of a larger, collective endeavor that is the concern of all scholars in the field, irrespective of their personal research interests. Activities furthering the control of access to bibliographic data or other categories of information ought to be subject to greater rationalization and organization than is the conduct of research into substantive topics in the field. This explains why such projects as the preparation and distribution of annual bibliographies are among the chief activities of organizations such as the History of Science Society.

This essay explores aspects of the history of science in America from the viewpoint of access to sources. Considered are a number of specific bibliographic and other reference works, as well as important recent projects and developments that aim to preserve and to improve access to archival and other documentation and historical data for future use. The intention is not only to gauge what has been done, but also to suggest needed directions for collective social action in the future. Finally, the essay suggests that the character of reference aids and the nature and use of primary documentation should themselves be objects of study if a fully realized historical methodology and historiography for the field is to emerge.

USE OF HISTORICAL SOURCES

At the outset of the period under review, that is, 1971 to the present, Walter Rundell published the results of a study of the condition of historical research

* University Archives, Harvard University, Cambridge, Massachusetts 02138.

on American history in general. He concluded that greater and more effective training in historical research methodology was needed. A more recent study explored the reference and research strategies of historians. Its author, Margaret Stieg, proclaimed that the overall outcome was an embarrassment for both historians and librarians. Historians were characterized as unsystematic in their research methods. They did not appear to make adequate use of available abstracts and indexes. (Historians of science were noted to be exceptions in their strong reliance on the *Isis Critical and Cumulative Bibliographies.*) Conversely, librarians and archivists failed to perform the function of educating historians about sources. The inadequacy of guides and indexes to manuscripts was cited as one reason why unpublished sources rank third, behind books and articles, in the use patterns of historians.[1]

A few works have appeared that consider the sources used by historians in writing the history of science, but there is no indication that this has become a self-conscious or sustained area of consideration or investigation. David Knight published a work that examined the character and use of various categories of sources for the history of the early modern period, especially the history of British science. A study of examples of the usefulness of documentary sources from the more recent period, and in the more limited context of the history of biochemistry, has been reported by John Edsall and David Bearman.[2]

Stieg's study of historical research methodology relied on questionnaires and was concerned with the strategy for finding relevant sources as well as gauging their relative value in actual use. Another approach to the question of use is the examination of footnote citations. The one study in this area that relates specifically to the history of science showed that even within this historical specialty there are significant differences determined by approach. The internalist history of science literature makes much less use of manuscript material than does the externalist.[3] One lesson that can be drawn from this and other studies is that the design of reference aids, and perhaps the appraisal of primary materials, must consider more closely the likely patterns of historical methodology ordinarily employed by different groups of potential users.

REFERENCE AIDS AND DOCUMENTS: GENERAL NOTE

The history of American science is not an isolated field. It is closely related to aspects of both the history of science in general and the history of the United States. The corpus of relevant reference works and archival sources is large indeed, but all cannot be noted here. This article generally concerns itself with works relating more or less specifically to American science, and selectively

---

[1] Walter Rundell, Jr., *In Pursuit of American History: Research and Training in the United States* (Norman: Univ. Oklahoma Press, 1970), p. 36; and Margaret F. Stieg, "The Information of (sic) Needs of Historians," *College and Research Libraries,* 1981, 42:549–560, on pp. 551–552, 554–555, 558–559.

[2] David M. Knight, *Sources for the History of Science, 1660–1914* (Sources of History: Studies in the Uses of Historical Evidence) (Ithaca, N.Y.: Cornell Univ. Press, 1975); and John T. Edsall and David Bearman, "Historical Records of Scientific Activity: The Survey of Sources for the History of Biochemistry and Molecular Biology," *Proceedings of the American Philosophical Society,* 1979, 123:279–292.

[3] Clark A. Elliott, "Citation Patterns and Documentation for the History of Science: Some Methodologic Considerations," *American Archivist,* 1981, 44:131–142.

even within that limit. These are the scholarly parameters within which Americanist historians of science have been and will continue to be most active, and where they can contribute directly to the solution to access and documentary problems.

Among the more general sources available are the bibliographies, indexes, and abstracts designed initially to serve scientists and also useful to historians using scientific publications as primary sources. Significant retrospective bibliographies of American books on science, technology, and medicine have been published recently by the Bowker Company. Also of interest are other specialized historical-scientific bibliographies; that on geology by Robert and Margaret Hazen includes over 11,000 alphabetically arranged entries, and demonstrates not only the content but also the quantity and varied publicaton format of the early American literature.[4]

In 1967, Robert Downs and Frances Jenkins edited an evaluative, state-of-the-art report on bibliographic works in several fields that touch the subject of this article. Especially relevant are the chapters on manuscripts and archives (Frank G. Burke), American history (Gerald D. McDonald), and history of science (John Neu). Read together, these chapters give a helpful if somewhat fragmented view of the available catalogues and bibliographies relating to the history of American science at that time. Robert Lovett's 1971 guide to information sources for business and economic history has a brief chapter on the history of science and technology. More recently, a general and international bibliography of reference works for the historian of science has been prepared by S. A. Jayawardene. That publication lists not only items relating directly to the history of science, but also includes references to works in some related areas and to general reference sources.[5]

ESSAYS AND RESOURCE REVIEWS

In recent years the general history of science field has witnessed two collective publications that review secondary works in the context of larger historiographic considerations. Both the work edited by Paul Durbin and that edited by Pietro Corsi and Paul Weindling include references to American output, although only the Corsi-Weindling guide has a separate chapter on the history of American science and medicine (by Edward H. Beardsley). The sections in the Durbin book on the sociology of science and science policy studies and on the history of technology relate particularly to American interests.[6]

[4] *Pure and Applied Science Books 1876–1982,* 6 vols. (New York: Bowker, 1982); *Health Science Books 1876–1982,* 4 vols. (New York: Bowker, 1982); and Robert M. Hazen and Margaret Hindle Hazen, *American Geological Literature, 1669–1850* (Stroudsburg, Pa.: Dowden, Hutchinson & Ross, 1980).

[5] Robert B. Downs and Frances B. Jenkins, eds., *Bibliography: Current State and Future Trends* (Illinois Contributions to Librarianship, 8) (Urbana/Chicago/London: Univ. Illinois Press, 1967); Robert W. Lovett, *American Economic and Business History Information Sources* (Management Information Guide, 23) (Detroit: Gale Research, 1971), pp. 211–221; and S. A. Jayawardene, *Reference Books for the Historian of Science: A Handlist* (Occasional Publication, 2) (London: Science Museum, 1982).

[6] Paul T. Durbin, ed., *A Guide to the Culture of Science, Technology, and Medicine* (New York: Free Press; London: Collier Macmillan, 1980); and Pietro Corsi and Paul Weindling, eds., *Information Sources in the History of Science and Medicine* (London/Boston: Butterworth Scientific, 1983).

The bibliographic essays or notes on sources at the end of many books have special utility insofar as they are evaluative, and because they discuss the sources in relation to a topic. Usually, however, such essays simply refer to publications or archival collections consulted, without examining or explaining the ways in which they were useful specifically for the particular study—mentioning, for example, the value of a certain manuscript collection as a whole but not explaining what parts were used or how. One of the things historians as well as archivists need to know is the relative worth of correspondence of various kinds (that is, family, professional, institutional), published articles and their drafts, laboratory notes, data sources (of various kinds), minutes of meetings, annual reports, and other types of records, considered in relation to categories of historical research interest. Not only would these reflections aid other historians in developing an effective methodological and evaluative approach to sources, they would also assist archivists in the appraisal of categories of material. Moreover, they could aid in the construction of finding aids and guides that would direct historians to sources appropriate to their interests.

While essays on methodology of the type outlined are rare, bibliographic notes and essays to a monograph, inevitably interjecting an element of evaluation and selectivity, are published relatively frequently. An example of a somewhat specialized recent work on American science that includes a bibliographic essay, incorporating both print and manuscript sources, is Hamilton Craven's *The Triumph of Evolution*. Daniel Kevles's *The Physicists* contains one of the best general bibliographic introductions to American science for the period since the Civil War. It, too, refers to manuscript collections as well as published works.[7]

A further type of essay on sources surveys a topic before research is undertaken, reviewing the general historical context and the possible sources of information. These are likely to refer to archival and manuscript sources, and in fact may be devoted to the holdings of particular repositories. Two examples for federal records are Meyer Fishbein's paper on World War II research and development activities and John Jameson's article on the National Park System.[8]

BIBLIOGRAPHIES OF PRINTED WORKS

Marc Rothenberg's 1982 bibliography of the history of American science represents a significant advance toward the control of access to the secondary literature. Intended for the nonspecialist or beginning student, its annotations not only summarize the works but also give a good sense of the interests and points of view of historians who wrote during the period 1940–1980. Some 800 books, articles, and dissertations are listed. Ironically, however, it includes relatively

[7] Hamilton Cravens, *The Triumph of Evolution: American Scientists and the Heredity-Environment Controversy 1900–1941* (Philadelphia: Univ. Pennsylvania Press, 1978); and Daniel J. Kevles, *The Physicists: The History of a Scientific Community in Modern America* (New York: Knopf, 1978).

[8] Meyer H. Fishbein, "Archival Remains of Research and Development During the Second World War," in *World War II: An Account of Its Documents,* ed. James E. O'Neill and Robert W. Krauskopf (National Archives Conferences, 8) (Washington, D.C.: Howard Univ. Press, 1976), pp. 163–179; and John R. Jameson, "The National Park System in the United States: An Overview with a Survey of Selected Government Documents and Archival Materials," *Government Publications Review,* 1980, 7A:145–158.

few bibliographies and catalogues, and therefore Rothenberg missed an opportunity to open up a wider range of literature for more advanced study. For that
purpose, the Americanist must fall back upon the cumulations of the *Isis Critical
Bibliography,* works on science in other bibliographies of American history, or
specialized listings. Henry Beers's bibliography of bibliographies on American
history, published in 1982, most certainly is destined to stand for some time as
the classic work in its category. Beers's work includes nearly 12,000 entries
(including guides and catalogues for manuscripts and archives). The entries are
arranged by subject, and there are sections for the sciences, making this the
best source for references to bibliographies of interest to historians of American
science as well as for other specialties.[9]

Philip Mitterling's annotated bibliography of books and articles on American
cultural history published since about 1950 has a chapter on science and medicine. Scientists are listed in a separate chapter on biography. Useful guidance
sometimes shows up in rather unexpected places, such as Robin Higham's guide
to military history. The format consists of a series of brief essays on special
topis with an appended bibliography. Edward C. Ezell prepared the section
"Science and Technology in the 19th Century," and Carroll Pursell compiled
the section "Science and Technology in the 20th Century" (in both the basic
and the supplementary volumes). There also is a chapter on medicine. The combination of succinct discussion of subjects and evaluative references would seem
a model means of offering orientation to an area of study in the context of a
reference work.[10]

Bibliographies relating to the history of certain desciplines have also been appearing, some of them international in scope, that provide access to segments
of the secondary literature on the history of science and collateral areas in the
United States. For example, Ronald Fahl and the Forest History Society have
prepared an alphabetically arranged bibliography for forestry and conservation,
and William Sarjeant has compiled a nearly definitive work for the history of
geology. The Sarjeant work is truly masterful. Topics are arranged or indexed
by nationality and are therefore easily accessible to historians of American science. The entries are not annotated. There are references to associations and
industrial and commercial firms, and extensive biographical listings that incor-
porate succinct biographical information along with the bibliographic citations.
References to the histories of institutions and to various subject fields make this
a work of potential value for the historians of American science in general. The
Office of Science and Technology at the University of California at Berkeley has
been active in bibliographic endeavors. For example, John Heilbron and Bruce

---

[9] Marc Rothenberg, *The History of Science and Technology in the United States: A Critical and
Selective Bibliography* (Bibliographies of the History of Science and Technology, 2) (New York/
London: Garland, 1982); *Isis Cumulative Bibliography: A Bibliography of the History of Science
Formed from Isis Critical Bibliographies 1–90, 1913–1965,* ed. Magda Whitrow, 5 vols. (London:
Mansell, 1971–1982); and *Isis Cumulative Bibliography 1966–1975,* ed. John Neu (London: Mansell,
1980–); and Henry P. Beers, comp., *Bibliographies in American History, 1942–1978: Guide to
Materials for Research,* 2 vols. (Woodbridge, Conn.: Research Publications, 1982).

[10] Philip I. Mitterling, *United States Cultural History: A Guide to Information Sources* (American
Government and History Information Guide Series, 5) (Detroit: Gale Research, 1979); Robin D. R.
Higham, ed., *A Guide to the Sources of United States Military History* (Hamden, Conn.: Archon,
1975); and Higham and Donald J. Mrozek, *A Guide to the Sources of United States Military History:
Supplement I* (Hamden, Conn.: Shoe String, 1981).

Wheaton have produced a work on recent physics that is international in coverage and arranged by topic; it is part of a larger project at Berkeley to control the literature and unpublished correspondence for the history of modern physics. Also to be noted is the Garland Publishing Company's current series Bibliographies of the History of Science and Technology, edited by Robert Multhauf and Ellen Wells. Rothenberg's bibliography is a part of this series.[11]

On occasion somewhat more specialized bibliographies appear. Some are devoted to the writings of individuals, such as Joan Gordon's bibliography on Margaret Mead. Jeffrey Stine and Michael Robinson's work on the U.S. Army Corps of Engineers represents a somewhat less common type of bibliography, that of writings relating to a particular institution or agency. There is need for more such works on scientific institutions, particularly as historians of science in America come to appreciate more fully the role of institutions in the shared community of science. Another specialized publication is R. Reginald's listing of works of science fiction. The second volume is a biographical directory of contemporary science fiction authors. However, there is no way of extracting American authors alone if their names are not known in advance, except by scanning the entries.[12]

BIOGRAPHICAL AND OTHER REFERENCE WORKS

If the function of bibliographies is to bring under control an ever-growing output of primary and secondary published works in the history of American science, the role of other reference works is to control the substantive data that form the foundation for historical studies. None of these undertakings can be complete, and therefore they not only codify but select the elements that constitute the starting point for historical investigation.

For American science, as for the history of science in general, a major event in recent years has been the completion of the *Dictionary of Scientific Biography*. Other works of biographical reference have appeared lately. Clark A. Elliott's *Biographical Dictionary of American Science* summarizes the major events in the lives of a number of scientists active before this century and gives references to sources of additional information, including some manuscripts. The chief source of ready information on the vast number of scientists in this century has been *American Men and Women of Science*. A recently published index to the first fourteen editions (1906–1979) allows access to sketches of some 270,000 individuals. Another notable index to biographical information now available is that by Earnest Barr, which covers obituaries and other biographical sketches

[11] Ronald J. Fahl, *North American Forest and Conservation History: A Bibliography* (Santa Barbara, Calif.: Clio Press for the Forest History Society, 1977); William A. S. Sarjeant, *Geologists and the History of Geology: An International Bibliography from the Origins to 1978*, 5 vols. (New York: Arno, 1980); and J. L. Heilbron and Bruce R. Wheaton, *Literature on the History of Physics in the 20th Century* (Berkeley Papers in History of Science, 5) (Berkeley: Office of Science and Technology, Univ. California, 1981).

[12] Joan Gordon, ed., *Margaret Mead: The Complete Bibliography, 1925–1975*, (The Hague: Mouton, 1977); Jeffrey K. Stine and Michael C. Robinson, *The U.S. Army Corps of Engineers and Environmental Issues in the Twentieth Century: A Bibliography* (Environmental History Series) (Fort Belvoir, Va.: Historical Division, Office of Chief of Engineers, U.S. Army Corps of Engineers, 1984); and R. Reginald, *Science Fiction and Fantasy Literature: A Checklist 1700–1974*, 2 vols. (Detroit: Gale Research, 1979).

in English-language journals up to about 1920. The entries identify the subjects by field, nationality, and birth and death dates.[13]

Among recent specialized biographical works, several cover engineering. For example, Christine Roysdon and Linda Kharti have prepared a biographical index to nineteenth-century technical journals and to several general biographical works such as the *Dictionary of American Biography*. The publication includes very brief biographical identifications of the engineers, with their birth and death dates. American chemists and chemical engineers are well served by the work edited by Wyndham Miles. The short, signed sketches cite published sources and sometimes manuscripts; some entries are based on personal recollections of the biographers. On occasion special biographical sources appear that have a place not filled by more general works, of which Ronald Stuckey's index to publications of the Ohio Academy is an example.[14]

The study of scientific institutions does not appear to have been as well served by American historians or bibliographers. The *Isis Cumulative Bibliography* has a special section on institutions, and those that are American are reasonably easily identified by scanning the pages. The historical-descriptive work on independent learned and research institutions edited by Joseph Kiger provides access to information on selected scientific as well as other organizations. The articles are written by various persons and include bibliographies.[15]

Several special dictionaries and encyclopedias of some value to historians of science have emerged, especially in applied science. These include Richard Davis's encyclopedia on forestry and conservation. The revised *Dictionary of American History* has articles on science, for example, scientific education (Robert C. Davis), periodicals (Michele Aldrich), and societies (Sally Gregory Kohlstedt). Richard Morris's *Encyclopedia of American History* has a chronology of science, invention, and technology (including medicine and public health).[16]

GUIDES TO MANUSCRIPTS AND ARCHIVES

This section considers especially the progress made in recent years to improve access to archival resources through the publication of guides and catalogues.

[13] *Dictionary of Scientific Biography*, Charles C. Gillespie, ed.-in-chief, 16 vols. (New York: Scribners, 1970–1980); Clark A. Elliott, *Biographical Dictionary of American Science: The Seventeenth Through the Nineteenth Centuries* (Westport, Conn.: Greenwood Press, 1979); *American Men and Women of Science Editions 1–14 Cumulative Index,* (New York: Bowker, 1983); and Earnest Scott Barr, *An Index to Biographical Fragments in Unspecialized Scientific Journals* (University: Univ. Alabama Press, 1973).

[14] Christine Roysdon and Linda A. Kharti, *American Engineers of the Nineteenth Century: A Biographical Index* (New York: Garland, 1978); Wyndham D. Miles, ed., *American Chemists and Chemical Engineers* (Washington, D.C.: American Chemical Society, 1976); and Ronald Stuckey, "Index to Biographical Sketches and Obituaries in Publications of the Ohio Academy of Sciences, 1900–1970," *Ohio Journal of Science,* 1970, 70:246–255.

[15] Joseph C. Kiger, ed., *Research Institutes and Learned Societies* (Greenwood Encyclopedia of American Institutions, 5) (Westport, Conn.: Greenwood, 1982). This work can be considered as complementary to the overall historical account given in Ralph S. Bates, *Scientific Societies in the United States,* 3rd ed. (Cambridge, Mass.: MIT Press, 1965), which has an extensive bibliography on pp. 245–293.

[16] Richard C. Davis, ed.-in-chief, *Encyclopedia of American Forest and Conservation History,* 2 vols. (Forest History Society) (Riverside, N.J.: Macmillan, 1983); *Dictionary of American History,* rev. ed., 8 vols. (New York: Scribners, 1976); and Richard B. Morris, ed., *Encyclopedia of American History,* 6th ed. (New York: Harper & Row, 1982), pp. 777–818.

(The term archival or archives is used here generally to mean both institutional records and the personal and professional papers of individuals.) These access tools fall naturally into those that describe the holdings of particular repositories or the contents of specific collections, and those that are union lists for a number of repositories. Henry Beers's *Bibliographies in American History* (see note 9) is important as a means of locating available catalogues and guides to archival repositories and collections. Robert Downs's recent supplement to *American Library Resources* updates an established compilation of references to published catalogues and other guides to American library holdings.[17]

The number of published repository and individual collection guides produced since the early 1970s is extensive, but relatively few relate directly to the history of science. An examination of many of the guides to state historical societies and the like has suggested how little of American science is reflected in their holdings. This perhaps indicates a prejudice against or lack of interest in science on the part of many builders of those collections, although it also may indicate the inadequacy of manuscript and archives descriptions. While many state and local historical societies do not appear promising from an inspection of their published catalogues, undoubtedly they hold treasures that can be uncovered by the persistent, the ingenious, or the fortunate. They will be particularly relevant to those in pursuit of the history of science on a local or regional level. Given the nature of scientific work in this country, the records of science are more likely to show up in academic archives and government repositories. Unfortunately, the state of business and industrial archives suggests that a substantial part of the record of research and development in that sector has been lost.

Because of the nature of the collections they represent, some published repository guides are especially promising for the historian of science. Of the general repositories of special significance for science history that have published guides to their holdings, the U.S. National Archives and the American Philosophical Society are notable. The APS has both a general guide prepared by Whitfield Bell, Jr., and Murphy Smith and a published version of its card catalogue. Examples of different kinds of guides to academic and research repositories having material relating to science are an article on the Harvard University Archives and the index to the Lick Observatory Archives based on its card catalogue.[18]

In recent years, the Smithsonian Institution has produced three editions of an exemplary guide to its archives and special collections which displays the rich resources there for the history of American science. The holdings include not only the Smithsonian's institutional records but the papers of individual scientists as well. Important collections for the history of American science and tech-

[17] Robert B. Downs, *American Library Resources: A Bibliographic Guide, Supplement 1971–1980* (Chicago: American Library Association, 1981).

[18] National Archives and Records Service, *Guide to the National Archives of the United States* (Washington, D.C.: GPO, 1974); American Philosophical Society, *Guide to the Archives and Manuscript Collections of the American Philosophical Society*, comp. Whitfield J. Bell, Jr., and Murphy Smith ((Memoirs, 66) (Philadelphia: American Philosophical Society, 1966); and American Philosophical Society Library, *Catalog of Manuscripts in the American Philosophical Society Library*, 10 vols. (Westport, Conn.: Greenwood, 1970); Clark A. Elliott, "Sources for the History of Science in the Harvard University Archives," *Harvard Library Bulletin*, 1974, 22:49–71; and Lick Observatory, *Preliminary Finding Aid to the Archives of the Lick Observatory, from the Card Catalogs Maintained by the Lick Observatory Staff, University of California at Santa Cruz* (A.I.P. National Catalog of Sources for History of Physics, Report 5) (New York: American Institute of Physics, 1980).

nology are held at the Eleutherian Mills Historical Library, and these have been delineated in a published guide by John Riggs. Useful aids to research in somewhat more restricted areas have been issued, especially by several government agencies. These include Alex Roland's booklet for NASA history and the guide by Maizie Johnson and William Heynen for records of the U.S. Hydrographic Office in the National Archives. A guide to a special type of material is Barbara Hughes's catalogue of instruments at the College of Charleston.[19]

Along with their general guide, the Smithsonian has begun to publish model finding aids for individual collections, beginning with William Cox's guide for the W. L. Schmitt papers. The series of registers of collections from the Library of Congress sometimes relate to scientific topics, as in that for the Oppenheimer papers.[20] Some of the most useful and insightful guides to individual collections have appeared in special newsletters of which the *History of Anthropology Newsletter* (issued by the University of Chicago) and the *Mendel Newsletter* (issued by the American Philosophical Society, for the history of genetics) are noteworthy. These types of publications frequently include descriptions or analyses of collections written by historians, based on their research experiences.

The repository and collection guides discussed above represent a sampling of the available aids for research in archives and manuscript collections. Localized aids, however, cannot suffice for many areas of research, especially in a field such as the history of science, which recognizes no closed geographical boundaries. Collections are not always where one expects to find them, and exchanges of letters and other documents crucial to a research topic can appear in unexpected places. Following the well-established lead of librarians, archivists now are beginning increasingly to consider means of sharing information on their holdings, and efforts are under way in professional, governmental, and commercial circles to promote the means of access to archival material.

The National Historical Publications and Records Commission has published a directory of American archives and manuscript repositories. Through automation of its data base, there is provision for periodic update and the capability of expansion of the information content.[21] The 1978 *Directory* established the cornerstone for national control of archival materials, beginning with identification and characterization of some 3,000 repositories of archival and manuscript material. While the entries are brief, they give the researcher a basic

---

[19] Smithsonian Institution, *Guide to the Smithsonian Archives* (Archives and Special Collections of the Smithsonian Institution, 4) (Washington, D.C., 1983). John Beverley Riggs, *A Guide to the Eleutherian Mills Historical Library: Accessions through the Year 1965* (Greenville, Del.: Eleutherian Mills Historical Library, 1970), with a supplement published in 1978, covering accessions in years 1966–1975; Alex Roland, *A Guide to Research in NASA History*, 7th ed. (Washington, D.C.: History Office, National Aeronautics & Space Administration, 1984); Maizie Johnson and William J. Heynen, *Inventory of the Records of the Hydrographic Office, Record Group 37* (National Archives Inventory, 4) (Washington, D.C.: National Archives and Records Service, 1971); Barbara Hughes, *Catalog of the Scientific Apparatus at the College of Charleston, 1800–1940*, ed. with additional material by Ralph Melnick (Charleston, S.C.: College of Charleston Library Associates, 1980).

[20] William E. Cox, *Guide to the Papers of Waldo Leslie Schmitt* (Guides to Collections, 1) (Washington, D.C.: Archives and Special Collections, Smithsonian Institution, 1983); and Library of Congress, Manuscript Division, *J. Robert Oppenheimer* (Registers of Papers in the Manuscript Division of the Library of Congress, 43) (Washington, D.C.: Library of Congress, 1974).

[21] National Historical Publications and Records Commission, *Directory of Archives and Manuscript Repositories in the United States* (Washington, D.C.: National Archives and Records Service, 1978).

orientation to materials in various subject and other categories and include references to published guides.

For more detailed information on the contents of many individual collections,
there is the *National Union Catalog of Manuscript Collections (NUCMC)*, prepared at the Library of Congress. It is difficult to gauge the degree of success
that *NUCMC* has achieved as a shared guide to collections for the history of
science, or to use *NUCMC* in turn as a means to evaluate the overall success
of archivists and historians in preserving significant collections in the first place.
Unfortunately, *NUCMC* systematically omits institutional records still in the
custody of the creating institution, and therefore it is fruitless to evaluate it as
a guide to archives of the records of universities, industry, research foundations,
and the like. Something of a tentative nature can be said about personal and
professional papers. Of a small sample of major twentieth-century American scientists listed in the *Dictionary of Scientific Biography,* fewer than half had their
papers listed in *NUCMC* as of 1981.[22] It appears from this evidence that the
papers of many of the country's most significant scientists either are not being
saved or at least are not being made available through the major national catalogue of manuscript collections. Additionally, the *DSB* also must be criticized
in that its entries for three-quarters of the scientists in the sample referred to
above make no reference at all to manuscript collections. This is a significant
oversight for a work that is recognized as the chief source of information on the
first echelon of scientists.

While not addressing the questions raised above in regard to preservation of
an adequate portion of the total historical record of American science, the
project undertaken by Chadwyck-Healey, Inc., to distribute in microform a large
number of the previously unpublished finding aids to manuscript and archival
collections is an important step forward for national access to archival materials.[23] The printed name and subject indexes that will accompany the microform
copies of the finding aids will further facilitate use of unpublished resources.

Several specialized union lists have appeared that may have value for historians of science. Particularly to be noted is Andrea Hinding's guide to archives
on women's history, the result of a wide-ranging and thorough survey project.
Its index indicates materials for the history of science. The entries on collections
not only indicate the content but also have brief biographical identifications of
women who are the subjects of the collections, and some of this information
may not exist in convenient form elsewhere.[24]

Other special projects have led to the publication of guides to manuscripts in
the history of science specifically. Edwin Layton's guide to materials in the
upper Midwest is a regional approach. Most of the guides, however, have been
discipline-based. They include Richard Davis on forest history and Michael

[22] Library of Congress, *National Union Catalog of Manuscript Collections, 1959–* (Ann Arbor,
Mich.: Edwards, 1962; Hamden, Conn.: Shoe String, 1964; Washington, D.C.: Library of Congress,
1965–present). Of 26 Americans listed in the *Dictionary of Scientific Biography* who died between
1930 and 1960 and whose surnames fall in the range A–Bo, 10 did not appear in *NUCMC* as of
1981. The papers of 11 were listed, and papers of others were indexed as being among those of
another person or agency.

[23] *National Inventory of Documentary Sources in the United States* (Teaneck, N.J.: Chadwyck-
Healey, 1983–).

[24] Andrea Hinding, ed., *Women's History: A Guide to Archives and Manuscript Collections in
the United States,* 2 vols. (New York: Bowker, 1980).

Sokal and Patrice Rafail on psychology. The catalogue of archival sources relating to biochemistry edited by David Bearman and John Edsall is likely to be
the most influential as an aid and incentive for similar undertakings in other
fields. The guide resulted from an extensive survey of sources, and the project developed an innovative computer system for access and control of its
findings.[25]

DOCUMENTARY PUBLISHING AND REPRINT PROJECTS

One way to make historical materials more readily accessible is to publish them,
whether by issuing documents in letterpress or microform, or by reprinting
classic published works. This field has been an active one for all historical areas
in recent years, and science has been affected as well. Among the reprint ventures, those by the Arno Press made a significant contribution to the history of
science. One of the most ambitious was that announced by Arno in 1980 under
the editorship of I. Bernard Cohen, entitled "Three Centuries of Science in
America." The series consisted of sixty-six books, with one original historical
work and several collected anthologies relating to certain individuals or themes.
Other reprints of published items have been carried out by individual scholars,
for example, Robert Kargon's edition of fifteen of the addresses by presidents
of the American Association for the Advancement of Science.[26]

The publication of manuscript materials in letterpress and microform has become a major scholarly enterprise in this country, with a separate organization,
the Association for Documentary Editing, which issues the journal *Documentary
Editing*. While this is not the place for a general review of the rationale and
practices of these editing projects, it is important for historians and archivists
concerned with the history of science in America to examine the projects'
meaning and value for the field. Here collective consideration and planning may
reap significant benefits and prevent wastage of funds. Also to be considered
are the ways and the degree to which such publishing projects affect the scholarly enterprise itself by making particular bodies of documentation readily available at many locations.

The National Historical Publications and Records Commission is the chief
sponsor of documentary editing projects in this country. The NHPRC Annual
Report for 1982 included a list of editing projects that were endorsed or financed
by NHPRC since its grants program began in 1964.[27] Science-related projects

---

[25] Edwin T. Layton, Jr., *A Regional Union Catalogue of Manuscripts Relating to the History of
Science and Technology Located in Indiana, Michigan, and Ohio* (Program in the History of Science
and Technology, Publication 1) (Cleveland: Case Western Reserve Univ., 1971); Richard C. Davis,
*North American Forest History: A Guide to Archives and Manuscripts in the United States and
Canada* (Santa Barbara, Calif.: Clio Books for the Forest History Society, 1977); Michael M. Sokal
and Patrice A. Rafail, *A Guide to Manuscript Collections in the History of Psychology and Related
Areas* (Millwood, N.Y.: Kraus International, 1982); and David Bearman and John T. Edsall, *Ar-
chival Sources for the History of Biochemistry and Molecular Biology: A Reference Guide and
Report* (Boston: American Academy of Arts and Sciences; Philadelphia: American Philosophical
Society, 1980).

[26] Robert H. Kargon, *The Maturing of American Science: A Portrait of Science in Public Life
Drawn from the Presidential Addresses of the American Association for the Advancement of Science, 1920–1970* (Washington, D.C.: American Association for the Advancement of Science, 1974).

[27] National Historical Publications and Records Commission, *Annual Report for 1982* (Washington, D.C.), pp. 16–21.

were relatively few and included some that relate to Western exploration (e.g., the journals of Major Stephen H. Long and the expeditions of John Charles Fremont). Letterpress editions in progress, as reported by NHPRC, included the papers of Thomas A. Edison, Joseph Henry, Constantine Rafinesque, and Leo Szilard. Microform editions completed included minutes and correspondence of the Academy of Natural Sciences of Philadelphia (1812–1924) and the papers of George Ellery Hale; those in progress included the papers of Charles E. Bessey and John Muir and records relating to the Franklin Institute. Of all NHPRC editions in progress in 1982, seventy-two were letterpress, and only eighteen were microform.

Whatever the long-term judgment on definitive editions of manuscript collections, increased attention should be given to the production of single-volume works of original documents that are intended to be read as monographs, for whatever reason—context, introduction to major issues in a historical period, or a sense of time or personality. Nathan and Ida Reingold have produced a selection of twentieth-century documents that is an effective general introduction to the science of the period. Other single-volume works of lesser scope have appeared, such as Roy Lokken's edition of the papers of James Logan. James Watson and John Tooze's scrapbook selection of printed, pictorial, and manuscript material relating to the DNA controversy is itself an illustration of the diversity and dispersal of sources on the history of science in modern society.[28]

SOCIAL ORGANIZATION AND PRESERVATION AND ACCESS TO SOURCES

The foregoing review discloses a wide variety of activities and an impressive, if somewhat uneven, display of products created for the bibliographic control of resources for the history of science in America. The work accomplished in recent years has involved efforts in a variety of quarters: by individual scholars, libraries and archives (alone and in concert), government agencies, commercial organizations, publishers, professional and learned societies, and private foundations.

In addition to efforts resulting in bibliographic products, other activities and projects have had significance for long-term preservation and access to historical information. In some instances, these efforts have been directed toward the more effective preservation of local historical sources, while in other cases the intent has been to promote the coordination of efforts to preserve and make accessible sources on a larger scale (often within particular scientific disciplines). The development of an increasing number of effective archival programs within particular institutions, the conduct of short-term projects, and the establishment of permanent efforts such as the discipline history centers suggest the confluence of multiple concerns. There is every reason to believe that joint or coordinated efforts will come more and more to characterize the future, in response to intellectual and social needs as well as to a technological imperative in the form of computer-based information systems.

A few examples of the events and projects of the last decade or so will suggest

---

[28] Nathan Reingold and Ida H. Reingold, *Science in American: A Documentary History 1900–1939* (Chicago: Univ. Chicago Press, 1981); Roy M. Lokken, "The Scientific Papers of James Logan," *Transactions of the American Philosophical Society,* 1972, *62*(6); and James D. Watson and John Tooze, *The DNA Story: A Documentary History,* (San Francisco: Freeman, 1981).

the degree of accomplishment. A significant commitment was made for the support of archival programs at the Massachusetts Institute of Technology and the Scripps Institution of Oceanography. The American Association for the Advancement of Science began efforts to preserve its historical records. The Rockefeller Archive Center, opened at Pocantico Hills in Tarrytown, New York, in 1975, exerted an important influence on studies of funding and other aspects of twentieth-century science. The Bancroft Library at the University of California at Berkeley established a program in 1973 to collect source materials in the history of science and technology in the San Francisco Bay area.

A survey of sources for the history of biochemistry and molecular biology was carried out under the joint sponsorship of the American Academy of Arts and Sciences and the American Philosophical Society, while the Society of American Archivists' Committee on Archives of Science surveyed plans of members of the National Academy of Sciences for preservation of their papers. A Laser History Project is currently in progress, under the sponsorship of several scientific societies.

Discipline-history centers constitute more permanent agencies for the promotion of the history of science and for the preservation and coordination of access to the sources. Progress on that front is exemplified by the establishment of the Center for History of Electrical Engineering by the Institute of Electrical and Electronics Engineers in 1980, the permanent location of the Charles Babbage Institute for the History of Information Processing at the University of Minnesota in the same year, and the organization in 1982 of the Center for History of Chemistry at the University of Pennsylvania. The well-established American Institute of Physics Center for History of Physics has engaged during recent years in special programs for the preservation of material relating to the history of astrophysics and of solid state physics.

As individual repositories, short-term survey projects, and discipline history centers have multiplied, other projects have addressed the complexity of issues involved in the appraisal and preservation of the records of modern science. The examination of these questions and the promulgation of a program for future action was the mission of the Joint Committee on Archives of Science and Technology, which carried out its work under a grant from the National Historical Publications and Records Commission. JCAST issued a report in 1983, subsequently summarized in *Isis*.[29] The A.I.P. Center for History of Physics in 1981 completed a study of documentation in U.S. Department of Energy laboratories that gave special attention to survey procedures and methods for archival appraisal. The Massachusetts Institute of Technology, operating under a grant from the Mellon Foundation, has engaged in a study of appraisal procedures and guidelines for scientific and technological records and has developed an appraisal guide.

The dissemination of information on various aspects of the history of science is now carried out through several special newsletters. *History of Science in*

---

[29] Joint Committee on Archives of Science and Technology, *Understanding Progress as Process: Documentation of the History of Post-War Science and Technology in the United States* (Final Report of the Joint Committee on Archives of Science and Technology [HSS–SHOT–SAA–ARMA]), ed. Clark A. Elliott (Chicago: Distributed by the Society of American Archivists, 1983); Clark A. Elliott, "Joint Committee on Archives of Science and Technology (JCAST): Summary from the Final Report," *Isis*, 1984, 75:158–162.

America: News and Views began issuance in 1980. In 1982 the history committees of both the American Geophysical Union and the American Physical
Society began to publish newsletters, joining earlier newsletters for special disciplines, including that of the history division of the Geological Society of
America. The discipline history centers referred to above also issue newsletters.

   Some events have occurred that caution against an overly optimistic attitude.
The U.S. National Archives established a science and technology archives division, but later diminished the program for budgetary reasons. Backward steps
were taken when the history program in the United States Geological Survey
was discontinued and an established archival program for mathematics at the
University of Texas was disrupted.[30] On the whole, however, progress has been
made. There is even some small evidence that a historical consciousness and
awareness of archival values is beginning to influence leaders of industry, a
sector of American science in which documentation has hitherto been largely
neglected. Among the encouraging signs are a project to write the history of
research and development efforts within the Du Pont Company (under auspices
of the Eleutherian Mills—Hagley Foundation, and with the support of Du Pont)
and recent efforts to encourage use of the corporate archives of the American
Telephone and Telegraph Company. The disciplinary history centers can be expected to play a particularly significant part in the promotion of historical and
archival interest within the industrial sector.

THE FUTURE

The concerns of this paper should be considered along with those examined in
the JCAST report mentioned above. Within the parameters of the current essay,
several needs and opportunities are apparent. A regular means by which historians, archivists, and librarians can distribute short or specialized bibliographies, methodological and historiographical papers, essays on sources (including
their actual use by historians engaged in research projects), and announcements
of new archival collections would fill an important need. Insofar as the history
of American science is a coherent field of investigation, such a serial publication
should be limited to American science in order to make it maximally relevant
to researchers sharing that common interest.

   There is also an obvious need for systematic and exhaustive bibliographies of
secondary works on American science that are intended for the specialist. These
need not necessarily be annotated except in cases of possible ambiguity. Such
reference aids could be issued in the form of bibliographies of biographies, institutional histories, disciplines, and topics such as education, regions, or chronological divisions. Bibliographies of primary scientific publications (books, articles, pamphlets, reports) also would be useful, for example, definitive lists of
the writings of selected scientists (perhaps conceived as supplements to existing
sources, such as the bibliographies that appear with the *Biographical Memoirs*
of the National Academy of Sciences) or bibliographies of the output of particular scientific laboratories or other institutions.

   A number of needs exist in the area of general reference works. Collections

---

[30] Gina Kolata, "Math Archive in Disarray," *Science,* 1983 (25 Feb.), *219*:940.

of short evaluative biographies written by experts, for all periods, and a biographical dictionary or directory for twentieth-century scientists are two important gaps in the existing literature. Also useful would be a reference work on the histories of scientific institutions. Both biographical and institutional works should devote special effort to listing relevant publications (primary and secondary) and to indicating the availability (or not) of archival sources. The history of science in America still lacks a general text, and that should be a high priority, as a collective effort if necessary. Development and publication of chronologies and an encyclopedia of American science would be of significant value to students and advanced researchers.

It is obvious that not only is current-day documentation for the history of science frequently in danger of disappearing, but access to that already in repositories often is inadequate. These two aspects of the problem of documentation sometimes compete for attention. Insofar as economic resources and knowledgeable personnel are limited, highest priority should be given to identifying or locating sources in danger of permanent loss, wherever they may be situated at present. Effort should be exerted to assure that records-creating institutions have adequate archival programs or that documents deemed to have permanent value are placed in other established repositories for preservation and future use. Such an emphasis is necessary in light of the enormous bulk of postwar documentation and the attendant difficulties of appraisal for historical use. Given such priorities, effort cannot be expended on preparing descriptions or indexes of greater and greater detail for collections now in repositories, while other documentary sources disappear for lack of attention. Historians and archivists need to continue to work toward a standard of adequate archival description that is useful for the one and feasible for the other. In regard to national or shared archival access programs, it is tempting to put efforts into the production of specialized, perhaps detailed name- and subject-access, catalogues serving the history of science alone. It is likely, however, that more is to be gained by improving the quality and coverage of general tools such as the *National Union Catalog of Manuscript Collections*. Alternatives may emerge in the form of on-line union catalogues of manuscript and archival collections, and this is much to be desired. Discipline-history centers and other organizations can maintain union catalogues for the location of collections and perhaps issue special-interest checklists for individual scholars.

The overall intent of this paper has been to demonstrate the diversity and accomplishments but also the gaps in efforts to control and make accessible the great body of primary and secondary materials relating to the history of American science. In years past, the body of quality secondary literature was limited and generally manageable. Now the corpus of professional and respectable amateur output is increasing. At the same time, the multiple channels in which the history of science moves makes documentary sources on many aspects of American life relevant. No longer can the history of science field depend—if in fact it ever could—on reprints of journal articles and the personal papers of a few great men. Investigations of the more distant past now can delve below the upper layer of elitism that was inevitably the concern of the first generations of American historians. Moreover, the locus of research interest is moving closer to World War II, which for science and its documentation is the great dividing line for modern times. On this side of the line the enterprise of science and its record have a size and character out of all proportion to those of its parent.

Not only better but perhaps different approaches to the evaluation and control of documentation will be necessary in order to meet the needs of historians in the future. To do so effectively, reference aids and documentation must themselves become objects of study. For example, the character of documents in their historical context must be understood more clearly. The reason is twofold. First, the proper evaluation of, as well as design of access to, historical sources (both printed and manuscript) requires an understanding of their social, intellectual, and physical qualities. Second, the understanding of the past, which depends on the evidence borne by documents, requires that historians comprehend the true functions and qualities of the documents themselves. Besides needing reference works of all kinds, therefore, the history of science in America in the future will require serious consideration both of the nature of those reference aids and their relations to scholarly inquiry and of the character and significance of primary documents as sources for understanding the past.

Notes on Contributors

Hamilton Cravens is Professor of History at Iowa State University. He has published *The Triumph of Evolution* (1978) and *Ideas in America's Cultures* (1982) as well as numerous articles on the natural and social sciences. He is now working on science and children in modern America.

Clark A. Elliott is Associate Curator of the Harvard University Archives. He is the author of *Biographical Dictionary of American Science: The Seventeenth Through the Nineteenth Centuries* (1979) and papers on American science and archival topics. He edits the newsletter *History of Science in America: News and Views*.

Mott Greene is in the second year of a MacArthur Fellowship. He is the author of *Geology in the Nineteenth Century* and is at work on a biography of Alfred Wegener and a collection of essays on science in preclassical antiquity.

Clara Sue Kidwell is an Associate Professor in Native American Studies at the University of California at Berkeley. Her research interests are in native American religions and medicine, and she is working on a book on concepts of power in native American societies.

Sally Gregory Kohlstedt is Associate Professor of History at Syracuse University. Her book *The Formation of the American Scientific Community* analyzes the early years of the American Association for the Advancement of Science. She is now writing on museums of natural history in nineteenth-century America.

Jane Maienschein, Assistant Professor of Philosophy, teaches history and philosophy of science at Arizona State University. She is editing a volume of selections from the Marine Biological Laboratory's *Biological Lectures* of the 1890s and completing a study of the establishment of American biology.

Albert E. Moyer is Associate Professor of History at Virginia Polytechnic Institute and an adjunct member of the university's Center for the Study of Science in Society. He is the author of *American Physics in Transition: A History of Conceptual Change in the Late Nineteenth Century* (1983).

Ronald L. Numbers is Professor of the History of Medicine and the History of Science at the University of Wisconsin–Madison. He is the editor, with David C. Lindberg, of *God and Nature: Historical Essays on the Encounter between Christianity and Science* (forthcoming) and is writing a history of creationism in the twentieth century.

Alex Roland is Associate Professor of History at Duke University, where he teaches military history and history of technology and directs the Program in Science, Technology, and Human Values. Formerly a historian at NASA, he has written *The National Advisory Committee for Aeronautics, 1915–1958* (NASA, 1984).

Margaret W. Rossiter is a research associate at the American Academy of Arts and Sciences in Cambridge. Her book *Women Scientists in America: Struggles and Strategies to 1940* (Johns Hopkins, 1982) won the 1983 Berkshire Prize for the best book in history by an American woman. She is now at work on a sequel, covering to the 1970s.

Marc Rothenberg is Associate Editor of the Joseph Henry Papers Project at the Smithsonian Institution. He is studying the interaction between amateur and professional astronomers in the United States during the early twentieth century, with special interest in the American Association of Variable Star Observers.

John Servos is Associate Professor of History at Amherst College, where he teaches the history of science. His articles on the history of science in modern America have appeared in *Isis*, *Historical Studies in the Physical Sciences*, *Ambix*, and the *Journal of Chemical Education*.

Sharon Gibbs Thibodeau, an archivist in the National Archives, appraises and describes the records of civilian federal agencies engaged in scientific activity. She represented the History of Science Society on the Joint Committee on the Archives of Science and Technology and assisted in the preparation of its final report.

John Harley Warner is a Fellow at the Wellcome Institute for the History of Medicine in London. He has published several articles, is completing a book on medical practice, knowledge, and professional identity in America, 1820–1885, and is studying the comparative transmission of nineteenth-century Parisian clinical medicine.

George Wise is a historian with the General Electric Research and Development Center, Schenectady, New York. He is the author of *Willis R. Whitney and the Origins of American Industrial Research* (Columbia, 1985).

Index

ISIS CUMULATIVE INDEX, 1953-1982

The new *Cumulative Index* provides direct access to nearly 1,000 major articles and 4,000 authoritative book reviews from 30 years of *Isis: An International Review Devoted to the History of Science and its Cultural Influences.*

AUTHOR-SUBJECT INDEX

Subject classifications and chronological divisions based on the widely used *Isis Critical Bibliography* lead to articles directly related to your interests. Among topics available are

Alchemy

Computer sciences and artificial intelligence

Historiography and historical method

Medicine and medical sciences

Physics

Women in science

BOOK REVIEW INDEX

Complete author-title citations place at your fingertips authoritative evaluations of 4,000 books central to the history of science.

SPECIFICATIONS

Uniform in format with *Isis*, cloth and paperback editions printed on acid-free paper, 168 pages. Each article indexed by author, subject, geographical locus, cultural focus, major individuals discussed, and institution covered. Each book indexed by author and title.

ORDER NOW

Send your orders to

History of Science Society Publications Office
215 South 34th Street/D6
Philadelphia, PA 19104, U.S.A.

or call TOLL FREE
(orders only, please)

1-800-341-1522

DATATEL-800™

IS YOUR *ISIS* FILE COMPLETE?

BACK ISSUES, listed below, are currently available for $12 each, $15 for Critical Bibliographies (*), Postpaid. A 25% discount is applied to the purchase of 5 or more issues.

TO ORDER, circle the desired issue(s), indicate the quantity in the space provided, and fill in the cost and mailing information.

CLIP AND MAIL the coupon to Isis Publication Office, University of Pennsylvania, 215 South 34th Street/D6, Philadelphia, PA 19104.

Selected out-of-print issues of Isis are available from
Johnson Reprint Corporation, 111 Fifth Avenue, New York, NY 10003; (212) 741-6800.

Isis is also available on microfilm from
University Microfilms, Inc., 300 North Zeeb Road, Ann Arbor, MI 48107; (313) 761-4700.

Volume	Year	Issue Number	Volume	Year	Issue Number
___ 33/34	1942	91 92 93	___ 58	1967	192 193 194 195*
___ 34	1943	96 97	___ 59	1968	196 197 198 199 200*
___ 39	1948	113/114	___ 60	1969	202 203 204
___ 44	1953	135/136 138	___ 61	1970	207 208 209 210*
___ 45	1954	139 140 141 142	___ 62	1971	214
___ 46	1955	144 145	___ 63	1972	219
___ 47	1956	149	___ 64	1973	221 222 223 224 225*
___ 48	1957	151 152 154	___ 65	1974	226 227 228 229 230*
___ 49	1958	155 156 157 158	___ 66	1975	231 232 233 234 235*
___ 50	1959	160 161 162	___ 67	1976	236 237 238 239 240*
___ 51	1960	165	___ 68	1977	242 243 245*
___ 52	1961	167 168 169 170	___ 69	1978	246 248 249 250*
___ 53	1962	171 172 173 174	___ 70	1979	251 252 253 254 255*
___ 54	1963	175 176 177 178	___ 71	1980	257 258 259 260*
___ 55	1964	179 180 181 182	___ 72	1981	262 263 264 265*
___ 56	1965	183 186	___ 73	1982	266 267 268 269 270*
___ 57	1966	187 188 189 190	___ 74	1983	271 272 273 274 275

$_____ for ____ regular issues at $12 each

$_____ for ____ CB (*) issues at $15 each

$_____ Subtotal

$_____ Less 25% discount, if applicable

$_____ TOTAL

_____ Check enclosed   _____ Bill me
(issues sent on receipt of payment)

NAME _____

ADDRESS _____

_____

CITY _____

STATE _____ ZIP _____